The author, who is 80 this year, in retirement at Quarr Abbey.

George Temple, F.R.S., is Sedleian Professor Emeritus of Natural Philosophy at Oxford.

100 Years of Mathematics

100 YEARS OF MATHEMATICS

A personal viewpoint

George Temple, F.R.S.

Sedleian Professor of Mathematics Emeritus
in the University of Oxford

Springer-Verlag

New York

Published by
Gerald Duckworth & Co. Ltd.,
The Old Piano Factory,
43 Gloucester Crescent,
London N.W.1

Published simultaneously by
Springer-Verlag New York (175 Fifth Avenue,
New York 10010, U.S.A.), for distribution in
the United States and its possessions,
Canada and Mexico

ISBN 0-387-91192-8   Springer-Verlag   New York

Text set in 10/11½ pt Linotron 202 Century, printed and bound
in Great Britain at The Pitman Press, Bath

Contents

PART II. SPACE

PART III. ANALYSIS

Preface

Ninety per cent of all the mathematics we know has been discovered (or invented), in the last hundred years, and this account is, of necessity, a personal selection of its major developments. It offers a continuous record of the advances made in each of some dozen directions of research, and indicates how these directions are converging into one single discipline uniting algebra, topology and analysis.

The theory of number is explored from the work of Méray, Weierstrass, Cantor and Dedekind to Peano, Frege and Zermelo, with an excursion into non-standard analysis from Veronese and Hahn to Skolem, Luxembourg and A. Robinson. Vectors and tensors are traced from Hamilton, Grassmann and Gibbs to Pauli, Dirac and the theory of spinors. The theory of 'distance' in geometry is treated fully from Cayley, Riemann and von Staudt to Klein, Christoffel, Ricci and Levi-Civita, with an account of the work of Helmholtz, Lie, Weyl and A. N. Whitehead. The origins of modern algebraic geometry are examined in the researches of Sylvester, Dedekind and Kronecker, with some reference to the work of Hilbert, Lasker and Emmy Noether.

The primitive notions of topology are traced in the theories of convergence from Fréchet to H. Cartan, Weil and J. W. Alexander, in the beginnings of functional analysis from Volterra to Banach, and in the fixed point theorems of Kronecker and Brouwer.

The classical subject of derivatives and integrals is expounded with the contributions of Cauchy, Riemann, Darboux and Borel; the theory of sets of points from Riemann, Hankel and Cantor, Peano, Jordan and Borel; measure theory from Lebesgue and W. H. Young to Dini; and the theory of distributions (or generalized functions) from Kirchhoff and Heaviside to Hadamard, Bochner, Dirac and Sobolev, L. Schwartz, Mikusiński and van der Corput.

Analysis is represented in a long chapter on differential equations from Briot and Bouquet to Lefschetz and M. Cartwright; in a chapter on the calculus of variations from the commentators on Jacobi to Morse and L. C. Young; and in an extensive exposition of potential theory from Green and Kelvin to de Rham and Hodge.

A concluding chapter examines mathematical logic from Boole to Frege, Russell, Hilbert and Brouwer.

The book is addressed primarily to the working mathematician, and to those embarking on this vast ocean of study and research. The author was a Professor of Mathematics at King's College, London, from 1932 to 1953, and Sedleian Professor of Natural Philosophy at Oxford from 1953 to 1968.

I must express my indebtedness to many librarians who have helped me. I am particularly grateful to Colin Haycraft who suggested that I should write this book and who has patiently waited for the typescript, to Mrs P. Raven who has meticulously revised the references and carried out the copy-editing, and to Dominic Welsh who has given invaluable help in correcting the proofs and considerably improving the final version of this work.

I acknowledge with gratitude the award, for two years, of a Research Fellowship from the Leverhulme Trust Fund. Finally I must express my indebtedness to the Provost and Fellows of the Queen's College, Oxford for all the help, sympathy and understanding they have given me while writing this book.

1981   G. T.

1. Introduction

1.1. 'A Century of Mathematics'

Almost the whole of mathematics, as it is studied and taught at the present time, is the work of mathematicians during the last hundred years. During this century mathematics has been transformed, its subject matter has been enormously widened and its methods have been profoundly deepened. This book has no pretensions to give an encyclopaedic history of the mathematics of this period, nor anecdotal biographies of mathematicians who flouished in that era. Its object is to present the work of some of the mathematicians who have carried out this transformation of mathematics, to describe their ideals, their concepts, their methods and their achievements.

It is a daunting and tremendous task, but it is an enterprise well worth the attempt. The professional mathematician can scarcely avoid specialization and needs to transcend his private interests and take a wide, synoptic view of the whole landscape of contemporary mathematics. His scientific colleagues are continually seeking enlightenment on the relevance of mathematical abstractions. The undergraduate needs a guidebook to the topography of the immense and expanding world of mathematics. There seems to be only one way to satisfy these varied interests—namely to offer a concise historical account of the main currents of recent mathematical thought. Only by a study of the development of mathematics can its contemporary significance be understood.

For the professional mathematician such an historical survey is provided, in certain branches of the subject, by the brilliant summaries by Bourbaki (1969) and Weyl (1951). These essays set a high standard of exposition, but they do not pretend to cover the whole century nor all the main divisions of the subject. There is room for a history of mathematical ideas which will demand less mathematical expertise and offer a more detailed account of the motivation of research.

This book has been written therefore to appeal to those who desire a broad survey of the main currents of mathematical thought. It is primarily and essentially an account of the discovery or invention of mathematical concepts, and the historical material is therefore divided and classified, neither chronologically nor biographically, but philosophically. Thus, for example, we shall study the theory of limiting processes, tracing them from the convergent sequences of Cauchy to the filters of H. Cartan.

1

1.2. Definitions of Mathematics

The non-mathematician who has been taught to regard mathematics as a closely woven tapestry of definitions and theorems may well expect a history of mathematics to open with a definition of the nature of mathematics as it was understood in the century under review. From Pythagoras (ca. 550 BC) to Boethius (ca. AD 480–524), when pure mathematics consisted of arithmetic and geometry while applied mathematics consisted of music and astronomy, mathematics could be characterized as the deductive study of 'such abstractions as quantities and their consequences, namely figures and so forth' (Aquinas ca. 1260). But since the emergence of abstract algebra it has become increasingly difficult to formulate a definition to cover the whole of the rich, complex and expanding domain of mathematics. In fact the subject matter discussed by mathematics has increased so rapidly and so extensively that there is some element of truth in maintaining that mathematics is not so much a subject as a way of studying any subject, not so much a science as a way of life.

We turn, then, from the attempt to characterize the material object of mathematics to an attempt to determine its formal object, i.e. its methodology. However, it appears to be impossible, and is certainly difficult, to begin with a description of mathematical methods that will be both comprehensive and intelligible. A simple example will illustrate the problem.

The concept of 'number' in its most elementary sense as the signless integer appears to be an immediate abstraction from quantitative reality subjected to processes of counting or measurement. Vulgar fractions arise from division of a quantity into equal parts. But in what sense is zero a number? Are there negative numbers? Are there numbers corresponding to incommensurable ratios?

Each question requires for its solution a fresh exercise of that kind of creative imagination which we call mathematical abstraction. There is no general paradigm or algorithm to direct and guide our thought. At each stage in the advance of mathematical thought the outstanding characteristics are novelty and originality. This is why mathematics is such a delight to study, such a challenge to practise and such a puzzle to define.

There are, however, two attempts to define mathematics which must be given serious consideration. They express two extreme positions, which should perhaps be regarded rather as sighting shots straddling the target.

First there is the definition, boldly proposed by Peirce (1881) that 'Mathematics is the science which draws necessary conclusions', and more explicitly formulated by Russell (1903, p. 3) that 'Pure Mathematics is the class of all propositions of the form "p implies q", p and q being carefully specified. It was indeed the purpose of Russell's treatise to provide a complete, exact and convincing justification of this definition, and the author employed all his encyclopaedic knowledge of the philosophy of mathematics, all his powers of acute and rigorous reasoning, and all the charm of his lucid and persuasive style to achieve this object. The second volume of Russell's treatise was never written, but, instead, he and A. N. Whitehead collaborated to give a magisterial account of the reduction of pure mathematics to logic in the three volumes of their *Principia Mathematica* (Whitehead and Russell, 1910).

We do not belittle the considerable influence this great work has had (even on those who have not read it) by maintaining that its fundamental conception of pure mathematics is too narrow and restrictive to be universally accepted. Logic undoubtedly has a role to play in mathematics, but logic is also indispensable in philosophy, science, law and indeed every intellectual activity. Mathematics has no monopoly of logical reasoning and the function of logic in mathematics is critical rather than constructive. Logical analysis is indispensable for an examination of the strength of a mathematical structure, but it is useless for its conception and design. The great advances in mathematics have not been made by logic but by creative imagination. For example, the title of mathematician can scarcely be denied to Ramanajan who hardly gave any proofs of the many theorems which he enumerated (Hardy 1927).

We turn therefore to a second definition, one which is implicitly held by many applied mathematicians, that mathematics is the language of experimental physics. Now there is no doubt that the language of physical science is mathematics, but the converse of this statement must be repudiated. The applied mathematician is happy to employ the methods and results of pure mathematics in formulating concise and economical descriptions of events in the world of physics, chemistry and biology. But pure mathematics is much more than an armoury of tools and techniques for the applied mathematician. On the other hand, the pure mathematician has ever been grateful to applied mathematics for stimulus and inspiration. From the vibrations of the violin string they have drawn the enchanting harmonies of Fourier series, and to study the triode valve they have invented the whole theory of non–linear oscillations. Moreover there are great subjects in pure mathematics which have arisen in complete and serene independence of the necessities of physical science. The 'language theory' is inadequate as a description of the nature of mathematics.

1.3. The Three-fold World of Mathematics

In default of a definition of mathematics we can offer a preliminary survey of three different types of intellectual activity which we can discriminate in the mathematics of the last hundred years. Mathematical activity has taken the forms of a *science*, a *philosophy* and an *art*.

As a *science* mathematics has been adapted to the description of natural phenomena, and the great practitioners in this field, such as von Kármán, Taylor and Lighthill, have never concerned themselves with the logical foundations of mathematics, but have boldly taken a pragmatic view of mathematics as an intellectual machine which works successfully. Description has been verified by further observation, still more strikingly by prediction, and sometimes, more ominously, by control of natural forces. Happily, unresolved problems, such as turbulence and elementary particles, still remain as challenges.

In great contrast mathematics has also been developed as a *philosophy*, in the sense in which this term is defined by A. N. Whitehead (1929, p. 3) as 'the endeavour to frame a coherent, logical and necessary system of general ideas in terms of which every element of our experience can be interpreted'. Substitute 'mathematics' for 'experience' and we have an admirable description of its speculative and philosophic development. The careful analysis of mathematical concepts, the formulation of rigorous canons of deduction and proof, the search for satisfactory evidence of self-consistency and the development of logical analysis are examples of the work achieved in this field. The philosophy of mathematics, like metaphysics, has its paradoxes and antinomies, and also diverse schools of thought—the formalists, the logicists and the intuitionists, as exemplified by Hilbert, Russell and Brouwer.

But for the great majority of mathematicians, mathematics is not just scientific description nor philosophic speculation but a whole world of invention and discovery—an *art*. The construction of a new theorem, the intuition of some new principle, or the creation of a new branch of mathematics is the triumph of the creative imagination of the mathematician, which can be compared with that of the poet, the painter and the sculptor. It was this quality in which Hardy was preeminent and which he described with such charm and precision (Hardy 1969) that Grahame Greene was impelled to say 'I know no writing—except perhaps Henry James' introductory essays—which conveys so clearly and with such an absence of fuss the excitement of the creative artist' (Greene 1940). As to whether the works of art exhibited by mathematicians are to be reckoned as inventions or discoveries, it must be admitted that most mathematicians are by nature Platonists who cheerfully, unreflectingly and habitually employ such loaded phrases as 'We assume that there exists . . .' or 'Therefore there exists . . .' an entity with such and such characteristics. Challenged by the realist they would probably reply that since the truths of mathematics are absolute, universal and eternal it is hard indeed to deny them an existence independent of human intelligence.

1.4. The Unity of Contemporary Mathematics

The preceding section, with its rapid survey of the different kinds of activity in which mathematicians have been engaged, may serve as an excuse for the absence of any formal definition of the subject. There is, however, one outstanding characteristic of mathematical activity in the last hundred years which the historian must emphasize at the very beginning of his work. This characteristic is the growing unification of the diverse mathematical disciplines.

It is commonly remarked that this is an era of specialization and it is inferred that the growth of mathematics means its proliferation into an ever-increasing diversity of distinct and specialized topics. It is of course true that few if any mathematicians could now claim acquaintance with every branch of the subject. The days of the great universalists, the polymaths such as George D. Birkhoff, Poincaré, E. T. Whittaker, Hilbert and Weyl, seem to have gone. But whereas in 1870 mathematics appeared to be diversified into distinct 'branches' spreading out in different directions, at the present time the side-shoots of these branches are growing together and reuniting in a single trunk. This process is exemplified in pure mathematics by the creation of such new subjects as algebraic topology and topological algebra which exhibit the fruitful interactions of two great disciplines. In applied mathematics, the themes of electricity, hydrodynamics and gas dynamics are being subsumed under the general title of continuum dynamics. This is the explanation of the delightfully unpredictable sequence of the successive fascicules of *Eléments de mathématiques* (Bourbaki 1969).

This characteristic of contemporary mathematics determines the structure of the present historical survey. It is no longer possible to discuss separately the developments in algebra, analysis, geometry, etc. Nor is it convenient to consider one by one the work of the great mathematicians, for each of these has enriched different branches by his researches. The historian is therefore driven to attempt a history of mathematical concepts and ideas.

It remains for him to decide whether he should describe each change in mathematics as it was seen by the contemporary mathematicians or whether he should view these changes from the standpoint of the present day. The first alternative lends a certain vividness to the narrative but seems to be more appropriate to the biographies of mathematicians. The second alternative gives an interpretation of the past which should make the present picture more intelligible. Accordingly we have adopted the second alternative and have attempted to see the last hundred years of mathematics from the vantage-point of the 1970s.

1.5. An 'Historical Introduction'

The purpose of this book is therefore not only to present an account of the development of mathematics during the last hundred years but also to provide an introduction to some of the various branches of mathematics which have grown during this period. In such an exploratory survey full and rigorous proofs, minute and careful definitions are out of place. What is needed is an appeal to the imagination and intuition of the reader which shall awaken his interest and encourage deeper and more serious study. Thus the reading of the earlier part of each chapter should require only a modicum of mathematics, although, as each chapter progresses, the exposition becomes inevitably more technical and demanding.

In order to make the descriptions of the various branches of mathematics intelligible it has been necessary to pass in review the pioneer work accomplished before the period 1870–1970. It is scarcely necessary to apologise for this enlargement of our field of study. What would geometry be without Gauss, mathematical logic without Boole, algebra without Hamilton, analysis without Cauchy? A 'century' of mathematics must therefore be interpreted in a free and generous manner. At once there arises a serious problem, admirably stated by Chesterton (1925):

A long and splendid literature[1] can be most conveniently treated in one of two ways. It can be divided as one cuts a currant cake or a Gruyère cheese, taking the currants (or the holes) as they come. Or it can be divided as one cuts wood—along the grain: if one thinks that there is a grain. But the two are never the same: the names never come in the same order in actual time as they come in any serious study of a spirit or a tendency. The critic who wishes to move onward with the life of an epoch must be always running backwards and forwards among its mere dates; just as a branch bends back and forth continually; yet the grain in the branch runs true like an unbroken river.

There can be no doubt that the convenience of the reader is best achieved by treating the history of mathematics organically and philosophically rather than in strict chronological order. For example, the subject of potential theory begins with Laplace and is still in active growth. It would be monstrous to expect the student of this subject to pursue its development fragmented into paragraphs distributed among some ten chapters for ten decades of research. We have therefore discussed our history, subject by subject, hoping that repetition and overlapping may be positive advantages making for clarity and intelligibility.

This book is far from an encyclopaedic account of a century of mathematics and it is abundantly clear that rigorous selection is as inevitable as it is difficult and uncongenial. As a general canon of selection we have concentrated on the beginnings of each subject and have endeavoured to select such topics as have proved to be fruitful and to become essentially involved in contemporary mathematics. Rightly or wrongly, this rules out the historical study of the 'special functions' of mathematics, for which we must refer the reader to such standard treatises as those of Watson (1922) on Bessel functions and Hobson (1931) on spherical harmonics.

Like Kelley (1955) I should have been proud to give this book the sub-title of 'What every young mathematician should know' of the history of mathematics, but I fear that there are many omissions which would falsify the claim. However, I take comfort in the thought that some of the great treatises can be criticized for some startling omissions. Forsyth (1890–1906) wrote six volumes on differential equations without a mention of Poincaré's geometric theory of ordinary differential equations. Bourbaki (1969) has summarized the foundations of topology without a mention of Kuratowski, J. W. Alexander or Veblen.

It must be confessed that there are great tracts of mathematics which are not explored in this book. The writer has had to choose between intelligibility and comprehensiveness, and it seemed better to treat some subjects in detail, rather than to attempt to include all subjects in one volume with tantalizing brevity. This must be his apology for the omission of such great developments as the theory of functions of a complex variable, the theory of modern abstract algebra, the later developments of algebraic geometry and topology, statistics, probability and numerical analysis.

---

[1] Here we apply this passage to mathematical writings!

An historical introductory study of mathematics has three great advantages:

(1) It has a permanent value as compared with the ephemeral interest of expositions contructed in accordance with what appeared (at the time of writing) to be the natural, essential, logical order of presentation. Merely as a record of facts, of the dating of discoveries, of the indebtedness of one mathematician to another, historical accounts cannot be faulted or become unfashionable. The magnificent achievement of Bourbaki (1939–) planned to present in an orderly logical sequence the whole of mathematics is already dated, and a new edition appears to need a complete revision based perhaps on category theory rather than on sets and logic; but the dull and pedestrian researches of Todhunter (1861, 1865, 1886–93) on the histories of probability, the calculus of variations and the theory of elasticity will always preserve their value for the historiographer.

(2) Secondly, I would maintain that the best way to learn mathematics (and this means of course the best way for the best students) is to attempt to solve each problem as it arises without recourse to any book and then, and only then, to compare the solution so painfully and partially achieved with the accepted and official answer. This is a counsel of perfection. A less exacting, but almost equally valuable approach, is to retrace the paths taken by the great pioneers and to appreciate the various shifts and artifices they were obliged to adopt for each question before stumbling on the right solution.

(3) Thirdly, the historical approach introduces us personally to the great mathematicians and infuses a humane and genial spirit into what can be a most arid and abstract study. Why did Kelvin initiate the study of perfect fluids, why did Heaviside invent generalized functions? The Marxist may regret that we have not discovered any economic pressures to account for these discoveries, but the historically minded mathematician will rejoice to establish the genealogical tree of discovery and research.

Unfortunately the history of mathematics throws little light on the psychology of mathematical invention. Hadamard (1945) and Cartwright (1955) have each explored this problem without reaching any satisfying solution. Poincaré (Hadamard 1945, pp. 11–15) as is well known, has recalled in detail how he saw, in a flash of insight, the elucidation of a problem in the theory of Fuchsian functions, while setting foot on a country bus. Disciples of Jung may have recourse to the theory of the Universal Unconsciousness as a great well from which a genius can draw his original discoveries. Others may prefer the more theological confession of that great aeronautical engineer, Barnes Wallis (the 'dam buster'), who confided in me that he could attribute his greatest discoveries only to some species of divine inspiration.

PART I. NUMBER

2. Real Numbers

2.1. Introduction

The arbitrary chronology, which restricts the present work to the last hundred years, is completely satisfactory in a history of the theory of the concept of 'number', for the year 1859 marks the date of Weierstrass's lectures on analytic functions which inaugurated a mathematical revolution.

Four striking features characterize this revolution. It was, in the literal sense of the words, reactionary and complete, it transferred the concept of number from geometry to arithmetic, and it has ever since dominated mathematical analysis.

The revolution was literally a reaction and a 'complete revolution' since it was a return to the Greek theory of ratio and proportion expounded in the Fifth Book of Euclid, and to the rigorous analysis of Eudoxus and Archimedes. This theory was forgotten or negelected in the first half of the nineteenth century, when it was replaced by a quasi-geometric theory of continuous magnitudes. The modern arithmetical theory of number lay concealed in some unpublished pages by Gauss (1800), and in a neglected work by Bolzano (1817), and was first publicly proclaimed in the unpublished course of lectures delivered by Weierstrass in Berlin in 1859. There has been no subsequent reaction. The arithmetical theory of real numbers has been the fertile source of all modern analysis. It has been followed by the logical analysis of the concepts of rational numbers and of the integers themselves. Finally, in recent years we have even seen the rehabilitation of 'infinitesimals'.

2.2. The Geometric Theory of Real Numbers

The geometric concept of a real number found classical expression in the researches of Cauchy (1821). For Cauchy, what we should now call a 'real number' was something undefined but envisaged as a symbol of some magnitude or quantity actually or potentially occurring in the real physical world. Thus the real numbers were thought to form an ordered and continuous system, to be subject to the algebraical operations of addition, subtraction, multiplication and division, and to furnish the material for approximation of any desired accuracy.

Cauchy's great contribution was to express the theory of limits as a theory of approximation. The mathematicians of the eighteenth century were familiar with the use of infinite series and of infinite sequences to represent such 'transcendental' numbers as π and e, but it was left to

11

Cauchy to give a precise analysis of the concept of the convergence of an infinite series or sequence, granted the validity of the concept of a real number. His 'general principle of convergence' defines a sequence of real numbers,

$$S_1, S_2, \ldots, S_n, S_{n+1}, \ldots$$

to be convergent if to any prescribed degree of approximation, specified by a (small!) positive number δ, there corresponds a term S_n, such that all the successive numbers of the sequence,

$$S_{n+1}, S_{n+2}, \ldots, S_{n+p}, \ldots$$

are clustered about S_n in such a way that

$$|S_{n+p} - S_n| < \delta \quad \text{for} \quad p = 1, 2, \ldots.$$

If this criterion were satisfied it was assumed to be obvious that the sequence $\{S_n\}$ converged to a unique limiting value, a real number, l, such that corresponding to a prescribed degree of approximation δ there corresponded a number S_n, such that all the successive numbers were clustered about l in such a way that

$$|l - S_{n+p}| < \delta \quad \text{for} \quad p = 1, 2, \ldots.$$

The defect in Cauchy's thought was that it furnished no proof for the existence of a limit of a convergent sequence, and gave no warning that such a proof was necessary. The great and permanent value of Cauchy's 'general principle of convergence' is that it provided an 'internal' criterion of the convergence of a sequence $\{S_n\}$, which was independent of any knowledge of the numerical value of the limit of the sequence.

By one of the ironies of history a much deeper, but incomplete, analysis of the problem of convergence had been given by Bolzano (1817). In modern terminology Bolzano set out to prove that if any infinite set of numbers S has *an* upper bound B, then it possesses a supremum or a least upper bound M such that no member of S exceeds M and that some member of S exceeds any number $M - \varepsilon$ less than M. Bolzano's proof was defective because he had no definition of the concept of a real number, but his method of proof introduced the new and powerful technique of the 'successive dimidiation' of an interval (A_n, B_n), A_n being less than some number of the set S and B_n being some number already identified at the nth stage of approximation as *an* upper bound of S. Then S must possess an upper bound, either in the interval (A_n, C_n) or in the interval (C_n, B_n) where $C_n = (A_n + B_n)/2$. Thus there is an upper bound B_{n+1} in one of these intervals, and $B_{n+1} \leqslant B_n$. All that Bolzano achieved, and it was a substantial achievement, was the construction of a monotone descending sequence of upper bounds B_n.

The need for some clarification of the concept of real *numbers* became apparent when Abel (1826) corrected Cauchy's belief that a sequence of continuous functions must have a continuous limit function, and thus demonstrated that intuition can be a misleading guide in theories of convergence.

2.3. Méray and Infinitesimal 'Variants'

The first satisfactory theory of real numbers was given by Méray (1869) and was achieved, in effect, by defining an 'infinitely small number' to be, not a number, but a *sequence* of rational numbers, $\{v_m\}$, converging to zero. In general Méray considered a 'variant', i.e. a family of rational numbers $\{v_i\}$, indexed by a set of integers, such that, 'however small one may take the positive quantity ε, one ends by having numerically $v_i < \varepsilon$'.

Any sequence of rational numbers $\{v_i\}$ is said to tend to the rational number V as a limit if the difference $\{V - v_i\}$ is infinitely small.

Finally Méray (1887) notes that a convergent variant may not tend to a rational limit, and in this case he says 'we assign to such a variant an *ideal* which we call an incommensurable number or quantity'.

In fact Méray's theory may be summarized by defining a real number to be a family of rational numbers $\{v_i\}$, such that for any tolerance $\varepsilon > 0$ and almost all indices i and j, $|v_i - v_j| < \varepsilon$.

Méray's theory is free from the logical defects which Russell (1903, Chap. XXIV) has noted in the later theories of Weierstrass, Cantor and Dedekind, and it is unfortunate that these writers were ignorant of his pioneer investigation.

2.4. Weierstrass and Monotone Sequences

It appears that Méray's theory did not become very widely known and the arithmetical reform of the geometric theory of real numbers is usually attributed to the theories of Weierstrass, Cantor and Dedekind. The task of submitting these theories to a minute logical analysis has been carried out with devastating results by Russell (1903, Chap. XXIV)[1], and the best service which an historian can now render is to describe the fundamental principles of analysis which a benign interpretation can discover in the wrecks of these theories.

The theory of Weierstrass was given in his unpublished lectures at Berlin in the years 1877–8 and was rapidly disseminated by his students. According to Pincherle (1880) the essence of Weierstrass' theory is the specification of a (positive) real number by the elements of which it is composed, the elements being unity and its aliquot parts. For example, a number expressed in decimal notation as 3·1416 is regarded as composed of 3 units, 1 tenth, 4 hundredths, 1 thousandth etc. In general a positive real number x is specified as composed of elements

$$\frac{1}{m_1}, \frac{1}{m_2}, \ldots, \frac{1}{m_n}, \ldots,$$

---

[1] Russell's account is itself far from satisfactory.

where $m_1, m_2, \ldots, m_n, \ldots$ are equal or unequal integers; the rational approximations to x are the numbers

$$x_n = \frac{1}{m_1} + \frac{1}{m_2} + \ldots + \frac{1}{m_n},$$

and the real number x is defined to be greater than or equal to a rational number c if $x_n \geq c$ for some integer n.

It appears that Weierstrass carefully refrained from *defining* x as the *sum*

$$\sum_{n=1}^{\infty} 1/m_n,$$

but his theory can be compactly expressed in modern terms by saying that the real number x *is* the bounded monotone sequence of rational numbers $(x_1, x_2, \ldots, x_n, x_{n+1}, \ldots)$.

The definitions of fractions and negative numbers required the existence of the elements ε_n and ε' such that $n\varepsilon_n = 1$ and $1 + \varepsilon' = 0$, and complex numbers required the existence of a further element i such that $i^2 + 1 = 0$.

In a careful and painstaking way Weierstrass elaborated a sound theory of the convergence of sequences of real numbers, but two critical observations must be made:

(1) What the theory really defines is not a unique real number, such as π or e, but a number of equivalent representations of a real number, such as

$$e = 1 + 1 + \frac{1}{2!} + \frac{1}{3!} + \ldots \quad \text{and} \quad e = \lim_{n \to \infty} \left(1 + \frac{1}{n}\right)^n.$$

This criterion is easily countered by defining a real number to be a class of equivalent representations, each of which is a monotone sequence, but the resulting theory begins to look somewhat clumsy.

(2) A more searching (but completely unjust) criticism is that the use of monotone sequences is crucially dependent on the fact that the rational numbers form a completely ordered system, i.e. if a and b are any two rational numbers, then either

$$a = b \quad \text{or} \quad a < b \quad \text{or} \quad a > b.$$

The Weierstrassian method of 'completing' the system of rational numbers by the system of real numbers cannot therefore be applied to systems of elements, such as the points of a topological space, which cannot be ordered in this way.

Finally we must mention explicitly that one of the triumphs of the Weierstrassian theory was the formal and rigorous proof of Bolzano's theorem that any bounded set S of real numbers, infinite in number, but not necessarily enumerably infinite, must possess a supremum μ and an infimum λ, real numbers, such that, if a is any number of S, then

$$\lambda \leqslant a \leqslant \mu$$

and if ε is any positive number, then there are members of the set S say x and y, such that

$$\lambda \leqslant x \leqslant \lambda + \varepsilon \quad \text{and} \quad \mu - \varepsilon \leqslant y \leqslant \mu.$$

A critical account of Weierstrass' theory has been given by Cantor (1889a)[1], with which we may fittingly pass on to Cantor's own theory.

2.5. Cantor and Convergent Sequences

The theory which we have attributed to Weierstrass requires the manipulation of both non-increasing and non-decreasing sequences and thereby becomes somewhat cumbrous. A considerable simplification is effected in Cantor's theory (Cantor 1872) at the cost of making Cauchy's concept of convergence fundamental, rather than the concept of bounded monotonic sequences.

Cantor's definition of a 'fundamental sequence' of rational numbers $\{S_n\}$ may be briefly expressed by saying that almost all numbers of the sequence are almost equal, i.e. for any assigned tolerance $\varepsilon > 0$

$$|S_m - S_n| < \varepsilon,$$

for almost all integers m and n, i.e. $|S_m - S_n| < \varepsilon$, except for a finite number of integers m and n. In the customary jargon we say that, given any positive number ε, there exists an integer $\mu(\varepsilon)$ (depending on ε), such that

$$|S_m - S_n| < \varepsilon \quad \text{if} \quad m \text{ and } n > \mu(\varepsilon).$$

A real number is then defined to be a fundamental sequence (Cantor 1883, p. 567) of rational numbers, and the subsequent arithmetic of real numbers follows the same general line as in the theory of Weierstrass. It is then possible to define a convergent sequence of real numbers by Cauchy's criterion and to give a rigorous proof that such a sequence has a limit which is also a real number.

As in the case of Weierstrass' theory, Cantor's definition of a real number is really no more than a definition of the representations of the same real number by different convergent sequences of rational numbers with the same limit. Strictly speaking, we should define a Cantorian real number as a class of equivalent representations, each with the same limit.

Cantor's definition has one great advantage over the definition which we have attributed to Weierstrass: it can easily be generalized to define the convergence of an enumerable sequence $\{P_n\}$ of points in any 'metric space' (Section 9.3), by Cauchy's criterion in which $|P_m - P_n|$ now means the *distance* between the points P_m and P_n.

---

[1] See also Jourdain (1915), Illigens (1889) and Cantor (1889b).

2.6. Dedekind Sections

We have it on the authority of Dedekind himself (Dedekind 1888) that the source of his theory of real numbers was the celebrated definition given in Book V of Euclid's *Elements* on the equality of two ratios. Other mathematicians, such as Bertrand (1867) and Stolz (1885) had recognized the significance of Euclid's definition in geometry, where incommeasurable ratios are presented as objectively existing, as for example, in the ratios of the hypotenuse to the other sides of an isosceles right-angled triangle. But it was left to Dedekind (1872) to provide a purely arithmetical theory of irrational numbers, in which the existence of these numbers is not assumed but demonstrated by actual logical construction. It seems that the same discovery was made independently by Dini (1878), Pasch (1882) and Tannery (1886), but undoubtedly Dedekind enjoys the priority.

From the standpoint of the computer what we need to know about an irrational number x is whether it is greater or less than an assigned rational number r. By meditating on this elementary fact Dedekind was led to formulate the concept of a 'section' (*Schnitt*) of the rational numbers as a pair of classes (A and B) of the rational numbers such that

(1) any rational number belongs to either A or B, but not to both, and
(2) any rational number in A is less than any rational number in B.

If the class A has a supremum which is a rational number r, or if B has an infimum which is a rational number r, then the section (A,B) is a rational real number; otherwise it is an irrational real number.

Algebraical operations on the Dedekind sections can now be defined in a natural and straightforward way, together with a rigorous definition of the convergence of a sequence of real numbers and a proof that a convergent sequence has a unique limit which is also a real number.

The definition of a real number as a Dedekind section has the advantage that, with the trivial exception of rational real numbers, each real number is a unique section, and we thus avoid the multiplicity of representations of a real number which arises in the theories of Weierstrass and Cantor.

It only remains to say that there is complete equivalence between the theories of Weierstrass, Cantor and Dedekind, in the sense that any bounded monotone sequence of rational numbers is convergent, and that any convergent sequence of rational numbers specifies a unique Dedekind section.

When we are working in the field of rational numbers a convergent sequence does not necessarily have a limit, e.g. the sequence of decimals, 1·0, 1·4, 1·41, 1·414, ... which converges to $\sqrt{2}$, has no rational limit. But when we are working in the field of real numbers, then a convergent sequence has a unique limit, itself a real number. Similarly if (A,B) is a Dedekind section of the *real* numbers (with similar properties to a Dedekind section of the rational numbers) then there is a unique real number r, which is either the greatest number in A or the least number in B.

Thus the operations of constructing a convergent sequence or a

Dedekind section are 'closure' operations. No further repetition of these operations can enlarge the field of real numbers.

The real numbers, like the rational numbers, form an ordered system, in the sense that, if x and y are any two real numbers, then either $x > y$ or $x < y$ or $x = y$. Moreover, the real numbers, like the rational numbers, form an 'Archimedean' system, in the sense that if x and y are any two positive real numbers, then, for some positive integer n,

$$nx > y.$$

Hence there is no (infinite) real number greater than all (finite) real numbers and zero is the unique number not greater in absolute value than any real number.

2.7. *Kronecker's Programme*

All the problems of the nature and existence of irrational, fractional and negative numbers were completely swept away in the radical programme of 'arithmetization' inaugurated by L. Kronecker and proclaimed in his notorious paper on the concept of number (Kronecker 1887). The object of Kronecker's reform was to express the whole of analysis explicitly in terms of the natural numbers, 0, 1, 2, ..., and to eliminate all reference to real and complex numbers.

Kronecker's theory is expressed in terms of the ideas of 'congruence' and 'modulus' developed by Gauss (1801). In general, if a, b, m are non-negative integers, then

$$a \equiv b \quad (modulo \ m),$$

if the absolute difference, $|a - b|$, is exactly divisible by m, Similarly

$$a \equiv b \quad (mod \ x, \ y, \ z),$$

if

$$|a - b| = \lambda x + \mu y + \nu z,$$

where λ, μ, ν are non-negative integers.

Kronecker eliminated negative integers by replacing an equation such as

$$7 - 9 = 3 - 5$$

by the congruence

$$7 + 9x = 3 + 5x \quad (mod \ x + 1).$$

The rule for the addition of fractions

$$\frac{a}{m} + \frac{b}{n} = \frac{an + bm}{mn}$$

was transformed into the congruence

$$ax + by = (an + bm)z \quad (mod \ mx - 1, \ ny - 1, \ mnz - 1).$$

Similarly relations connecting complex numbers were purified from the imaginary unit $i = \sqrt{-1}$ by the use of congruences modulo $x^2 + 1$. In all these congruences the 'indeterminates' x, y or z are variables which can take all positive integral values.

By a further display of ingenuity Kronecker was able to evade the explicit introduction of the irrational or complex zeros of a polynomial

$$F(x) = x^n - c_1 x^{n-1} + \ldots \pm c_n,$$

with integral coefficients. The functions $f_1, f_2, \ldots, f_n$ of the n indeterminate integral variables $x_1, x_2, \ldots, x_n$ are defined by the equation

$$G(x) = (x - x_1)(x - x_2) \ldots (x - x_n) = x^n - f_1 x^{n-1} + \ldots \pm f_n,$$

whence it follows that

$$F(x) = G(x) \qquad (\mathrm{mod}\ f_1 - c_1, f_2 - c_2, \ldots, f_n - c_n)$$

and thus $F(x)$ is factorized without any explicit reference to its zeros!

Kronecker's programme of arithmetization remained a mathematical curiosity until it was resuscitated in a somewhat different form by Brouwer (Heyting 1956). But in estimating its significance it must be remembered that he was undoubtedly right in demanding that all analysis should be shown to be reducible to propositions about integers. So much would be conceded by Weierstrass, Cantor and Dedekind. Moreover, in his insistence on the explicit expression of analytical results in terms of integers, he was anticipating the requirements of the electronic computer engaged in numerical analysis. For example, the computer is not interested in the 'exact' real zeros of a function $f(x)$ but only in the determination of sufficiently small intervals, say $h < x < k$ such that $f(h)$ and $f(k)$ have opposite signs, while $f(h)$ and $f(k)$ are each sufficiently small.

But, although analysis must begin and end with integers, the intervening arguments are enormously facilitated by the use of real numbers, complex numbers, convergent series, and the like. Kronecker demanded that throughout the whole process only relations between integers should be explicitly employed. This demand could be met only by cumbrous arguments, whose ingenuity was equalled only by their length.

3. Infinitesimals

3.1. Indivisibles

The possibility of non-Euclidean geometries arises from the fact that the famous 'parallel postulate' of Euclid is independent of the other Euclidean axioms. Now another questionable postulate is that of Archimedes according to which any two segments OA and OB on a straight line are comparable, in the sense that however large is OB and however small is OA, nevertheless, for some integer n,

$$n|OA| > |OB|,$$

where $|OA|$ and $|OB|$ denote the lengths of OA and OB.

The question arises, are there 'infinitesimal' segments OE, such that for all integers n,

$$n|OE| < |OB| \quad ?$$

The 'infinitesimal calculus', proudly so named with its apparatus of 'indivisibles' and 'ultimate ratios', implicitly assumed the existence of such infinitesimal segments and of the corresponding infinitesimal numbers which measured their lengths in terms of some unit. But the absence of any rational theory of infinitesimals rendered even Newton's theory of fluxions vulnerable to the criticism of Berkeley (1734)—criticism which was violently rejected by contemporary mathematicians on the ground that the infinitesimal calculus gave the 'right results'.

With the development of the theory of limiting processes the uncritical use of infinitesimals gradually ceased, although vestiges remain in the notorious 'Calculus Made Easy' S. P. Thompson (1910a).

Applied mathematicians have never ceased to use the convenient symbolism of infinitesimals which eliminates the careful, if tedious, reference to sequences converging to zero, and which seems to find a ready response in uncritical audiences. The very notation for the derivative

$$\mathrm{d}f(x)/\mathrm{d}x$$

or the integral

$$\int_a^b f(x) \, \mathrm{d}x$$

of a function $f(x)$ suggests the practical utility of the 'infinitesimal' $\mathrm{d}x$.

19

3.2. Veronese and Infinitesimal Units

In the system of real numbers, as constructed by Weierstrass, Dedekind
or Cantor, there is no room for infinitesimals. If x and y are any two
positive real numbers and $x < y$, then for some integer n,

$$nx > y.$$

Nevertheless, the question still persists, is there some other system of
numbers in which there are infinitesimals?

An affirmative answer to this question was first given by Veronese
(1891) who constructed a geometric model in which the Archimedean
postulate is not valid.

Consider a set of points $(a, -2)$, $(b, -1)$, $(c, -0)$ on the lines $y = -2$,
$y = -1$, $y = 0$ in the Euclidean plane with Cartesian coordinates (x, y).
Let us specify a relation of *order* such that the point (x_1, y_1) precedes the
point (x_2, y_2) if either

$$y_1 < y_2,$$

or $\qquad\qquad\qquad y_1 = y_2 \quad \text{and} \quad x_1 < x_2,$

and let us define multiplication by the rule

$$n(x, y) = (nx, y)$$

Then $(x, 0)$ is a Veronesean real number, and $(x, -1)$, $(x, -2)$ are
Veronesean infinitesimals, such that

$$n(x, -1) = (nx, -1) < (\xi, 0)$$

and $\qquad\qquad n(x, -2) = (nx, -2) < (\xi, -1)$

for all real ξ and positive n.

Holder (1910) and Schoenflies (1906) have given this theory an algebraic form. In effect they take a Veronesean number (an infinitesimal or
infinity) to be a Laurent power series

$$\sum_{n=+1}^{\infty} c_n \eta^n + \sum_{m=0}^{M} k_m \eta^{-m} = f(\eta)$$

where the coefficients c_n and k_m are real numbers, and the ordering
relation is

$$f(\eta) < g(\eta),$$

if $g(\eta) - f(\eta)$ is positive for all values of η greater than some value $\eta_0 > 0$.

Levi-Civita (1892/3) generalized these numbers by allowing the indices
n and m to be fractional numbers, and a similar extension was given by
Hilbert (1899, p. 22).

In discussing the independence of the axioms of geometry Hilbert was
inevitably interested in non-Archimedean geometries which allow for
infinitesimal and infinite segments. He considered (Hilbert 1899, p. 65)
the domain of all the functions of the real variable t which are obtained
from t by addition, subtraction, multiplication, division and the operation

$f(t) \rightarrow |(1 + t^2)^{\frac{1}{2}}|$. Then as t increases indefinitely the function $f(t)$ is either always positive or always negative or always zero. Hence these functions form an ordered set, in the sense that $f(t) - g(t)$ is either, ultimately, positive, negative or zero. But if n is any positive rational number there are functions $f(t)$ such that $f(t) - n$ is ultimately positive, and so also is $\lambda f(t) - n$ where λ is any positive integer. Hence this system is non-Archimedean.

Hilbert gave a simple construction for infinitesimals when considering the equality of the base angles of an isosceles triangle (Hilbert 1899, Appendix II, p. 194; 1903). Functions of the form

$$f(t) = a_0 t^n + a_1 t^{n+1} + \ldots$$

where the variable t and the coefficients a_0, a_1, ... are real, can be classified as positive or negative according as the leading coefficient a_0 is positive or negative. If

$$f(t) = At \quad \text{and} \quad g(t) = 1 \qquad (A > 0),$$

then
$$f(t) - g(t) = -1 + At$$

and therefore $f(t) - g(t)$ is classified as negative and in this sense $At < 1$, however large A may be. We thus have a non-Archimedean set.

3.3. Orders of Infinity

Non-Archimedean numbers also figure in the various schemes for measuring and comparing the rate of growth of functions of a real variable $f(x)$ as the argument tends to infinity.

Thus Thomae (1880) represents the 'order of infinity' of

$$f(x) = x^\alpha (\log x)^{\alpha_1} (\log \log x)^{\alpha_2} \ldots$$

by the symbol

$$\alpha + \alpha_1 l_1 + \alpha_2 l_2 + \ldots$$

where the letters l_1, l_2, ... are new (infinite) units of a non-Archimedean algebra, such that

$$n l_2 < l_1, \quad n l_1 < 1, \quad \text{etc.}$$

for all positive integers n.

A systematic calculus of orders of infinity was developed by du Bois-Reymond (du Bois-Reymond 1882, Hardy 1910) based on a sequence of positive, continuous and 'steadily increasing' functions $\{\varphi_n(x)\}$, $n = 1$, $2, \ldots$, such that

$$\varphi_{n+1}(x)/\varphi_n(x) \rightarrow \infty \quad \text{as} \quad x \rightarrow \infty.$$

The main theorem in this calculus proves the existence of a function $f(x)$ which grows more rapidly than any function $\varphi_n(x)$ in the sequence.

3.4. Hahn's Transfinite System of Non-Archimedean Numbers

The most exhaustive investigation of infinitesimals and infinite numbers is that due to Hahn (1907), who generalized and extended the researches of Levi-Civita by showing that a system of non-Archimedean numbers can be associated with any aggregate which is simply ordered in accordance with Cantor's definition, and that such a system is the most general non-Archimedean system possible.

In Hahn's system any non-Archimedean number can be expressed in the form

$$A = a_0 e_0 + a_1 e_1 + \ldots + a_\omega e_\omega + a_{\omega+1} e_{\omega+1} + \ldots$$

where the suffixes $0, 1, \ldots, \omega, \omega + 1, \ldots$ form a complete set of finite and transfinite integers, the coefficients $a_0, a_1, \ldots$ being real numbers and the symbols $e_0, e_1 \ldots$ representing the non-Archimedean units such that

$$(a_\alpha e_\alpha)(a_\beta e_\beta) = (a_\alpha a_\beta)\, e_{\alpha+\beta},$$

with the natural law for addition as in linear algebra, and the ordering relation that, if $m\beta > n\alpha$ for all positive integers m and n, then $m e_\beta > n e_\alpha$.

3.5. Skolem and the Proper Extension of the Natural Number System

The topic of infinitesimals remained a mathematical curiosity after the work of Hahn (1907) until 1929 when Thorald Skolem showed that the system of natural numbers was incomplete in the sense that it has a 'proper extension' into a non-Archimedean system. This result is implicit in the theory of infinitesimals developed by Veronese and Levi-Civita, which can be regarded for the present purpose as an Abelian group with one generator j and the order relation generated by the laws

$$x_1 + y_1 j < (x_1 + 1) + y_2 j,$$

$$x_1 + y_1 j < x_1 + (y_1 + 1)j,$$

$$j^n < j^{n+1},$$

for all integers x_1, y_1, y_2, n.

The natural numbers correspond to the sub-set of 'Veronesean integers', $x + 0j$, between which are intercalated the non-standard integers $x + yj$.

3.6. A. Robinson and Non-standard Analysis

A very general method of constructing a 'non-standard analysis' containing infinitesimals and infinite numbers was developed by A. Robinson (1966). If K is the set of all sentences in a formal language L which can express statements about the field of real numbers R, then it follows from standard results of Predicate Logic that there exists a proper extension $^\dagger R$ of R, such that all sentences of K are true also in $^\dagger R$, in a certain restricted sense. Corresponding to the set N of natural numbers in R there is set $^\dagger N$ in $^\dagger R$ with the 'same properties' as N. $^\dagger R$ is a non-Archimedean ordered field with non-zero infinitesimals η such that $|\eta| < r$ for *all* standard positive r, however small. If x is any finite non-standard number such that $|x| < r$ for some standard r, then there is a unique standard real number n, called the standard part of x and such that $n - x$ is infinitesimal.

3.7. Luxembourg and Ultrapowers

The attempts made by Schmieden and Laugwitz (Schmieden and Laugwitz 1958; Laugwitz 1959, 1961) to construct a calculus of infinite numbers in which the elements were sequences of rational numbers $x = \{x_1, x_2, \ldots\}$ suffered from the existence of 'divisions of zero', such as

$$a = \{1, 0, 1, 0, \ldots\} \quad \text{and} \quad b = \{0, 1, 0, 1, \ldots\},$$

for which $ab = \{a_1 b_1, a_2 b_2, \ldots\} = \{0, 0, \ldots\}$. This defect was remedied in the non-standard model of analysis constructed by Luxembourg (1962) in the form of an 'ultrapower'.

As in the theory of Schmieden and Laugwitz, Luxembourg operates in the space R^∞, the points of which consist of sequences $\{x_n\}$ of real numbers x_n, and in which a sequence $\{c_n\}$ in which $c_n = c$ for all n represents the real number c. But whereas the German mathematicians defined relations between the elements of the space R^∞ relative to the Frechet filter on N (the space of the positive integers), Luxembourg employs a free ultrafilter U on N, and defines two elements $A = \{a_n\}$ and $B = \{b_n\}$ of R^∞ to be equivalent modulo U if the set of integers $\{n\}$ for which $a_n = b_n$ belongs to U. The elements of the non-standard system $^\dagger R$ are then defined to be classes of elements of R^∞ which are equivalent modulo U.

The system $^\dagger R$ is then proved to be a totally ordered non-Archimedean extension of R with no divisors of zero.

3.8. The Rehabilitation of Infinitesimals

The naive eighteenth-century use of infinitesimals can now be given an exact and rigorous formulation. Thus a function $f(x)$ of the non-standard variable x is continuous at $x = x_0$ if $f(x_0 + \eta) - f(x_0)$ is infinitesimal for all infinitesimal η, and $f(x)$ is differentiable with derivative $f'(x)$ if $f'(x)$ is the standard part of

$$[f(x + \eta) - f(x)]/\eta.$$

Again the Riemann integral of $f(x)$ over $[0, 1]$ is the standard part of

$$\frac{1}{\omega} \sum_{k=0}^{\omega-1} f(t_k)$$

where ω is a positive infinite non-standard number and

$$\frac{k-1}{\omega} \le t_k \le \frac{k}{\omega}.$$

If $\gamma(x)$ is a non-negative, continuous, standard function such that

$$\int_{-\infty}^{\infty} \gamma(x)\ \mathrm{d}x = 1$$

then $\gamma(\omega x)$ is the Dirac delta function $\delta(x)$.

Luxembourg also gives interesting proofs of the Heine–Borel covering theorem and the Hahn–Banach extension theorem (Luxembourg 1962).

4. Cantor and Transfinite Numbers

4.1. Cantor and the Hierarchy of Convergent Sequences

In the theory of Fourier series we may consider the series

$$(F) \quad \frac{1}{2}a_0 + \sum_{n=1}^{\infty} (a_n \cos nx + b_n \sin x)$$

generated by a prescribed function $f(x)$, with coefficients

$$a_n = \frac{1}{\pi} \int_0^{2\pi} f(x) \cos nx \, dx, \qquad b_n = \frac{1}{\pi} \int_0^{2\pi} f(x) \sin nx \, dx.$$

Alternatively we may consider a series, such as F, with any prescribed coefficients, and study its convergence to a limit function. If two such series converge to the same function at each point x in $(0, 2\pi)$ then their difference must converge to the null function $f(x) = 0$. Cantor's researches on Fourier series in 1870–2 considered series such as F which converged to the null function at each point x in $(0, 2\pi)$ apart from a set S of exceptional points (Jourdain 1915, Cavaillès 1932).

This investigation forced upon Cantor the necessity for a rigorous theory of irrational numbers, especially when he came to deal with the case in which the set S contains infinitely many points.

Cantor proceeds by establishing a hierarchy of sequences. The collection (or set) A of fundamental sequences are those sequences of rational numbers $a = \{a_n\}$ ($n = 1, 2, \ldots$), which converge in accordance with Cauchy's criterion that, for each assigned tolerance $\varepsilon > 0$, $|a_p - a_q| < \varepsilon$ for almost all integers p and q. The next rank B of sequences are convergent sequences whose elements are the fundamental sequences $\{a\}$ of the collection A, convergence being now defined by a suitable extension of Cauchy's criterion. Similarly Cantor then constructs a higher rank of sequences C from the family B of sequences, and continues to define yet higher ranking families D, E, ...

Any two fundamental sequences $x = \{x_n\}$, $y = \{y_n\}$ are equivalent, in modern terminology, if, for each assigned tolerance $\varepsilon > 0$, $|x_n - y_n| < \varepsilon$ for 'almost all'[1] integers n. A collection of equivalent fundamental sequences is an element of the set of sequences B, and these elements are what we should now call real numbers, i.e. a real number is a collection of equivalent convergent fundamental sequences of rational numbers.

---

[1] See section 2.5.

The fact that this definition was not given explicitly by Cantor seems to be due to his obsession with the unending character of the successive ranks of the hierarchy of sequences A, B, ...

4.2. Cantor and Derived Sets

By analogy with his generation of this system of sequences A, B, . . . Cantor showed how to generate a system of derived sets of points from any given set.

The fundamental concept is that any sequence in any one of the Cantorean hierarchies A, B, . . ., determines a point on a straight line, and, conversely, that any such point is determined by such a sequence.

By the 'neighbourhood' of a point P Cantor means any interval I of which P is an *inner* point, i.e. does not coincide with either of the extremities of I. This is the germ of the later definition (Sections 9.4, 9.5) of a neighbourhood of P as an 'open' set containing P.

A limit point of a set of points $\{P\}$ is a point Q such that each neighbourhood of Q contains an infinite number of points of the set $\{P\}$. By the Bolzano–Weierstrass theorem, any bounded infinite set of points must possess at least one limit point.

Hence any given set of points $\{P\}$ possesses a possibly empty set of limit points $\{Q\}$, which Cantor (1872) called the first 'derived set' of $\{P\}$. And just as he was able to generate a hierarchy of sequences A, B, . . . so now by analogy he was able to define inductively the nth derived set of $\{P\}$.

It must be sadly admitted that this analogy which guided Cantor is in fact unnecessary. The definitions of a neighbourhood, of limit points and of successive derived sets are independent of the concept of the Cantorean hierarchy, but they are of permanent value and form the basis of point-set topology.

Cantor generalized these concepts for points in n-dimensional space, in which a point is represented by an ordered set of n real numbers $(x_1 \ x_2 \ . . ., x_n)$, and in which an interval is replaced by a ball or solid sphere so that a neighbourhood of the point $(a_1 \ a_2 \ . . ., a_n)$ is the set of points $(x_1, x_2 \ . . ., x_n)$ such that

$$(x_1 - a_1)^2 + (x_2 - a_2)^2 + \ldots + (x_n - a_n)^2 < \varepsilon,$$

for some positive number ε.

4.3. The Anatomy of Point Sets

The limit points of a given set $\{P\}$ may, or may not, be points of this set, and, if the set contains only a finite number of points, the limit points do not exist.

By comparing a given set of points E with its first derived set E_1, Cantor introduced a number of fertile concepts which are fundamental in modern analysis and topology. It is convenient to use the symbol $\subset$, invented by Peano (1889)[1], and the notation $A \subset B$, to denote that the set A is contained in the set B. Then Cantor describes the set E as

(1) closed, if $E_1 \subset E$,
(2) dense in itself, if $E \subset E_1$,
(3) perfect if $E_1 \subset E \subset E_1$, i.e. if $E_1 = E$ (Cantor 1872),
(4) dense in another set F, if $F \subset E_1$.

4.4. *Derived Sets of Transfinite Rank*

From any given set of points E_0 Cantor constructed in succession, the derived set E_1, the derived set E_2 of the points in E_1, and so on, and hence was led to classify sets of points into two classes:

(1) Sets E_0 whose derivative E_n of some finite rank n consists of a finite number of points, and for which all the succeeding derivatives E_{n+1}, E_{n+2} ... are empty.
(2) Sets E_0 such that each derived set E_n consists of an infinite number of points.

In considering the hierarchy of convergent sequences Cantor had envisaged the existence of convergent sequences of infinite rank, but in considering successive derived sets he went far beyond this infinitely distant horizon, and considered the set of points which are common to all the derived sets $\{E_n\}$ for all values of n. This set he denoted by E_ω and called it the first derived set of transfinite order. By a repetition of this infinite process he introduced the first transfinite derived set of E_ω, denoted by $E_{2\omega}$, the second derived set $E_{3\omega}$, and then, in succession the set E_{ω^2} which consists of all the points common to $\{E_{n\omega}\}$ for all values of n, and the set E_{ω^ω} with the points common to E_ω, E_{ω^2}, E_{ω^3} ... .

Cantor himself left open the question whether this process terminated in a derived set which was empty, but this question was answered by Bendixson (1883) who showed that if the first derived set of E_0 is not finite nor enumerably infinite, then the first transfinite derived set E_ω is perfect and the complement of E_ω with respect to E_0 is enumerable.

4.5. *The Lebesgue Chain*

In his researches on the definition of the length of a curve, H. Lebesgue (1928) was led to consider a method of covering an interval which illustrates the concept of transfinite numbers.

If the curve is specified by the parametric equations

$$x = x(t), \; y = y(t), \; z = z(t),$$

where the functions $x(t)$, $y(t)$, $z(t)$ have derivatives $x'(t)$, $y'(t)$, $z'(t)$, then, starting at any point with parameter t_0, we can determine an interval (t_0, t_1) such that the chord length

$$\{[x(t_1) - x(t_0)]^2 + [y(t_1) - y(t_0)]^2 + [z(t_1) - z(t_0)]^2\}^{\frac{1}{2}}$$

---

[1] To be accurate Peano used the symbolism $A \subset B$ to denote that A is deduced from B, and the symbol $\in$ to denote the relation of an individual to its class.

differs from

$$(t_1 - t_0) \, [x'^2(t_0) + y'^2(t_0) + z'^2(t_0)]^{\frac{1}{2}}$$

by less than any prescribed tolerance $\varepsilon(t_1 - t_0)$. Suppose we require the arc length from $t = a$ to $t = b$, we can trace such a chord from the point $t_0 = a$ to a point t_1, and thence trace another chord from t_1 to t_2, and so on, until we reach a given point $t = b$ or until we reach a limit point t_ω, corresponding to the first transfinite number ω. We can then start again and proceed from t_ω to $t_{2\omega}$, and, if necessary, from $t_{2\omega}$ to $t_{3\omega}$, and thus in succession through the sequence of transfinite numbers, until, it is hoped, we reach the point b.

Cantor had already noted that if a finite (or indeed an infinite) interval is covered by a non-overlapping collection of intervals, then such a collection must be enumerable, and this result was asserted by H. Lebesgue to prove that for some transfinite number α we shall have $t_\alpha = b$ and thus succeed in covering the interval $[a, b]$.

The collection of intervals considered by Lebesgue is such that to each point s of a closed interval $[a, b]$ there corresponds a sub-interval $[s, t)$ such that $s \leqslant t < b$. A covering of the interval $[a, b]$ by a collection of such non-overlapping sub-intervals is now called a 'Lebesgue chain'.

A formal proof that a finite interval can be covered by a Lebesgue chain of intervals using only transfinite numbers of Cantor's second rank has been given by W. H. Young (1910).

Pal (1912) has shown that the existence of a Lebesque chain can be established without the use of transfinite numbers.

W. H. Young and G. C. Young (1915) have also established a number of related covering theorems without the introduction of Cantor's numbers. These theorems seem adequate to prove all that has been effected by Lebesgue's original theorem and even to yield further results.

4.6. The Abstract Theory of Transfinite Ordinal Numbers

Although the Lebesgue chain was devised some twenty years after Cantor had invented his theory of transfinite ordinal numbers, we have described H. Lebesgue's concept first as it gives an objective reality to the more shadowy concepts of Cantor's fertile imagination.

The essential feature of the Lebesgue chain is a mapping f of the unit interval $0 \leqslant x < 1$ into itself such that

$$0 < f(x) \leqslant 1 \quad \text{and} \quad x < f(x)$$

so that any point s is the left-hand end-point of a half-open interval

$$s \leqslant x < f(x)$$

Cantor's ordinal numbers provide a convenient notation to indicate the order in which the intervals of the Lebesgue chain are successively generated.

The three principles enunciated by Cantor for the construction of ordinal numbers correspond to:

(1) The simple iteration of the mapping process f, which successively generates the points

$$f_1 = f(0), f_2 = f(f_1), f_3 = f(f_2), \ldots, f_{n+1} = f(f_n), \ldots.$$

corresponding to the natural numbers

$$1, 2, 3, \ldots, n, n+1, \ldots$$

(2) The introduction of the first limit point

$$f_\omega = \lim f_n \quad \text{as} \quad n \to \infty,$$

of the monotonic, increasing sequence $\{f_n\}$.

By a further application of the first principle Cantor generates the sequence

$$f_{\omega+1} = f(f_\omega), f_{\omega+2} = f(f_{\omega+1}), \ldots, f_{\omega+n+1} = f(f_{\omega+n}), \ldots$$

The second principle then generates the second limit point

$$f_{2\omega} = \lim f_{\omega+n} \quad \text{as} \quad n \to \infty.$$

Successive applications of these two principles yield a whole hierarchy of points

$$f_{3\omega} = \lim f_{2\omega+n}, \quad \text{as} \quad n \to \infty$$

and so on;

and

$$f_{\omega^2} = \lim (f_{n\omega}) \quad \text{as} \quad n \to \infty.$$

Further iteration also yields

$$f_{\omega^3} = \lim (f_{n\omega^2}), f_{\omega^4} = \lim (f_{n\omega^3}), \ldots,$$

and so on, and even

$$f_{\omega^\omega} = \lim (f_{\omega^n}).$$

However, at this stage Cantor applies a guillotine—his third principle—which is really a theorem to the effect that the aggregate of points x generated by applications of the first and second principles, and such that $0 \leqslant n < f_{\omega^\omega}$ is enumerably infinite. It is upon this theorem that Lebesgue relied in asserting that the unit interval could be covered by a chain of intervals identified by transfinite numbers of the first class (1, 2, ..., n, ...) or of the second class (ω, $\omega + 1$, ..., ω^2, ..., ω^ω).

The symbols ω, ω^2, ..., ω^n, ..., ω^ω, ... are the transfinite ordinal numbers, and the process of generating the Lebesgue chain of intervals provides an existence theorem guaranteeing the absence of any contradiction in the invention of these numbers, provided that we are satisfied that there is no contradiction in the definition of the real numbers, and, of course, of the rational numbers and the integers.

A critical examination of the definition of the integers had to await the researches of Frege, Peano and Russell, for Cantor relied on our intuitive concepts of the 'natural numbers' and of the idea of an 'aggregate', 'set' or 'collection'. The theory of ordinal numbers could then be abstracted from the theory of sets of points in an interval by introducing the concept of a 'simply ordered aggregate' M as a set of distinct elements $\{x, y, \ldots\}$, between any pair of which there existed a relation (denoted by the symbol $<$) such that, either $x < y$ or $y < x$. The use of the usual symbol, $<$, associated with the idea of 'less than', is a concession to human weakness. All that is implied by this symbol now is that the corresponding relation is asymmetrical and transitive, i.e. that either

(1) $x < y$ or $y < x$ for any pair x, y, and
(2) if $x < y$ and $y < z$, then $x < z$.

Two 'simply ordered aggregates' M and N were described as 'similar' if

(1) there was a 1–1 correspondence between the elements $\{m\}$ of M and $\{n\}$ of N such that
(2) if m_1, $m_2 \in M$ and $m_1 < m_2$ and if m_1, m_2 correspond to n_1, $n_2 \in N$ then $n_1 < n_2$.

The relation of similarity is clearly reflexive, symmetrical and transitive. The most important species of simply ordered aggregates are those which are 'well-ordered'. An aggregate M is 'well-ordered' if every sub-aggregate of M has a first term, when ordered by the same order relation as M itself. Cantor defined the 'order-type' or ordinal number of a well-ordered aggregate M as the characteristic of the class of similar aggregates—an evocative description rather than a definition as the word 'characteristic' requires clarification.

There followed an arithmetic of ordinal numbers, in which addition and multiplication are not necessarily commutative. But of much greater interest are the three distinct order types described by Cantor:

(1) The order type η, of the rational numbers r such that $0 < r < 1$ in which there is no lowest nor highest term, but the set is everywhere 'dense'. Cantor showed that the aggregate of all algebraic numbers can be well-ordered and that it has the order type η.
(2) The order type θ of the real numbers x, such that $0 \leqslant x \leqslant 1$. This aggregate is perfect and between any two terms of θ there always lies a rational number.
(3) The order type π of the negative and positive integers (and zero) in their natural order.

Cantor proved that these three aggregates are distinct, i.e. they are not 'similar'. He also thought that he had proved that any aggregate can be well ordered but, in fact, this theorem requires and is equivalent to the 'multiplicative axiom', which asserts that if M is an infinite collection of aggregates A, B, ..., then there is an aggregate which contains one term from each of the aggregates A, B, ...

4.7. The Theory of Cardinal Numbers

The theory of ordinal numbers is applicable only to aggregates in which a relation of order has been established, but there is also a theory of numeration applicable to any aggregate. This theory, which is more abstract and therefore simpler than the theory of ordinal numbers, was in fact developed by Cantor after this latter (Cantor 1895, 1897) and is called the theory of 'cardinal numbers'.

Cantor did not give a formal definition of cardinal numbers but described a cardinal number, or 'power', as a characteristic of a class of 'equivalent' aggregates, two aggregates M and N being equivalent if there is a 1–1 correspondence between their elements, irrespective of any concept of order. He was then able to give the first positive definition of an *infinite* aggregate M as an aggregate which is equivalent to a part of itself. Thus the aggregate of natural numbers, 1, 2, 3, ..., is equivalent to the aggregate of odd numbers 1, 3, 5, ...

From the description of cardinal numbers Cantor was able to develop their arithmetic, describing, with no pedantic care for precision, the notions of addition, multiplication and exponentiation. The cardinal number of the aggregate of natural numbers Cantor denoted by the Hebrew letter 'Aleph', with suffix zero, $\aleph_0$, and easily established the startling results that

$$\aleph_0 + 1 = \aleph_0,\ 2\aleph_0 = \aleph_0,$$

and

$$\aleph_0 \times \aleph_0 = \aleph_0.$$

He showed that the class of rational numbers, and even the class of algebraic numbers has the cardinal number $\aleph_0$. But more remarkable was his proof that the class of all real numbers has a cardinal number c such that

$$c = 2^{\aleph_0} = n^{\aleph_0} \quad \text{for} \quad n = 2, 3, \ldots$$

and

$$c^{\aleph_0} = c^n = \aleph_0^{\aleph_0} = nc = c,$$

whence it follows that the class of points with real coordinates in Euclidean space of n dimensions has the cardinal number c.

4.8. Well-ordered Sets

A collection of elements, x, y, z, ..., is 'simply ordered' if there is a binary relation, denoted, say, by $<$, such that

(1) if x and y are distinct, then either $x < y$ or $y < x$,
(2) if $x < y$ and $y < z$, then $x < z$.

Then, if $x < y$, we say that x precedes y.

A simply-ordered collection is 'well-ordered' if every non-empty subset S has a first element, which precedes all other elements of S.

The most important application of well-ordered sets is to 'transfinite induction'. Well-ordered sets possess a first term which precedes all the others. A property P of the terms of a sets is 'inductive', if it is true of a term x when it is true of every term preceding x. If P is an inductive property, and if P is true of the first term, then it is true of every term of the sets. This is proved by a *reductio ad absurdum* and the proof is therefore rejected by the intuitionists (Chapter 16).

But for Cantor the main interest of well-ordered sets was that they provided the material for constructing a hierarchy of cardinal numbers.

In developing the arithmetic of ordinal numbers Cantor had established their comparability, i.e. he had proved that if α and β are any two ordinals, then either $\alpha > \beta$, or $\alpha = \beta$, or $\alpha < \beta$. From this he deduced that the ordinals, in order of magnitude, themselves form a well-ordered series.

We have already (Section 4.7) introduced $\aleph_0$, the cardinal of the aggregate of natural numbers 1, 2, ... Now consider the aggregate of all ordinals β such that all the ordinals less than β form a set of ordinal number $\aleph_0$. This set is well ordered and its cardinal is denoted by $\aleph_1$. Similarly Cantor introduced the sequence of Alephs, $\aleph_n$, $n = 1, 2, 3, \ldots$.

When, however, we consider cardinal numbers in general, i.e. without restricting them to the characterization of well-ordered sets, we enter a different region of the transfinite theory, in which hypotheses take the place of theorems.

Cantor conjectured that the cardinal number, c, of the class of all real numbers, was equal to $\aleph_1$, but this plausible assertion has remained unproved. This is the famous 'continuum hypothesis'.

Cantor asserted the comparability of any two cardinal numbers, but he gave no proof of this assertion. This gap in his theory was partly filled by Schröder (1896) and Bernstein (1905)[1] who proved independently that if A and B are any two cardinals and if $A \leqslant B$ and if $B \leqslant A$, then $A = B$. The difficulty in the proof is that the relation $A \leqslant B$ merely signifies that there is a 1–1 correspondence between the elements of A and the elements of a part B_1 of B.

Cantor also considered that any aggregate can be made into a well-ordered aggregate by the construction of an appropriate ordering relation, but this hypothesis remains a challenge to posterity, perhaps more fruitful than many theorems of transfinite arithmetic.

The incompleteness of Cantor's theory should be regarded as a merit rather than a demerit since it has served as a stimulus to subsequent mathematicians, and the same must be asserted of an internal inconsistency in the theory—the famous paradox published to the world by Burali-Forti in 1897, but discovered by Cantor himself in 1895 and communicated privately by him to Hilbert in 1896 and to Dedekind in 1899.

4.9. The Paradoxes of Infinite Set Theory

For Hilbert the Cantorean theory of transfinite numbers was a mathematical paradise, from which he refused to be expelled. But the paradise contains serpents and tempters with which we must now engage in combat.

The first three adversaries encountered by paradisal mathematicians are the antinomies of Burali-Forti (1897), of Frege and Russell (1902),[2] and the ancient Greek fallacy called 'Liar' (Bochenski 1961).

---

[1] See also Hessenberg (1909).

[2] Russell's devastating discovery of the paradox of set theory and the 'consternation' it caused to Frege are conveniently accessible in Heinenoort's source book (pp. 125–128) (see Bibliography).

If, with Burali-Forti, we consider the totality of all ordinal numbers, then it appears that this aggregate is well-ordered and hence has itself an ordinal number N, which must have a successor $N + 1$, although all the ordinal numbers have already been enumerated by the ordinal number N.

Similarly, if with Zermelo and Russell, we consider the collection C of all classes which taken one by one are not members of themselves, then it appears that we can prove C both to be and not to be a member of itself.

And it is salutary to remember that Philetes (340–285 BC) died because he could not resolve the paradox of the man who said 'I am lying', for if this statement is true, it is false, and, if it is false, it is true.

Indeed there are a round dozen of genuinely different puzzles of this description (Fraenkel, Bar-Hillel and Levy 1958), some of them involving only terms of mathematics or logic, such as class and number, and some referring to non-logical terms, such as thought, language or symbolism. Among the second class are the paradoxes of the 'Liar', of the least integer not in fewer than nineteen syllables (Berry 1906). These paradoxes may well be due to faulty ideas concerning thought or language and not to faulty logic or mathematics, and we therefore do no more than make this brief reference to the semantic and linguistic antinomies.

According to Russell, all these paradoxes result from a certain kind of vicious circle (Russell 1908; Whitehead and Russell 1910, Vol. 1, Chap. 2) which arises from 'supposing that a collection of objects may contain members which can only be defined by means of the collection as a whole', as, for example, if we speak of the cardinal number of all cardinal numbers.

To eliminate such paradoxes Russel invented the 'theory of types'. This theory requires the most careful enunciation, and the exposition of Russell and Whitehead is ineluctably lengthy and complex. Furthermore, in order to legitimate a great mass of reasoning, it seems necessary to supplement the theory of types by an 'axiom of reducibility', which is not at all evident and which naturally is not proved.

Russell revised his theory of types many times, but the simplest account is undoubtedly the first, which is presented in a single paragraph of *The principles of mathematics* (Russell 1903, Section 104).

Certain sets may themselves become elements of a 'higher' class, and thus we may generate a hierarchy of sets of different types. But the detailed and systematic development of this naïve theory of types becomes so cumbersome and onerous that it required an 'axiom of reducibility' to legitimize what Bourbaki might call the 'abuse of language and logic' which characterizes ordinary mathematics.

5. Finite and Infinite Numbers

5.1. The Foundations of Arithmetic

The criticism of the foundations of geometry began early in the nineteenth century with the critical study of Euclid's axioms, and the examination of the nature of real numbers began in 1869 with the work of Méray, but the much more fundamental study of the rational numbers and the integers was commenced even later.

The serious critical study of the theory of the 'natural numbers', i.e. the integers, begins with a fragmentary writing of Dedekind not published until after his death, and therefore unknown until 1932 (Dedekind 1932). Priority of publication undoubtedly belongs to Frege (1884) and to Peano (1889), who was, however, ignorant of the writings of Frege. Russell (1903, p. xviii, lines 21–24) had not made a 'thorough study' of the work of Frege. The work of the historian in the genealogy of ideas is accordingly much simplified.

5.2. Frege and the Class Concept of Natural Numbers

In the naïve arithmetic, as it was taught in schools until recently, the concepts of number, of counting, of addition and multiplication were acquired experimentally, and, for university students, these basic ideas were supplemented by the principle of mathematical induction, usually presented as a self-evident truth.

Frege (1884) found this uncritical and dogmatic teaching of matters which should be accessible to reason unsatisfactory and in sharp contrast to contemporary work in mathematics which showed a tendency to rigour in proof and sharpness in definition. Nor did he find much help in the work of philosophers such as Mill and Kant or of logicians such as Schröder or Jevons. In a pleasantly discursive style he reviews earlier attempts to elucidate the fundamentals of arithmetic and finally proposes the following definition: 'the number which belongs to a concept F is the extension of the concept "equal to the concept F"', it being understood that the concept F is "equal" to the concept G when there is a one-to-one correlation of the objects which fall under F and under G'.

The number 0 is then defined as the number which belongs to the concept 'not identical with itself', and the number 1 to the concept 'identical with 0'. The subsequent numbers are defined by an analysis of counting and of the idea of succession expressed by the definition that, if n is the number of a concept F, if x is an instance of F, and if m is the

34

number of the concept 'falling under F but not identical with x', then n is the successor of m in the series of natural numbers. Finally the infinite number, ∞, is defined as the number which belongs to the concept 'finite number'.

Frege was justly proud of his discoveries and thought that he had demonstrated conclusively that the fundamental truths of arithmetic were universally valid logical principles. However, Frege's arithmetic does involve the notion of a class, or set, or aggregate, for in effect he defines a number to be the collection of classes equivalent (in the sense of one-to-one correlation) to one another.

But when Russell made his 'thorough study' of Frege's work there was just one point where he encountered a difficulty, namely in the concept of a class—for the idea of a collection of classes which do not belong to themselves seems to involve an inherent contradiction and to invalidate the concept of a class! Frege frankly accepted this criticism, and endeavoured to meet it by a restriction on the general concept of a class, but this attempt has not seemed to be satisfactory. The beginning of the correspondence between Frege and Russell is given in Heijenoort's valuable source book (1967) (see Bibliography).

5.3. Peano's Axioms

The contribution of Peano to the elucidation of the foundations of arithmetic is in sharp contrast to the work of Frege. Frege had attempted, with considerable success, to give a constructive definition of the natural numbers and a proof of the validity of mathematical induction, but Peano gave no such definition or proof. Instead he took the concept of number to be an indefinable—except in-so-far as it was characterized by a few axioms. This drastic simplification has contributed substantially to the popularity of his system and explains Zermelo's remark (Zermelo 1908, p. 115) that Peano was primarily interested in systematization rather than research.

In Peano's system (Peano 1889, 1895–1903) there is an undefined invertible operation which acting on any (natural) number x produces its (unique) successor $y = x + 1$. Unity is defined as the unique number which is not the successor of any other number. The principle of mathematical induction is expressed in the compact assertion that if C is any set of numbers such that

(1) $1 \in C$

and (2) *if* $x \in C$ *then* $x + 1 \in C$

then C contains *all* the natural numbers.

On the basis of his few axioms Peano was able to define addition and multiplication and thus to establish arithmetic on a simple and solid basis. Rational numbers were introduced in the form (p/q) where p and q are natural numbers, with the definitions

$$(p/1) = p \quad \text{and} \quad m(p/q) = (mp/q)$$

for all natural numbers m.

According to Jourdain (1912), who wrote with the approval of Peano, the latter made use of Grassmann's *Lehrbuch* (Grassmann 1861) in which the method of proof by recurrence was first used systematically, and was also indebted to Dedekind (1888). But in Dedekind's book the principle of mathematical induction is established by means of his theory of chains, whereas in Peano's work it has become an axiom.

But the most serious objection to Peano's system is that it provides no means of counting objects in the real world. It is true that the natural numbers which we employ in describing the size of a flock of sheep or the position of a house in a street do satisfy Peano's axioms—but so do the same numbers when each is increased by the same number, say one dozen. In fact Peano's axioms do not distinguish between the progression 1, 2, 3, ... and the progression 13, 14, 15, ... For a satisfactory definition of unity which does make such a distinction we must return to Frege or progress to Russell, whose work is described below (Section 16.4).

The same objection can also be argued against Dedekind's system, which fails similarly to identify the 'base-element' of a chain with the number 'one' of the real world—an objection of which Dedekind was well aware.

5.4. Dedekind's Theory of Chains

When, after many years of thought, Dedekind published (1888) his theory of ordinal numbers, in effect he stood the whole theory on its head by making the theory of finite ordinals subordinate to the theory of infinite ordinals. There was a good reason for this revolutionary change, for Cantor had already shown that it was possible to give a positive characterization of infinite numbers. An infinite set had been defined as a set which could be put into a one-to-one relation with a proper part, and inevitably a finite set could only be defined by the impossibility of such a relation. Thus the set of *all* natural numbers, 1, 2, 3, ..., is in one-to-one relation with the set of all even numbers, 2, 4, 6, ..., and also with the truncated set, 2, 3, 4, ..., but the set of numbers, 1, 2, 3, 4, 5, 6, 7 cannot be put into a one-to-one relation with any proper part such as 1, 3, 5, 7.

Dedekind's theory of ordinal numbers applies both to finite and to infinite ordinals, and is therefore wider in scope than Cantor's theory, which applies only to infinite ordinals and presupposes the properties of finite ordinals. For this reason Dedekind's theory is closely related to Peano's theory (Section 5.3) of finite integers, but it is conveniently described here because of its great influence on the mathematical logicians who have critically examined the axiom of choice.

Dedekind's theory is expounded in a lengthy memoir (Dedekind 1872, 1888) and in a letter to Keferstein in 1890[1], which reveals the train of thought which led Dedekind in the development of his ideas.

In this letter Dedekind analyses the familiar concept of natural numbers into the following properties. The natural numbers N form a

---

[1] Heijenoort's source book pp. 98–103 (see Bibliography).

system of elements 1, 2, 3, ..., n, ... in which there is an ordering relation $\varphi(n)$ which correlates each number n with its successor $\varphi(n)$ usually denoted by $n + 1$. Addition being yet undefined, the ordering relation is regarded simply as a mapping, $n \rightarrow \varphi(n)$, of the system N into itself, and, since the number 1 is not the successor of any other number, the set $\varphi(N)$ must be a proper part of the set N, i.e. $\varphi(N) \subset N$.

It would be tempting to define the natural numbers, one by one, as $2 = \varphi(1)$, $3 = \varphi(2)$ and so on, but Dedekind rejects this seductive invitation, precisely because the phrase 'and so on' clamours for analysis and definition, and introduces the concept of a 'chain'.

A set K of elements of a system S is called a 'chain' relative to a mapping φ if $\varphi(K) \subset K$, and we can consider all the chains which include a prescribed element a of S. The intersection of all these chains is called *the* chain of the element a. (Of course, what Dedekind had in mind was that any chain of elements which contains a should also contain $a + 1$, $a + 2$, ..., whence the intersection of all these chains—the chain of a— would be exactly the set a, $a + 1$, $a + 2$, ...)

The system of natural numbers S can now be characterized as consisting of a *basic* element a and of *the* chain of this element a.

This scheme provides a definition of unity as the *basic* element a and implies that the system S is infinite. Dedekind did not succeed in proving that such a system S does really exist, i.e. that his definitions would not lead to any contradiction; but he did succeed in establishing the famous method of mathematical induction for all finite natural numbers, and in defining the numerical operations of addition and multiplication (Section 16.11).

It should be emphasized that the Dedekind chain K is primarily one set of elements such that $\varphi(K) \subset K$. In fact K must contain the sets $L = \varphi(K)$, $M = \varphi(L)$, 'and so on', when this last phrase has been clarified.

5.5. Dedekind's Quasi-chains and Arcs

The Dedekind concept of a 'chain' can be reworded, and perhaps clarified, in terms of 'invariant' and 'semi-invariant' subsets of an ordinary relation which maps a set S into a subset $\varphi(S)$. An infinite set S is characterized by the existence of mappings such that $\varphi(S)$ is a *proper* subset S, i.e. there are elements of S not included in $\varphi(S)$, e.g.

$$S = \{1, 2, 3, \ldots\}, \quad \varphi(S) = \{2, 3, 4, \ldots\}.$$

A subset K of S is invariant under the mapping φ, if $\varphi(K) \equiv K$, and we venture to describe K as 'semi-invariant' if $\varphi(K)$ is a *proper* subset of K. Thus the subset (10, 11, 12, ...) is semi-invariant under the mapping, $n \rightarrow n + 1$. Dedekind characterized a finite set S by the non-existence of mappings φ such that under φ, S has a semi-invariant subset. This is the first known *positive* characterization of a finite set.

Since such a finite set possesses a mapping φ which has no proper semi-invariant set, or chain in Dedekind's terminology, he introduces the concept of a 'quasi-chain', H_b, such that each element x of H_b is mapped into another element of H_b, *except* a single element $x = b$. Thus, for example, if $H_b = \{\varphi^{-1}(b), b\}$, $\varphi(H_b) = \{b, \varphi(b)\}$.

Instead of the 'proper chains' appropriate to an infinite system, Dedekind introduced 'arcs'. The arc $C(a, b)$ is the intersection of all the quasi-chains H_b which contain the element a. With these powerful new concepts Dedekind was able to prove that if a, b, c are any three distinct elements, then, if $x' = f(x)$, then

either
$$C(b', c) = C(b', a) \cup C(a', c)$$

or
$$C(b', c) = C(b', a) \cap C(a', c).$$

This is sufficient to define an ordering relation among the system of finite elements, to establish the principle of mathematical induction, and to prove that any subset of a finite set is also finite, and that a finite set cannot be an infinite set.

Dedekind's theory of finite sets does seem to be more profound and more satisfactory than the theory of Peano and it is a matter of regret that it passed unnoticed until Cavaillès developed it still further (Cavaillès 1932). The concept of a quasi-chain was introduced independently by Zermelo (1908a), but he did not use the concept of an 'arc'.

Other definitions of finite sets due to Tarski (1925) and Janiszewski (1912) characterize a finite set in terms of 'irreducible' or 'saturated' elements.

Sierpinski (1918) characterized a finite set in terms of the collection K of all sets, each of which satisfies the following conditions:

(1) Any singleton, i.e. any set consisting of just one element, belongs to K, and
(2) if A and B are two sets which belong to K, so also does their union $A \cup B$.

A finite set is then defined as one which belongs to each such class K.

However, Kuratowski (1920) pointed out that collection K would possess the paradoxical properties of all such 'large' sets and he proposed to avoid this difficulty by the following definition:

(1) Its elements are (non-empty) subsets of M,
(2) it contains all the singleton sets, and
(3) if A and B are two subsets of M so also is their union $A \cup B$.

5.6. Zermelo and the Axiom of Choice

Cantor (1883, p. 550) bequeathed to mathematicians, not only the materials for the construction of logical antimonies, but also a problem the solution of which Hilbert (1900) considered to be one of the most important tasks confronting the mathematical world. The problem is to prove that 'every well-defined set can be brought into the form of a well-ordered set'.

To the astonishment of the International Congress of Mathematicians meeting in Heidelberg in 1904, J. König challenged the profound convictions of Hilbert by offering a proof that the continuum cannot be well-ordered, but almost immediately he perceived a flaw in his argument and withdrew his paper. Shortly afterwards Zermelo (1904) offered a proof of the well-ordering theorem after some conversations with E. Schmidt.

This proof was based on the assumption that from each subset S of a system of elements M we can choose a 'distinguished' element, $s = f(S)$. Zermelo's 'postulate of choice' had been challenged by Peano (1890a) and Hadamard *et al.* (1905) on the ground that he had provided no proof. This was readily admitted by Zermelo (1908a) who pointed out that his postulate was needed to establish a number of elementary and fundamental theorems and that it had already been used implicitly for this purpose. He might indeed have cited Peano's proof of the existence of solutions of the differential equation

$$\mathrm{d}y/\mathrm{d}x = f(x, y)$$

when the function f is continuous but not Lipschitzian in y (Peano, 1890a).

Indeed a more striking example is the proof that if the point ξ is a limit point of the set S then there exists a sequence of points $\{\xi_n\}$ which converge to ξ as n tends to infinity.

In this second paper Zermelo (1908a) deduced the well-ordering theorem from a more restricted version of the axiom of choice, viz. that there exists a set C which consists of exactly *one* element $s = f(S)$ from each set S of any *disjoint* collection of sets. The main technique used in this proof was a skilful adaptation of Dedekind's theory of chains (Section 5.4) in which the ordering relation is of the form

$$S' = \varphi(S) = S - f(S).$$

It was already known to Cantor that the well-ordering theorem implied the 'comparability' of any two well-ordered sets W, W', i.e. that either W and W' have the same order type or that one, W, has the same order type as a section of the other, W', i.e. a subset of W' consisting of the elements x such that $x < s \in W'$. Hartogs (1915) established the converse of this theorem and thus showed the logical equivalence of the axiom of choice, the well-ordering theorem and the theorem of comparability.

The axiom of choice had a very varied reception, especially in France, where it provoked a lively correspondence (Borel 1914) between Hadamard, Baire, Lebesgue and Borel. Indeed, as late as 1941, Lebesgue regarded this axiom primarily as a valuable instrument for suggesting lines of research, independently of its logical justification (Lebesgue 1941).

5.7. *Hausdorff's Maximal Principle*

Hausdorff (1914, pp. 140–1) showed that the well-ordering theorem could be used to establish a maximal principle. Any partially ordered set must contain completely ordered subsets $A, B, \ldots$ and these subsets are partially ordered by the relation of inclusion. Hausdorff introduced the concept of a *maximum* in the collection G of all completely ordered subsets as a subset M contained in G and such that, if N is any other subset of G which contains M, (i.e. $N \supset M$), then $N = M$. He then used the well-ordering theorem to prove that such a maximum necessarily exists.

It is the converse of this theorem which became so important, namely that the existence of a maximal completely ordered subset implies the well-ordering theorem. This showed that the axiom of choice and the maximal principle were equivalent axioms, and experience established the greater flexibility and wider range of the latter.

The maximal principle has subsequently been enunciated in many different forms and what is now known as 'Hausdorff's maximal principle' is often expressed in the following abstract form:

> A 'nest' is a collection N of sets, such that any two sets, A and B, of N are comparable, in the sense that either $A \subset B$ or $B \subset A$.
>
> If N is a nest in a certain collection of sets S, then there is a maximal nest in S which contains N.

5.8. Kuratowski and Maximal Chains

As noted by Kuratowski (1922a) the theory of transfinite numbers had found numerous applications in different branches of analysis and of topology, although the actual theorems established scarcely contained any reference to these numbers. Moreover, a number of mathematicians had succeeded in eliminating transfinite numbers from their theories by means of special, *ad hoc*, devices. Kuratowski therefore proposed a simple, general and uniform method which could be applied in all these applications without any reference to transfinite theory. His theory is a further development of the chain theory, invented by Dedekind (Section 5.4), and developed by Zermelo (1908a) and Hessenberg (1909).

The Dedekind chain σ was a single set of *elements* such that $\varphi(\sigma) \subset \sigma$ for a given ordering relation φ. The Kuratowski chain Z was a collection of *sets* such that, for a given ordering relation φ,

(1) Z contained a prescribed set α,
(2) if Z contained a set σ, it also contained $\varphi(\sigma)$,
(3) Z contained the intersection of any sets contained in Z.

Kuratowski assumed that among the collections Z there existed a minimum collection, $M(A)$, which would also be a chain. Similarly, starting with an ordering relation $\psi(\sigma)$ Kuratowski defined a maximum collection $N(A)$.

The existence of the minimum chain $M(A)$—or of the maximum chain $N(A)$—implies the existence of the intersection (or the union) of all the chains Z, and is now recognized to be a new axiom, with the same status as the axiom of choice, but with a more direct and flexible application to topology.

5.9. Zorn's Lemma for Nests

Zorn (1935) gave another method for the elimination of transfinite numbers and the well-ordering theorem by the use of 'nests', i.e.

collections of sets which are ordered by inclusion, so that if the sets σ and τ belong to the same collection, then either $\sigma \subset \tau$ or $\tau \subset \sigma$.

In the theory of the algebraic closure of fields there are fundamental theorems due to Steinitz, originally proved by means of the well-ordering theorem. Zorn proposed to make these theorems shorter and more algebraic by the use of a new maximal principle. This new principle he expressed in terms of the concepts of 'nests' and of a 'closed' collection of sets, i.e. a collection which contains the union of every chain in the collection. In its original form Zorn's maximum principle asserts that in a 'closed' collection of sets there exists at least one which is not contained as a proper subset in any other set of the collection, and which is therefore rightly called a maximum set.

This, the original form of Zorn's contribution, underwent a number of changes, and finally crystallized in its present form which is known as Zorn's lemma and which states that if every totally ordered set of elements of a partially ordered set S has an upper bound, then S itself has at least one maximal element. Zorn seems to have been the first to derive the axiom of choice from the maximal principle, and it appears that he was unaware of the work of Hausdorff and Kuratowski.

An exhaustive account of *Equivalents of the Axiom of Choice* has been given by H. Rubin and J. E. Rubin (1963) and by Jech (1973).

5.10. *The Continuum Hypothesis*

Among the many fruitful problems which Cantor bequeathed to posterity, was the conjecture made at the beginning of his investigations (1878) that, the cardinal number of the continuum, is the next cardinal number greater than $\aleph_0$, the cardinal number of the set of natural numbers. This implies that any uncountable set contains a subset which has the 'power of the continuum', i.e. the cardinal number of the continuum. This is the 'continuum hypothesis'. The 'generalized continuum hypothesis' is that, if c is any infinite cardinal, and if a is any cardinal strictly greater than c, then $a \geqslant 2^c$.

Many fruitless attempts were made to prove or to disprove the truth of the continuum hypothesis and its generalization, but the most remarkable result was obtained by Gödel and given in his lectures in 1938–9 and published in 1940 (Gödel 1940). The background of Gödel's investigation is the system Σ of axioms for set theory due to Zermelo, Fraenkel and Bernays (but excluding the axiom of choice). The foreground is the axiom of choice AC and the generalized continuum hypothesis CH. Gödel established that if a contradiction from AC and CH were derived in Σ, then it could be transformed into a contradiction arising from the axioms of Σ alone. Thus if the system Σ is consistent, it remains consistent when supplemented by AC and CH.

Gödel's result was completed in a course of lectures given by Cohen in 1965 and published in 1966 (Cohen 1966). Cohen established that neither can CH be proved from Σ and AC, nor AC from Σ and CH. This provides a complete proof of the independence of CH and AC.

PART II. SPACE

6. Vectors and Tensors

6.1. *The Beginning of Algebra*

The distinction between algebra and arithmetic, in their early development, was that arithmetic was concerned with the properties of specific, individual numbers, while algebra was concerned with the properties of all numbers. Thus the statement that 'seven is a prime number' is arithmetical, but the statement that

$$(x + y)(x - y) = x^2 - y^2$$

for all real numbers is algebraical.

The popular characterization of the elements, x, y, ..., of algebra as 'unknowns' emphasizes that the only properties of these elements employed in algebra were their ability to be combined by addition and subtraction or multiplication and division, and to be arranged in order of magnitude. In fact the elements of algebra were originally simply the real numbers, until the researches of Wessel in 1799, Argand in 1806 and Gauss in 1799 and 1831 (Crowe 1967) provided a satisfactory theory of complex numbers.

These numbers obeyed the same laws of addition and multiplication as the real numbers, but did not form a linear set ordered by the relations of 'greater' and 'less'. The geometrical representation of complex numbers in the Argand diagram not only gave them a certain physical reality but also provided a guarantee of the self-consistency of the rules of addition and multiplication.

Hamilton (1837) provided an autonomous theory of complex numbers as ordered couples (x, y) of real numbers and thus freed the subject from any appeal to geometry. In Hamilton's theory, which is now universally adopted, the imaginary unit i is the couple $(0, 1)$ and the real unit is $(1, 0)$, which by an abuse of notation was, and is still, represented by the same symbol as the unit of arithmetic: 1.

As it was gradually realized that the use of complex numbers not only greatly facilitated calculation but also led to the fascinating properties of 'analytic functions', numerous attempts were made to discover a similar theory which would operate in three-dimensional space and not be restricted to the complex plane of the variable $x + iy$.

As early as 1827, Möbius had invented his barycentric calculus (Möbius 1827), in which the letters A, B ... not only represented points in the manner of Euclid, but also became the elements of an algebra which could be multiplied by the real numbers, say a, b, ..., in such a way that the

45

centroid of S of masses $a, b, \ldots$, located at the points $A, B, \ldots$ was given by the formula

$$aA + bB + \ldots = (a + b + \ldots)S.$$

The immediate influence of the work of Möbius seems to have been negligible, but its remote influence is to be found in combinatorial topology, where the 'barycentric coordinates' of a point P in terms of the vertices $A, B, \ldots$ of a simplex are the numbers $a, b, \ldots$ such that

$$a + b + \ldots = 1$$

and $$aA + bB + \ldots = S.$$

Möbius did not consider the problem of multiplication until 1862 when it appears that he had independently discovered the 'geometric product', i.e. the vector product, and the 'projective product', i.e. the scalar product of two vectors (Möbius 1887).

6.2. Hamilton's Quaternions

In the algebra of complex numbers addition and multiplication are commutative, associative and distributive. It is impossible to preserve these characteristics in a three-unit algebra of space. However, Hamilton (1843) suddenly envisaged the possibility of a non-commutative four-unit algebra, in which each element is of the form

$$w\varepsilon + x\mathrm{i} + y\mathrm{j} + z\mathrm{k},$$

where w, x, y, z are real numbers, and ε, i, j, k are the four units, of which the first, ε, is usually denoted by the same symbol, 1, as the unit of the real number system. Multiplication is associative and distributive, but not commutative, for

$$\mathrm{i}^2 = \mathrm{j}^2 = \mathrm{k}^2 = \mathrm{ijk} = -\varepsilon = -1,$$

whence
$$\mathrm{jk} = -\mathrm{kj} = \mathrm{i},$$
$$\mathrm{ki} = -\mathrm{ik} = \mathrm{j},$$
$$\mathrm{ij} = -\mathrm{ji} = \mathrm{k}.$$

The elements of this algebra Hamilton called quaternions, after the Vulgate name for the squads of four soldiers who guarded St Peter in prison (*Acts* 12: 4).

The Hamilton quaternion, $q = w + x\mathrm{i} + y\mathrm{j} + z\mathrm{k}$, is the sum of a scalar part, $S_q = w$, and a vector part $V_q = x\mathrm{i} + y\mathrm{j} + z\mathrm{k}$, and the Hamilton product of two purely vectorial quaternions, $\alpha = a\mathrm{i} + b\mathrm{j} + c\mathrm{k}$ and $\xi = x\mathrm{i} + y\mathrm{j} + z\mathrm{k}$, is the quaternion, whose scalar part is *minus* the scalar product $(\alpha . \xi)$ of α and ξ and whose vector part $[\alpha \wedge \xi]$ is the vector product of α by ξ.

This combination of a scalar and a vector in one expression seemed unnatural to many mathematicians, and Hamilton devoted many pages of his *Lectures on quaternions* (1853) and his *Elements of quaternions* (1866) to a domestication of this strange animal. The difficulty in the visualization of a quaternion was partly removed by Hamilton's evaluation of the quotient of one vector α by another vector β. In the simplest case when β is a unit vector the equation

$$\alpha = q\beta$$

has a unique solution

$$q = (\beta \cdot \alpha) + [\beta \wedge \alpha].$$

This representation of a quaternion as the quotient of two vectors did provide a means of establishing the self-consistency of the quaternionic algebra on a geometric basis, but did little to recommend quaternions in mechanics except in the theory of tops and gyroscopes. Here quaternions furnished a simple and elegant means of representing rotations in space. A rotation through an angle ω about an axis with direction cosines l, m, n is represented by the quaternion

$$q = \cos \tfrac{1}{2}\omega + \sin \tfrac{1}{2}\omega \ (\mathrm{i}l + \mathrm{j}m + \mathrm{k}n)$$

and the vector $\varrho = \mathrm{i}x + \mathrm{j}y + \mathrm{k}z$ is then rotated into the vector

$$\varrho' = q\varrho q^{-1}$$

where

$$q^{-1} = \cos \tfrac{1}{2}\omega - \sin \tfrac{1}{2}\omega(\mathrm{i}l + \mathrm{j}m + \mathrm{k}n).$$

In the magisterial treatise *Über die Theorie des Kreisels* Klein and Sommerfeld (1897) give the motion of a top as most simply represented by the parameters

$$\alpha = \cos \tfrac{1}{2}\omega + \mathrm{i}n \sin \tfrac{1}{2}\omega$$
$$\beta = -\,m \sin \tfrac{1}{2}\omega + \mathrm{i}l$$
$$\gamma = m \sin \tfrac{1}{2}\omega$$
$$\text{and} \quad \delta = \cos \tfrac{1}{2}\omega$$

where i now stands for the ordinary imaginary unit $\sqrt{(-1)}$.

This representation of rotations, based on quaternions, is undoubtedly more convenient than the earlier representation due to Rodrigues and Darboux (Lamb 1920).

6.3. *Vector Algebra*

The parallelogram construction for the resultant of two forces was a commonplace in the early nineteenth century but it was only gradually realized that this construction was equivalent to the addition of two vectors. One of the clearest and earliest (and shortest) expositions of vector algebra and its application to Newtonian mechanics is that of Barré, better known as the Comte de Saint Venant (Barré 1845). It is remarkable that this author defines the 'geometric product' $\overrightarrow{ab}$ (i.e. the vector product now written as $a \wedge b$) but does not mention the scalar product—for the excellent reason that *his* vector algebra is closed under the operations of addition, subtraction and the geometric product.

But although the quaternion product of two vectors α and β had the vector product $[\alpha \wedge \beta]$ as its vectorial part, the scalar part was not the ordinary scalar product $(\alpha . \beta)$ but its negative. Moreover, mechanics found its natural expression in terms of vectors and Newtonian mechanics did not require quaternions.

No account of the history of vectors and quaternions would be complete without a reference to the antipathy of Lord Kelvin, who gave unambiguous expression to his opinion in his letters to Fitzgerald and Hayward (S. P. Thompson 1910b). To the former he wrote in 1896 '... "vector" is a useless survival, or off-shoot, from quaternions, and has never been of the slightest value'. To the latter he write in 1892: 'Quaternions came from Hamilton after his really good work had been done; and, though beautifully ingenious, have been an unmixed evil to those who have touched them in any way, including Clerk Maxwell'.

In spite of the spirited advocacy of Tait (1867) quaternions found no place in the *Treatise on Natural Philosophy* (Tait and Thomson 1867), nor in the famous *Treatise on Electricity and Magnetism* (Maxwell 1873). Interest in quaternions steadily declined and the publication of a series of textbooks on vector algebra marks the triumph of the latter in the field of Newtonian mechanics (Crowe 1967).

The last phase of the older school of quaternion enthusiasts is marked by the publication of McAulay's book in 1898 on octonians, i.e. quaternions, $q = w + ix + jy + kz$, where w,x,y,z are ordinary complex numbers; but in 1927 quaternions were rehabilitated as the 'spin variables' in Pauli's quantum theory.

6.4. The Linear Vector Function of Gibbs

In 1881 Gibbs privately printed an account of 'The elements of vector analysis' for the benefit of his students at Yale. This is the earliest and the best of the many treatises on vector algebra and vector analysis in which these admirable techniques for the physicist and engineer were freed from quaternionic shackles.

The most remarkable and original feature of Gibbs's system of vector algebra is his treatment of the linear vector function, i.e. an automorphism f of the space of vectors, such that, if ϱ and σ are any two vectors, and a and b are any two scalars, then

$$f(a\varrho + b\sigma) = af(\varrho) + bf(\sigma).$$

Hamilton (1862, 1864) had already discovered the linear vector function f and had given the cubic equation

$$0 = m + m'f + m''f^2 + f^3$$

satisfied by any such function. Hamilton's discovery of this equation is noteworthy as one of the first examples of what is now known as the 'Cayley–Hamilton' characteristic equation. This is the equation satisfied by any square matrix A. The characteristic equation of A is written in determinantal form as

$$\varphi(\lambda) = \det (A - \lambda I) = 0$$

where I is the unit matrix. Then the Cayley–Hamilton equation is

$$\varphi(A) = \|0\|$$

where the symbol $\|0\|$ denotes the zero matrix.

Gibbs gave a representation of the linear vector function in a form more easily intelligible to physicists in terms of the special linear vector functions, which he called 'dyads' and which had the form

$$f(\varrho) = \alpha(\lambda \cdot \varrho)$$

where α and λ are vectors and the dot represents the scalar product. He proved that, in general, any linear vector function has the form

$$f(\varrho) = \alpha(\lambda \cdot \varrho) + \beta(\mu \cdot \varrho) + \gamma(\nu \cdot \varrho)$$

(λ, μ and ν being vectors) i.e. is a trinomial dyad.

Dyads are well adapted to the representation of displacements, rotations and strains in a continuous medium, but do not seem to be capable of adaptation for the expression of stresses, or of general second-order tensors.

6.5. *The Definitions of a Vector and of a Tensor*

Initially a vector in Euclidean space was defined as a directed segment of a straight line specified by the ordered pair of its initial point A and terminal point B and was written as the vector $\vec{AB}$, or later, in a more sophisticated manner, as a set of such equivalent or 'equipollent' directed segments. For the purposes of numerical calculation this geometric definition was suplemented by an algebraic definition, in which a vector was specified by its components say (a,b,c), in a system of rectangular Cartesian coordinates.

The extension of this concept to n-dimensional Euclidean space was provided by the geometric definition of Grassmann (1844) and the algebraic definition of Cayley for whom a vector is specified in a given frame of reference by its n components $(a_1, a_2, \ldots, a_n)$.

This familiar representation is not completely satisfactory for it fails to answer the question, what is a vector, and only answers the question, how is a vector represented in a given frame of reference?

The question can be evaded by the abstract theory of vectors which characterizes the vectors in a Euclidean space of n dimensions. Properties of such a set S of elements, α, β, $\ldots$, include:

(1) There is a commutative and associative addition, $\alpha + \beta$;
(2) There is a distributive and associative multiplication by real numbers r, giving vectors $r\alpha$, $r\beta$, $\ldots$ etc.
(3) Any $n + 1$ vectors, α_1, α_2, $\ldots$, α_{n+1}, are connected by a relation linear

$$r_1\alpha_1 + r_2\alpha_2 + \ldots + r_{n+1}\alpha_{n+1} = 0, \ (r_i \in R).$$

Also

$$\alpha, \beta \in S, \; a, \; b, \in R \; \Rightarrow \; a\alpha + b\beta \in S,$$

$$a(\alpha + \beta) = a\alpha + a\beta,$$

$$(a + b) \, \alpha = a\alpha + b\alpha,$$

$$ab \, \alpha = a(b\alpha),$$

and $\quad 1\alpha = \alpha.$

If the set of elements S forms a vector space of n dimensions, and, if to each vector α there is associated a positive (or zero) real number $\|\alpha\|$, satisfying the usual axioms for the 'length' of a vector α, we then have a 'normed vector space'.

Another form of the abstract theory of vectors is given by Peano (1895–6), based on Grassmann's calculus of extension (Grassman 1844, p. 1–319).

The concept of a vector space admits of many generalizations:

The set of real numbers R can be replaced by the set of complex numbers, or, indeed, any other field, and the number of dimensions n increased to infinity, thus generating a Hilbert space.

The vectors can be taken to be the elements of any Abelian group, and the set R replaced by any commutative field.

Again the vectors α, β can be replaced by functions of one or more real variables x in an interval (a, b) with the norm

$$\|f(x)\| = \left\{ \int_a^b |f(x)|^2 \mathrm{d}x \right\}^{\frac{1}{2}},$$

thus giving the Fréchet space of L^2 functions.

There is another way of describing a vector, which exhibits explicitly the relation of a vector to its components in a given frame of reference. A direction is specified by its direction cosines $(l_1, l_2, \ldots, l_n)$, i.e. the cosines of the angles between the coordinate axes and the direction. The component of a vector v in the direction $(l_1, l_2, \ldots, l_n)$ depends upon the direction, i.e. a vector v determines a mapping $\Lambda \to R$ of the space Λ of the direction cosines (or, we may say, of the points of a unit sphere in n-dimensional space) into the real numbers R. If $v_1, v_2, \ldots, v_n$ are the components of the vector v along the axes then the component in the direction $(l_1, l_2, \ldots, l_n)$ is

$$v_1 l_1 + v_2 l_2 + \ldots + v_n l_n,$$

i.e. it is a *linear* function of direction. We may therefore, following Neville (1921), describe a vector to be a linear function of direction.

The stress in a continuous medium is specified by the components of force acting across an element of area. If $S_{\alpha\beta}$ is the α-component of the force per unit area of an element of area with normal in the direction of the β space coordinates, then the component of the force in the direction (l_1, l_2, l_3) for an element of area with normal in the direction (m_1, m_2, m_3) is

$$\sum S_{\alpha\beta} l_\alpha m_\beta$$

i.e. it is a linear function of the two directions (l_α) and (m_β). Thus the stress tensor may be regarded as a bilinear function of direction.

In general a tensor, even in n-dimensional Euclidean space, may be defined as a multilinear function of direction, i.e. a function of the form

$$f(l_1, l_2, \ldots, l_n) = \Sigma C_{\alpha_1 \alpha_2 \ldots \alpha_n} l_{\alpha_1} l_{\alpha_2} \cdots l_{\alpha_n},$$

where the ls are direction cosines, the Cs are real numbers—the 'components' of the tensor, and the suffixes each take the values $1, 2, \ldots, n$.

6.6. Quaternions in Relativity Theory

In Newtonian dynamics the equations of motion are invariant when the rectangular Cartesian coordinates of space, x,y,z, are subjected to a translation or rotation, and when the time coordinate t is subjected to a translation. But in the special theory of relativity the equations of dynamics and of electromagnetism remain invariant under rotations in space and under Lorentz transformations of space–time. These latter are most compactly described in the terminology of Minkowski (1909) as rotations in the E_4 with Cartesian coordinates $(x, y, z,$ and $w = ict)$, where c is the speed of light and $i = \sqrt{-1}$.

Now Hurwitz (1896) had shown that any rotation in four-dimensional space E_4 could be expressed in the form $q \to lqr^{-1}$ where q is the quarternion $w + ix + jy + kz$ and l, r are each unit quarternions. An account of this theory and of its applications to the theory of regular polytopes (i.e. the higher dimensional analogues of polygons and polyhedra) is given by du Val (1964).

However, the first use of quaternions in the special theory of relativity seems to be due to Conway (1911), and independently to Silberstein (1914) who expressed the general Lorentz transformation in the form

$$q' = QqQ^{-1}$$

where $q = ict + r$, $q' = ict' + r'$ are the quaternionic representations of the space vectors r, r' and the dates t, t' of the same event in two different 'inertial' systems, while

$$\sqrt{2}\,Q = (1 + \gamma)^{\frac{1}{2}} + u(1 - \gamma)^{\frac{1}{2}},$$

$$\gamma = (1 - v^2/c^2)^{-\frac{1}{2}},$$

$$u = v/|v|.$$

and v is the relative velocity of the two inertial systems.

6.7. Grassmann's Calculus of Extension

The first edition of Grassmann's *Die lineale Ausdehnungslehre*, or *Calculus of Extension*, was published in 1844. In 1862 he published a greatly extended version of this work under the title *Die Ausdehnungslehre: Vollständing und in strenger Form bearbeitet*, and in 1878 a second edition of the original work. Although these works are professedly treatises on the geometry of Euclidean space of n dimensions, their significance is primarily algebraical, and their influence has been mainly on differential geometry in the theory of 'exterior forms'.

In the realm of geometry the germs of Grassmann's calculus are to be found in the barycentric calculus of Möbius (1827) in which a clear distinction is made between the directed segment of a straight line from a point A to a point B, denoted by AB, and the directed segment from B to A, denoted by BA or by $-AB$. Möbius had denoted a triangle with vertices A,B,C as ABC, but Grassmann took the further step of distinguishing between the two orientations of the same triangle denoted by ABC and ACB, and of extending this concept to configurations in n-dimensional space.

Moreover Grassmann represented the points A,B,C by the elements a,b,c of an algebra in which the directed segment AB is represented by the product $a \wedge b$ and the oriented triangle ABC (or BCA or CAB) by the product $a \wedge b \wedge c$ (or $b \wedge c \wedge a$ or $c \wedge a \wedge b$). This implies that

$$a \wedge b = -b \wedge a$$

and that $\qquad (a \wedge b) \wedge c = a \wedge (b \wedge c)$

and, by an obvious extension to a point pair AA, that

$$a \wedge a = 0.$$

From Möbius, Grassmann took the idea of representing the mass centre of masses m_1, m_2, ... at the points a_1, a_2 ... by the algebraical element a, where

$$ma = \sum m_k a_k,$$

and $\qquad\qquad\qquad m = \sum m_k.$

Grassmann clearly envisaged addition and multiplication as abstract operations, addition being commutative and the multiplication of the algebraic elements being anti-commutative, but associative and distributive. He and Hamilton are therefore the founders of algebra emancipated from the theory of equations and independent of the theory of groups.

The original books by Grassmann are very scarce but they are reprinted in his collected works (1894–1911). A general account of the various publications dealing with the Calculus of Extension is given in Crowe (1967). The most accessible accounts of its mathematical context are in A. N. Whitehead (1897) and Forder (1941). We shall summarize here some elementary and fundamental concepts of Grassmann's theory.

In a space of n dimensions there are $n + 1$ unit elements

$$e_1, e_2, \ldots, e_{n+1}$$

which are independent in the sense that the linear relation

$$a_1 e_1 + a_2 e_2 + \ldots + a_{n+1} e_{n+1} = 0$$

where a_1, a_2, ..., a_{n+1} are real numbers, implies that

$$a_1 = 0, \; a_2 = 0, \; \ldots, \; a_{n+1} = 0.$$

These unit elements generate a linear vector algebra A_1 with elements of the form

$$\xi = \Sigma\, x_j e_j$$

where $x_1, x_2, \ldots, x_{n+1}$ are real numbers.

From these unit elements we can form the 'progressive', 'combinatorial' or 'outer' products $e_j \wedge e_k$ subject to the 'equations of condition',

$$e_j \wedge e_k + e_k \wedge e_j = 0 \quad \text{for all } j \text{ and } k,$$

which imply that

$$e_j \wedge e_j = 0 \quad \text{for all } j.$$

These $\frac{1}{2}(n+1)n$ independent unit elements form another linear vector algebra A_2.

Similarly we can form the progressive product $e_j \wedge e_k \wedge e_l$ of three unit elements e_j, e_k, e_l which is anti-symmetric in each pair of suffixed j, k, l, and thus generate a linear vector algebra A_3 with $\frac{1}{6}(n+1)n(n-1)$ independent units.

Again from any m units $(m \leqslant n+1)$, e_a, e_b, $\ldots$ we can form a progressive product $e_a \wedge e_b \wedge \ldots$ which is anti-symmetric in each pair of suffixes and thus generate a linear vector algebra A_m with $(n+1)!\,/m!\,(n+1-m)!$ independent units.

Furthermore we can form the progressive product of any two elements of the algebras A_ϱ and A_σ, but this product is zero if $\varrho + \sigma$ exceeds $n+1$. However, Grassmann triumphed over this restriction by introducing the 'supplement' of the units e_1, e_2, $\ldots$ to be the units

$$e_1{}^* = e_2 \wedge e_3 \wedge \ldots e_{n+1}, \; e_2{}^* = -e_1 \wedge e_3 \wedge \ldots e_{n+1}, \; \ldots$$

If E_ϱ is any unit of the algebra A_ϱ, its supplement $E_\varrho{}^*$ is defined to be the progressive product of those units e_1, e_2, $\ldots$, e_{n+1} which do *not* appear in E_ϱ, with a numerical factor chosen so that

$$E_\varrho \wedge E_\varrho{}^* = e_1 \wedge e_2 \wedge \ldots \wedge e_{n+1}$$

Then, if E_ϱ and E_σ are any units of the algebra, with supplements $E_\varrho{}^*$ and $E_\sigma{}^*$ respectively, then the 'regressive outer product' of E_ϱ and E_σ is defined to be the supplement of the progressive product of $E_\varrho{}^*$ and $E_\sigma{}^*$.

In all Grassmann defined some sixteen species of multiplication, but the preceding examples give some idea of the fertility of his creative powers.

6.8. Tensors and Functors

Perhaps the most abstract, the deepest and therefore to some eyes the simplest and most satisfactory definition of a tensor is in terms of categories and functors (Section 10.4).

We need not restrict ourselves to *rectangular* Cartesian systems. The applied mathematician thinks of a set of n independent directions or vectors issuing from a point O. The pure mathematician envisages an

n-dimensional vector space in which each vector x can be expressed uniquely in the form

$$x = x_1 a_1 + x_2 a_2 + \ldots + x_n a_n,$$

where $(x_1, x_2, \ldots, x_n)$ are real numbers, and $(a_1, a_2, \ldots, a_n)$ are independent elements of the vector space, which we will call a 'frame', A. The numbers $(x_1, x_2, \ldots, x_n)$ are the components of x in the frame A.

Any two such frames A and B, consisting of elements $(a_1, a_2, \ldots, a_n)$ and $(b_1, b_2, \ldots, b_n)$ respectively, are connected by relations of the form $b_p = \sum \lambda_{pq} a_q$ which define a transformation denoted by $A \to B$ or (A,B). Clearly there is a unique identical transformation (A,A); the product or composition of two transformations is given by

$$(A,B) . (B,C) = (A,C),$$

and the product is associative, that is,

$$(A,B)\{(B,C) . (C,D)\} = \{(A,B) . (B,C)\} . (C,D).$$

In technical language the transformations are the morphisms of a category consisting of the frames.

In the frame A a tensor T of rank r is a scalar multilinear function of r vectors $x^{(1)}, x^{(2)}, \ldots$ in A, i.e. it has the form

$$T = \sum t^{(A)}_{\alpha_1 \alpha_2 \ldots \alpha_r} x^{(1)}_{\alpha_1} x^{(2)}_{\alpha_2} \ldots x^{(r)}_{\alpha_r}$$

where $(x^{(p)}_{\alpha_1}, x^{(p)}_{\alpha_2}, \ldots, x^{(p)}_{\alpha_r})$ are the components of $x^{(p)}$, and $t^{(A)}_{\alpha_1 \alpha_2 \ldots \alpha_r}$ are real numbers called the components of the tensor T in the frame A.

The values of the components of the tensor T depend upon the frame A which is being used, and the transformation (A, B) from a frame A to a frame B induces a corresponding transformation from the components $T^A = (t^{(A)}_{\alpha_1 \alpha_2 \ldots \alpha_r})$ in the frame A to the components $T^B = (t^{(B)}_{\alpha_1 \alpha_2 \ldots \alpha_r})$ in the frame B. The only writer who explicitly includes a reference to the frame in his notation for tensor components is Whitehead (1922).

These sets of components form a second category T in the sense indicated above. To each frame A of the first category there corresponds a set of components T^A of the second category T, and to each transformation (A,B) in the category Φ there corresponds a transformation (T^A, T^B) of the components in the second category T, with similar properties for inversion, composition and associativity. Thus in the technical language of Eilenberg and MacLane (1945) the pairs of associated mappings,

$$A \to B, \quad T^A \to T^B$$

form a 'functor'. This functor corresponds to the abstract concept of a (contravariant) tensor (of rank r).

A tensor is almost a trivial example of a functor but it does provide the first and most elementary animal in the rich zoological garden of functors which constitutes the contemporary paradise of algebra.

6.9. *Tensors in the Ricci Calculus*

Although the 'Absolute Differential Calculus' (Ricci and Levi-Civita 1901) is essentially concerned with the tensor analysis, it opens with a new development in tensor algebra, in which the tensor theory is extended from rectangular Cartesian frames of reference to any local system of curvilinear coordinates in an n-dimensional manifold.

The neighbourhood of any point O in such a manifold is homomorphic with a neighbourhood of the origin in a Euclidean space X with coordinates $x^1, x^2, \ldots, x^n$, and therefore with a neighbourhood of the origin in any other Euclidean space Ξ with coordinates $\xi^1, \xi^2, \ldots, \xi^n$. Thus for any point on a two-dimentional surface we have representations of its neighbourhood in terms of (local) Cartesian coordinates (x, y) or (local) polar coordinates (r, θ).

A curve C through the point O is given by parametric equations

$$\left. \begin{array}{ll} x^\varkappa = f^\varkappa(s), & O = f^\varkappa(O), \\ \text{or} \quad \xi^\varkappa = \varphi^\varkappa(s), & O = \varphi^\varkappa(O), \end{array} \right\} \quad \varkappa = 1, 2, \ldots, n$$

The direction of the tangent to this curve at O is given by the set of numbers

$$t^\varkappa = [\mathrm{d} f^\varkappa(s)/\mathrm{d}s]_{s=0},$$

or

$$\tau^\varkappa = [\mathrm{d}\varphi^\varkappa(s)/\mathrm{d}s]_{s=0}.$$

The two coordinate systems X and Ξ are connected by the transformation scheme

$$\xi^\varkappa = \Lambda^\varkappa(x^1, \ldots x^n), \quad O = \Lambda^\varkappa(0, \ldots, O)$$

and its inverse

$$x^\varkappa = L^\varkappa(\xi^1, \ldots, \xi^n), \quad O = L^\varkappa(0, \ldots, O)$$

Hence

$$\tau^\varkappa = \sum_\alpha \Lambda^\varkappa_\alpha t^\alpha,$$

$$t^\varkappa = \sum_\beta L^\varkappa_\beta \tau^\beta,$$

where

$$\Lambda^\varkappa_\alpha = [\partial \Lambda^\varkappa / \partial x^\alpha]_O,$$

$$L^\varkappa_\beta = [\partial L^\varkappa / \partial \xi^\beta]_O,$$

the suffix meaning that the partial derivatives are to be evaluated at the point O.

Not the least of the innovations due to Ricci is the careful use of upper and lower suffixes (e.g. $\varkappa$ and α in $\Lambda^\varkappa_\alpha$) whose significance appears later. This notation was made more compact by Einstein (1916) with the convention that, when a suffix occurs both in a lower place and in an upper place, the sign of summation may be omitted. Thus we write

$$\tau^\varkappa = \Lambda^\varkappa_\alpha t^\alpha, \quad t^\varkappa = L^\varkappa_\beta \tau^\beta$$

when we have such a 'dummy' suffix, α or β.

It is easy to prove that

$$\Lambda^\varkappa_\alpha L^\alpha_\beta = \delta^\varkappa_\beta,$$

where the 'Kronecker delta' equals unity if $\varkappa = \beta$ and is otherwise zero.

By a natural generalization of the concept of a tensor in Euclidean space, Ricci defines a contravariant tensor, at O with components D^α in X or $\varDelta^\beta$ in $\varXi$, by the condition that these representations of the tensor are connected by the same law of transformation

$$\varDelta^\varkappa = \varLambda^\varkappa_\alpha D^\alpha, \ D^\varkappa = L^\varkappa_\beta \varDelta^\beta,$$

as the direction numbers t^α and τ^β.

The gradient of a scalar, $f(x^1, \ldots, x^n) = \varphi(\xi^1, \ldots, \xi^n)$ has components

$$f_\alpha = \partial f/\partial x^\alpha \text{ in } X,$$

or $\qquad\qquad \varphi_\beta = \partial\varphi/\partial\xi^\beta \text{ in } \varXi,$

which form representations of a 'covariant' tensor with the law of transformation

$$\varphi_\alpha = f_\beta \varLambda^\beta_\alpha, f_\beta = L^\alpha_\beta \varphi_\alpha.$$

There are also tensors of higher rank exemplified by forming products of, say μ, contravariant tensors of the first rank (such as t^α) and, say ν, covariant tensors of the first rank (such as φ_α). (As a mnemonic, 'co' = 'below', and 'contra' = up-against.)

Returning to the abstract functional theory of tensors we note that, while the pairs of associated mappings

$$X \to \varXi \quad \text{and} \quad D^\alpha \to \varDelta^\beta$$

form a contravariant functor, it is the mappings

$$\varXi \to X \quad \text{and} \quad D^\alpha \to \varDelta^\beta$$

which form a covariant functor.

6.10. *Tensors and the Linear Group*

The transformations

$$\tau^\varkappa = \varLambda^\varkappa_\alpha t^\alpha$$

which connect the direction numbers at O in X and $\varXi$ form a group G, as is easily verified by considering a third space Ω and direction numbers ω^α for the same curve C. This is the general linear group in n variables. These transformations induce corresponding transformations of the components of any tensor. Thus if $F^{\alpha\beta} = t^\alpha p^\beta$ and $\varphi^{\alpha\beta} = \tau^\alpha \pi^\beta$ are the components of a contravariant tensor of the second rank, we have

$$\varphi^{\alpha\beta} = (\varLambda^\alpha_\lambda t^\lambda)(\varLambda^\beta_\mu p^\mu) = (\varLambda^\alpha_\lambda \varLambda^\beta_\mu) F^{\lambda\mu},$$

and hence another group of transformations acting on these second-rank tensors.

This group of transformations is isomorphic with the original group G and similarly for mixed tensors of any order.

Just as Gibbs showed that the general linear vector function, or dyadic, in three-dimensional Euclidean space is the sum of three monomial dyads, so it can be shown that 'pure' covariant tensors, or 'mixed' tensors, can be formed as the sum of an appropriate number of monomial tensors, each of which is a product of the form

$$t^\alpha u^\beta v^\gamma \ldots p_\varkappa q_\lambda r_\mu \ldots,$$

Incidentally we note that tensors of the form $t^\alpha t^\beta$ play an important role in Eddington's *Fundamental Theory* (Eddington 1946) for, if t^α is a unit tensor,

$$(\text{i.e. } \sum_\alpha t^\alpha t^\alpha = 1)$$

the matrix $M = \|t^\alpha t^\beta\|$ is 'idempotent', i.e.

$$M^2 = M \times M = \|\sum t^\alpha t^\gamma t^\gamma t^\beta\| = \|t^\alpha t^\beta\| = M.$$

The situation when Ricci and Levi-Civita published a general account of their researches (Ricci and Levi-Civita, 1901) on the 'Absolute differential calculus' in 1901 may be summarized as follows.

(1) The theory of Cartesian tensors (referred to rectangular axes) employed in continuum mechanics and electromagnetic theory had been extended to curvilinear coordinates in n-dimensional Euclidean space.

(2) Christoffel's discovery of 'covariant differentiation' (Christoffel 1869) had enabled Ricci to extend the theory of tensor analysis to Riemannian space of n dimensions, and had unwittingly prepared the necessary machinery for Einstein's general theory of relativity.

(3) In this generalized tensor theory a tensor was represented in any local coordinate system, or frame, at a point O by a set of 'components' (such as the components $\partial\varphi/\partial n^\alpha$ of the gradient of a scalar φ), and the transformations connecting the components of the same tensor in different frames at O formed a group which was isomorphic with the general linear group in n dimensions.

(4) In fact Ricci and Levi-Civita were thus led to formulate an abstract definition of a tensor as an entity represented in any frame at a point O by a set of components which transformed according to the same law as the special tensors with components of the form

$$t^\alpha u^\beta v^\gamma \ldots p_\varkappa q_\lambda r_\mu \ldots,$$

where t^α, u^β, v^γ, $\ldots$ are components of contravariant tensors of the first rank and p_x, q_λ, $r_\mu \ldots$ are components of covariant tensors of the first rank.

This definition was accepted as the most general possible concept of a tensor, until the advent of the quantum theory of the spinning electron, when mathematicians realized that there could be another type of tensor, and that, in the words of Sir Charles Darwin, 'something had slipped through the net' of definition so tightly drawn by Ricci and Levi-Civita (Darwin 1928).

These new tensors are vectors referred to rectangular frames of reference and we shall proceed to describe their discovery.

6.11. Cartesian Tensors and the Rotation Group

Classical quantum theory (if that is not a contradiction in terms!) operated in the familiar three-dimensional space of Euclid and was concerned only with the familiar Cartesian tensors such as the linear and angular momentum of an electron.

According to the definition of Ricci and Levi-Civita, the components of these tensors at a given point should transform isomorphically with the corresponding transformations of the rotation group. In fact this isomorphism was even more in evidence in quantum mechanics than in Newtonian mechanics because the 'commutation laws' of angular momentum were precisely the same as the 'exchange relations' of the infinitesimal operators of the rotation group.

These latter, corresponding to infinitesimal rotations through an angle ε about an axis with direction cosines, l,m,n, had the form

$$R = \varepsilon(lX + mY + nZ) + O(\varepsilon^2)$$

where $\qquad\qquad X = y\partial/\partial z - z\partial/\partial y,$

and Y,Z are formed by cyclic permutation of the coordinates x,y,z. This implies that under this rotation any function $f(x,y,z)$ is transformed into

$$f + Rf + O(\varepsilon^2)$$

so that, for instance, x becomes $x + \varepsilon(mz - ny) + O(\varepsilon^2)$.

The 'infinitesimal operators of the rotation group', X, Y and Z, are connected by the commutation relations,

$$YZ - ZY = -X,$$

and two similar equations.

Now in classical quantum theory the Cartesian components of angular momentum L,M,N are operators which are connected by the exchange relations

$$MN - NM = i\hbar L,$$

and two similar relations, where $\hbar$ is Planck's constant divided by 2π and $i = \sqrt{-1}$. Manifestly the quantum operators $(iL/\hbar)$, $(iM/\hbar)$, $(iN/\hbar)$ satisfy the same exchange relations as the infinitesimal operators X, Y, Z.

In quantum theory the classification of atomic spectra and especially the elucidation of the Zeeman effect required the determination of the possible energy levels in the atom, of the possible values of the total angular momentum, and of the possible values of the component of the angular momentum in the direction of the applied magnetic field (which may be taken to coincide with the z axis). These quantities are given by the 'eigen-values' of the Hamiltonian H, of the operator $K = (L^2 + M^2 + N^2)^{\frac{1}{2}}$ and of the operator N respectively.

The formulation of quantum theory due to Heisenberg (1926) requires the construction of all the possible representations of L,M,N as matrices. It is found that in any representation K can be represented by a numerical multiple $\hbar(j^2 + j)^{\frac{1}{2}}$ of a unit matrix with $2j + 1$ rows and columns, while N is represented by the diagonal matrix with elements $(j, j - 1, j - 2, \ldots, -j)\hbar$ in the principal diagonal.

Since quantum theory was invented in order to elucidate the complexities of atomic spectra, physicists did not realize at once that the matrix representations of the angular momentum operators provided all the possible representations of the infinitesimal operators of the rotation group, and therefore yielded every possible class of tensor.

6.12. Spinors

Each representation of the rotation group (or of the angular momentum operators) is characterized by a 'quantum number', j, and the matrices in this representation have $2j + 1$ rows and columns. Hence j is either

(1) zero or a positive integer, or
(2) one of the numbers $\frac{1}{2}$, $\frac{3}{2}$, ..., $\frac{1}{2}(n - 1)$.

The zero or integral values of j correspond to the transformations of the classical tensors which are induced by rotations of three-dimensional Euclidean space. The half-integral values of j correspond to a new kind of tensor.

Pauli (1927) and Darwin (1927) were the first to realise that the exigencies of spectral theory required the point electron to possess, not only an angular momentum about the nucleus, but also an intrinsic angular momentum as if it were endowed with a spin and that the appropriate operator to represent the electronic spin was K with quantum number $j = \frac{1}{2}$.

In this case ($j = \frac{1}{2}$), N is represented by the diagonal matrix $\frac{1}{2}\hbar S_z$ where

$$S_z = \begin{Vmatrix} 1 & 0 \\ 0 & -1 \end{Vmatrix}$$

and $N^2 = (\frac{1}{2}\hbar)^2 I$ where I is the unit matrix of two rows and columns. It follows that, if

$$L = \tfrac{1}{2}\hbar S_x, \quad M = \tfrac{1}{2}\hbar S_y, \quad N = \tfrac{1}{2}\hbar S_z,$$

then

$$S_x S_y = -S_y S_x = iS_z, \quad S_y S_z = -S_z S_y = iS_x,$$

and

$$S_z S_x = -S_x S_z = iS_y,$$

i.e. S_x, S_y, S_z satisfy the same relation as the units, i, j, k of Hamilton's quaternion algebra.

S_x and S_y are of course also represented by matrices of two rows and columns, and the corresponding tensor has therefore only two components, although we are operating in three-dimensional space.

This two-component tensor is called a 'spinor', and the credit for its discovery really belongs to E. Cartan (1913, 1914) as a result of a much more general investigation.

The matrix representation of the rotation operators X, Y, Z corresponding to S_x, S_y, S_z is

$$X = \tfrac{1}{2}\begin{Vmatrix} 0 & i \\ i & 0 \end{Vmatrix}, \quad Y = \tfrac{1}{2}\begin{Vmatrix} 0 & 1 \\ -1 & 0 \end{Vmatrix}, \quad Z = \tfrac{1}{2}\begin{Vmatrix} i & 0 \\ 0 & -i \end{Vmatrix}.$$

Thus the corresponding infinitesimal transformations of the two-component spinor (ξ, η) are

$$X(\xi, \eta) = \tfrac{1}{2}(i\eta, i\xi), \quad Y(\xi, \eta) = \tfrac{1}{2}(\eta, -\xi),$$
$$Z(\xi, \eta) = \tfrac{1}{2}(i\xi - i\eta).$$

From this spinor we can generate a null vector with components

$$x = -\xi^2 + \eta^2, \quad y = -i(\xi^2 + \eta^2), \quad z = 2\xi\eta,$$

with the correct infinitesimal transformations of the rotation group in three-dimensional space

$$X(x,y,z) = (0, z, -y),$$
$$Y(x,y,z) = (-z, 0, x),$$
$$Z(x,y,z) = (y, -x, 0).$$

If we compare the finite transformations, and it is sufficient to compare the transformations generated by Z, we find that the rotation $x = -2iA^2 \sin \alpha t$, $y = -2iA^2 \cos \alpha t$, $z = 2A^2$ corresponds to $\xi = A \exp(\tfrac{1}{2}i\alpha t)$, $\eta = A \exp(-\tfrac{1}{2}i\alpha t)$.

Thus to the same rotation, with αt or $\alpha t + 2\pi$ as the angular parameter, there correspond two spinor transformations

$$\xi = A \exp(\tfrac{1}{2}i\alpha t), \quad \eta = A \exp(-\tfrac{1}{2}i\alpha t)$$

and
$$\xi = -A \exp(\tfrac{1}{2}i\alpha t), \quad \eta = -A \exp(-\tfrac{1}{2}i\alpha t),$$

i.e. a double-valued representation of the rotation group.

In Cartan's work (see above) spinors were defined with reference to infinitesimal rotations, but Brauer and Weyl (1935) give a full treatment with reference to finite rotations. Further developments of the theory have been given by E. Cartan (1937) and Chevalley (1954).

6.13. Dirac's Operators

The condition that the function $w(z) = u(x, y) + iv(x, y)$ of the complex variable $z = x + iy$ should be analytic (i.e. that the incrementary ratio

$$[w(z + \eta) - w(z)]/\eta$$

should tend to a unique limit as the complex increment η tends to zero in any direction), is expressed by the equations obtained by Cauchy and used with such effect by Riemann:

$$\partial u/\partial y + \partial v/\partial x = 0, \quad \partial u/\partial x = \partial v/\partial y.$$

These equations imply that

$$\Delta_2 u = 0, \quad \Delta_2 v = 0,$$

where Δ_2 is the Laplacian operator

$$\Delta_2 = \partial^2/\partial x^2 + \partial^2/\partial y^2$$

and they realize, in effect, a factorization of Δ_2 in the form

$$\Delta_2 = (\partial/\partial x + i\partial/\partial y)(\partial/\partial x - i\partial/\partial y) = \nabla_2\nabla_2{}^*,$$

for the first-order differential relation $\nabla_2 w = 0$ is equivalent to the Cauchy–Riemann equations.

Similarly Hamilton expressed the Laplacian operator

$$\Delta_3 = \partial^2/\partial x^2 + \partial^2/\partial y^2 + \partial^2/\partial z^2$$

in the form $\qquad\qquad\qquad \Delta_3 = -\nabla_3^2$

where $\qquad\qquad\qquad \nabla_3 = i\partial/\partial x + j\partial/\partial y + k\partial/\partial z,$

and i, j, k are the quaternionic units.

Dirac (1928) gave the first solution of the problem of expressing quantum theory in a form which was invariant under the Lorentz transformations of the special theory of relativity, by expressing d'Alembert's wave operator

$$\square = \partial^2/\partial x^2 + \partial^2/\partial y^2 + \partial^2/\partial z^2 - \partial^2/c^2\partial t^2$$

in the form

$$\square = \{\alpha_1\partial/\partial x + \alpha_2\partial/\partial y + \alpha_3\partial/\partial z + i\alpha_4\partial/c\partial t\}^2$$

where α_1, α_2, α_3, α_4 are operators which satisfy the relations

$$\tfrac{1}{2}(\alpha_j\alpha_k + \alpha_k\alpha_j) = 0 \quad \text{if} \quad j \neq k$$

$$\text{or} \quad +1 \quad \text{if} \quad j = k.$$

These four operators generate an algebra with sixteen linearly independent units

$$1, \quad \alpha_j(4), \quad \alpha_j\alpha_k(6), \quad \alpha_j\alpha_k\alpha_l(4), \quad \alpha_1\alpha_2\alpha_3\alpha_4,$$

and this algebra is the direct product of two quaternionic algebras with units

$$\sigma_1 = \alpha_1\alpha_4, \quad \sigma_2 = \alpha_2\alpha_4, \quad \sigma_3 = \alpha_3\alpha_4, \quad 1,$$

and

$$\tau_1 = \alpha_2\alpha_3, \quad \tau_2 = \alpha_3\alpha_1, \quad \tau_3 = \alpha_1\alpha_2, \quad i\alpha_4.$$

There is a corresponding theory in n-dimensional space which enables us to take the square root of the Laplacian operator in n dimensions, or, equivalently, the square root of the linear form

$$x_1^2 + x_2^2 + \ldots + x_n^2 = (p_1x_1 + p_2x_2 + \ldots + p_nx_n)^2.$$

In this theory we have n units, $p_1, p_2, \ldots, p_n$ which satisfy the relations

$$p_i^2 = 1, \; p_ip_k = p_kp_i = 0 \quad \text{if} \quad i \neq k,$$

and which generate an algebra with 2^n linearly independent units of the form

$$p_1^{\alpha_1}p_2^{\alpha_2} \ldots p_n{}^{\alpha_n}$$

where each α_k is either 0 or 1.

This algebra was discovered by Clifford (1878) in an endeavour to graft Hamilton's quaternions on to Grassmann's extensive algebra. The vital distinction between the two systems is that Grassmann's equations of condition

$$e_i e_k + e_k e_i = 0 \quad \text{for all } i \text{ and } k$$

are invariant when the units are transformed under the general linear group, whereas Clifford's relations

$$\tfrac{1}{2}(p_i p_k + p_k p_i) = 0 \quad \text{if} \quad i \neq k$$
$$= 1 \quad \text{if} \quad i = k$$

are invariant under the group of rotations in n-dimensional space.

7. Geometry and Measurement

7.1. The Foundations of Geometry

The history of geometry during the last hundred years begins, conveniently and inevitably, with Riemann's paper 'Über die Hypothesen, welche der Geometrie zu Grunde Legen', (Riemann 1867) which was written and read as his *Habilitationschrift* to the Faculty of Philosophy at Göttingen in 1854, but was not published until after his death in July 1866. To appreciate the influence of this epochmaking work, we must sketch the condition of geometrical studies in the first half of the nineteenth century.

For the history of geometry we are fortunate in possessing works by some great masters together with a fine bibliography (Somerville 1970, Pasch 1882b, Dehn 1926, Klein 1926–7, Bonola 1908).

7.2. The Parallel Postulate

The genius of Euclid is most clearly revealed in the famous 'parallel postulate' and in his use of congruence as a method of proof. In giving explicit statement to these two concepts, Euclid unconsciously issued to future geometers a challenge, to which the reply was the creation of Metageometry, i.e. the critical examination of the undefined notions and unproved axioms of Euclidean (or any other) geometry (Klein 1872).

The famous parallel postulate implies that if l is any straight line and if P is any point, not on the line l, then there is one and only one straight line which intersects the point P, is coplanar with l and does not intersect l. Is this axiom *independent* of the other axioms?

The answer would be in the negative if we could prove either (1) that this axiom can be deduced from the other axioms or (2) that the denial of this axiom (together with the affirmation of the others) leads to a contradiction. But all attempts to deduce this axiom have failed (and, as we shall see, are doomed to failure).

The axiom can be defined in two (or three) different ways.

(1) Lobatchewsky in the years 1826–1830 and Johann Bolyai in 1832 independently published a 'hyperbolic' geometry in which there are two distinct lines l_1 and l_2 which intersect the point P, are coplanar with l and do not intersect l, i.e. are two 'parallels'.

(2) Riemann in 1854 briefly considered a 'spherical' geometry in which every line through P coplanar with l meets l, i.e. there are no possible 'parallels'.

Both hyperbolic geometry and spherical geometry can be developed without any contradictions arising, but this is clearly not equivalent to a formal proof that contradictions will never arise. Such a formal proof was given by Beltrami (1868a) by providing a Euclidean 'model' for these non-Euclidean geometries, at least in the two-dimensional case when all the points and lines considered lie in the same plane. Beltrami showed that the axioms of elliptic geometry hold if 'straight lines' are taken to be great circles on a sphere, and that the axioms of hyperbolic geometry hold if 'straight lines' are taken to be geodesics (or lines of shortest length) on a 'pseudo-sphere', of which the simplest example is the 'tractoid'.

However, even Beltrami's proof is not complete for it assumes the consistency of the axioms of Euclidean geometry, and this requires an analysis of the concepts of congruence and of measurement which is one of the great achievements of the last hundred years.

7.3. *Curvature of Surfaces*

The work of Gauss in non-Euclidean geometry is known only from his private letters to Wolfgang Bolyai, the father of Johann Bolyai, and it is therefore difficult to assess its influence, but his work on the curvature of surfaces was of decisive importance (Gauss 1827).

All smooth curves in a plane, regarded as flexible and inextensible strings, are applicable to one another, but this is not true of smooth surfaces in space, regarded as flexible and inextensible membranes—a sheet of paper can be wrapped around a cylinder but not around a sphere. Gauss discovered the necessary and sufficient condition that two surfaces should be applicable to one another, without tearing or stretching, and expressed this condition in terms of the 'specific curvature K' of a surface at a point. The conditions for applicability are

(1) the obvious condition that there should exist a 'mapping' of one surface S_1 on to the other surface S_2 such that if neighbouring points P_1, Q_1 on S_1 correspond respectively to neighbouring points P_2, Q_2 on S_2, then the geodesic distances P_1Q_1 and P_2Q_2 are equal, and

(2) the far from obvious condition that the specific curvature of S_1 at P_1 should equal the specific curvature of S_2 at P_2.

The specific curvature is therefore an intrinsic property of a surface or rather of a class of developable surfaces. This intrinsic character is exemplified in Gauss' theorem that, if A,B,C are the internal angles of a triangle of area D formed by geodesics, then, as the triangle shrinks to a point P, the ratio $(A + B + C - \pi)/D$ tends to the specific curvature K at P. In a sphere of radius R, $K = 1/R^2$.

If the specific curvature is constant all over a surface S_1, and therefore all over any applicable surface S_2, then the surface S_1 can slide over itself in any manner without suffering any distortion by tearing or stretching, just as one piece of paper can slide freely over another, or as one spherical cap can slide freely over a sphere of equal radius. This property of surfaces of uniform specific curvature (whether it be positive, negative or zero) is called the property of 'free mobility' in such a surface.

Euclid's axioms of congruence express the property of free mobility in a plane. It seems that Gauss's investigation of specific curvature was undertaken with a view to establishing non-Euclidean geometry on an axiom of free mobility, which would guarantee the possibility of measurement by the transference from place to place of standards of length. This deliberately vague but evocative programme received precise and definite fulfilment later (see Sections 7.25–7.27).

7.4. *Projective Geometry*

Originally projective geometry was a subdivision of Euclidean geometry. The ellipse, the parabola and the hyperbola had been defined by Greek geometers as conic sections, i.e. as the projections of a circle (the base of the cone) from a point (the vertex of the cone) on a plane (a section of the cone). Fresh interest had been aroused by the geometrical theory of perspective drawing. The decisive influence, however, was the discovery of geometrical properties of figures which were unaltered by projection, such as the theorems of Pappus, Desargues and Pascal.

The great treatise of Poncelet (1822) expresses the hope that the haphazard procedures of Euclid may be replaced by a systematic development guided by the principles of projection, of continuity, and the introduction of imaginary points, such as the 'circular point at infinity', in which any circle meets the line at infinity in its own plane.

Most metrical properties disappear when geometrical figures are subjected to projection, but certain special metrical properties do survive.

The special metrical properties which persist after projection, or the projective invariants, are the cross ratios of two pairs of collinear points $(A,B; P,Q)$ i.e. the quantity

$$(A,B; P,Q) = (AP/BP) \div (AQ/BQ).$$

If the points A,B,P,Q are projected from any centre S onto any line l, and if the rays SA, SB, SP, SQ meet l in the points a,b,p,q then,

$$(AP/BP) \div (AQ/BQ) = (ap, bp) \div (aq, bq).$$

This property of the 'transversals' l was known to the great French geometers of the first half of the nineteenth century, e.g. Brianchon (1817), Poncelet (1822) and Chasles (1852). Indeed Laguerre-Verly (1853), as a student at the age of nineteen, generalized the result obtained by Chasles as follows.

Let $\alpha, \beta, \gamma, \ldots$ be a set of coplanar angles with vertices at $A,B,C, \ldots$. Let this figure be given a homographic transformation (i.e. any series of projective transformations) in which the points $A,B,C, \ldots$ become the points $A',B',C', \ldots$ and the circular points at infinity I,J become the points I',J'. Let the two sides of the angle at A' and the two lines $A'I'$, $A'J'$ intersect any transversal in a range of points with cross ratio a, and similarly for the angles at B', C' etc. Then any relation of the form $F(A, B, C \ldots) = 0$ is transformed into a relation of the form

$$F(-\tfrac{1}{2}i \log a, \ -\tfrac{1}{2}i \log b, \ -\tfrac{1}{2}i \log c, \ \ldots) = 0.$$

7.5. *The Problem of Measurement*

Euclid compares two triangles ABC and XYZ which are such that the angle at A equals the angle at X, while the sides AB, AC are respectively equal to the sides XY, XZ, and he roundly asserts that the triangle ABC can then be moved so it exactly coincides with the triangle XYZ. He appeals in fact to our everyday experience that the world does contain certain 'rigid' structures, which can be freely moved about, but which always exactly match whenever they are brought together, such as a carpenter's foot rule or a milliner's yard-stick.

The theory of measurement is an attempt to give mathematical expression to these concepts, and its history covers the period from von Staudt in 1847 until Whitehead in 1929. In order to include all the major contributions it seems necessary to abandon the schemes of classification proposed by Klein (1872) or Weyl (1923) and to divide our study into three parts, each characterized by certain presuppositions, which steadily become less specific and more general.

(1) The metrical theory, in global form due to Helmholtz and Lie, then in infinitesimal form due to Riemann, Levi-Civita, Weyl and Eddington.
(2) The projective theory, due to von Staudt, Cayley and Klein.
(3) The topological theory due to Menger and A. N. Whitehead.

The Work of Riemann

7.6. *Riemann's Concept of a Manifold*

It is one of the chief embarrassments of a historian that mathematical discoveries are not made in the right chronological order and, in particular, that some of the most profound and difficult concepts are adumbrated and discussed in one of the earliest texts that we have to cite on the problem of measurement, the *Habilitationschrift* of Riemann (Riemann 1867).

The verdict of Veronese (1894), quoted with approval by Russell (1897, p. 15) that 'Riemann in seiner Definition des Begriffs Grösse dunkel ist', may justify the following benign interpretation of Riemann's obscure and suggestive phrases.

(1) A one-dimensional manifold, M_1, is given by a continuous mapping

$$f : M_1 \to R$$

of the elements of M_1 (points p,q) into the real numbers, R, such that the function f is not constant in any open set in M_1. (This version of Riemann's definition implies that continuity is defined, topologically, in terms of open sets. The anachronistic use of this concept seems to be the simplest way of giving significance to Riemann's concepts of continuity and of a non-constant function.)

(2) An n-dimensional manifold, M_n, is defined inductively by a continuous mapping,

$$g : M_n \to R$$

of elements of M_n [$(n-1)$-dimensional manifolds M_{n-1}] into the real numbers R, such that the function g is not constant in any n-dimensional open set in M_n.

7.7. *Coordinate Systems*

From this concept of a continuous manifold M_n Riemann infers that, through any fixed point p in M_n, there passes an $(n-1)$-dimensional manifold M_{n-1}^* given by an equation of the form

$$g(M_{n-1}) = g(M_{n-1}^*)$$

and that, through the same point p^* in M_{n-1}^* there passes an $(n-2)$-dimensional manifold in M_{n-2}^*, given by an equation of the form

$$h(M_{n-2)} = h(M_{n-2}^*) \quad \text{etc.}$$

Thus the point p^* is characterized by the set of n numbers

$$g(M_{n-1}^*),\ h(M_{n-2}^*),\ \ldots$$

which are taken to be its coordinates in the system specified by the mapping functions, g, h .... Thus in three-dimensional Euclidean space M_3 we could take the manifolds M_2 to be concentric spheres, the manifold M_1 to be meridians of longitude, and the manifold M_0 to be points.

7.8. *Distance*

Having thus sketched the solutions of the problems of giving a definition of a manifold and its dimension and of coordinate systems, Riemann turns to the concept of 'distance'. Here he assumes that with any two points A, P of a manifold there is associated a number $\delta(A, P)$ to be called the 'distance' between A and P, and that the equation of a 'sphere' with centre A can be written in the form $f(P) = \text{constant}$, where f, regarded as a function of the coordinates $(x_1, x_2, \ldots, x_n)$ of P, has continuous second derivatives, and attains its minimum value, zero, at the point A with coordinates $(a_1, a_2, \ldots, a_n)$.

Then all the derivatives $\partial f / \partial x_n$ must vanish at A, and, unless all the second derivatives $\partial^2 f / \partial x_j \partial x_k \equiv g_{jk}(a_1, \ldots, a_n)$ vanish at A, the leading terms in the expression of f must form a positive definite quadratic function

$$Q = \sum_{j,k} g_{jk}(x_j - a_j)(x_k - a_k).$$

Assuming that the distance function $\delta(A,P)$ is approximately a homogeneous function of the coordinate differences $(x_k - a_k)$ when δ is small, Riemann concludes that to this approximation the quadratic expression Q is of the form

$$Q = \lambda(P)\delta^2(A,P).$$

The importance of this investigation lies not so much in the logic of the deduction, which is open to criticism, but in the conclusion that distance ds between two 'consecutive' points $(a_1, \ldots, a_n)$ and $(x_1, \ldots, x_n)$ has the form

$$ds = \{\textstyle\sum g_{jk}dx_jdx_k\}^{\frac{1}{2}}$$

when

$$dx_j = x_j - a_j.$$

Riemann himself noticed that in the exceptional case when all the second derivatives $\partial^2 f/\partial x_j\partial x_k$ vanish at A then the leading terms in the expression f must form a positive definite quartic function

$$\textstyle\sum g_{jklm}(x_j - a_j)(x_k - a_k)(x_l - a_l)(x_m - a_m).$$

This leads to manifolds with a line element of the form

$$ds = \{\textstyle\sum g_{jklm}dx_jdx_kdx_ldx_m\}^{\frac{1}{4}}.$$

7.9. Geodetic Coordinates

From the expression for the 'infinitesimal' line element ds, we can deduce the length l of any curve joining two points A and P, and hence deduce the equations of the geodesic, the curve which minimizes l for fixed points A and P. These equations have the form

$$\ddot{x}^i + \Gamma^i_{jk}\dot{x}^j\dot{x}^k = 0$$

where

$$x^i = x^i(t) \quad (i = 1, 2, \ldots, n)$$

are the parametric equations of the curve and the coefficients Γ^i_{jk} are defined by the equation

$$\Gamma^i_{jk} = \frac{1}{2}\sum_l g^{il}\left\{\frac{\partial g_{lk}}{\partial x_j} + \frac{\partial g_{lj}}{\partial x_k} - \frac{\partial g_{jk}}{\partial x_i}\right\}$$

We can now choose geodetic coordinates $y_1, y_2, \ldots, y_n$ such that the equations of the geodesics have the form

$$y^i = \eta^i s$$

where $\eta^1, \eta^2, \ldots, \eta^n$ are constants specifying the initial direction of the geodesic at A, and s is the arc length measured along the geodesic from A.

Riemann's brief discussion seems to imply that he knew that the terms of the second and fourth orders in the expression for ds^2 must now be of the form

$$ds^2 = \sum_i (dy^i)^2 - \frac{1}{3}\sum_{ij,\,rs} R_{ij,\,rs}\,(y^idy^j - y^jdy^i)(y^rdy^s - y^sdy^r)$$

and we may well believe that he identified the coefficients $R_{ij,\,rs}$ with the functions $(ij,\,rs)$ which he discovered in his researches on solutions of the heat equation (Riemann 1861).

In a Euclidean manifold all the coefficients $R_{ij,\,rs}$ vanish. On the surface of a sphere of radius a let us take as coordinates the longitude φ and the colatitude θ measured from the North Pole. Then the line element is given by

$$ds^2 = a^2 d\theta^2 + \sin^2\theta \; d\varphi^2$$

The geodesics drawn from the North Pole have the equations

$$s = a\theta, \; \varphi = \text{constant}$$

and the corresponding geodetic coordinates are ξ, η where

$$\xi = a\theta \cos\varphi \quad \text{and} \quad \eta = a\theta \sin\varphi.$$

Hence
$$ds^2 = (d\xi^2 + d\eta^2) - \tfrac{1}{3}a^2(\xi d\eta - \eta d\xi)^2 + \ldots$$

neglecting terms of higher orders in ξ, η.

In the general case the coefficients $R_{ij,rs}$ provide measures of the departure of the manifold from flatness, i.e. they are intrinsic measures of its curvature and are generalizations of the specific curvature of a two-dimensional surface introduced by Gauss.

Just as in the case of a surface $(n=2)$ the condition for the 'free mobility' is that the measures of curvature should be the same for all pairs of directions η^i, η^j and for all points on the surface. When this condition is satisfied we have a manifold of constant uniform curvature which may be positive, negative or zero, and in which new coordinates $(x_1 x_2, \ldots, x_n)$ can be introduced in such a way that the line element is given by

$$ds^2 = \sum_i dx_i^2 \div \left(1 + \tfrac{1}{4}K \sum_i x_i^2\right)^2$$

where K is the specific curvature.

For example on the sphere of radius a we can take

$$x = 2a \tan \tfrac{1}{2}\theta \cos\varphi$$
$$y = 2a \tan \tfrac{1}{2}\theta \sin\varphi$$

and
$$K = 1/a^2.$$

Theories of Distance

7.10. Cayley's Theory of Distance

Cayley's great discovery (Cayley 1859) in the metrical theory of distance is somewhat obscured in his famous paper, partly by the now antiquated language of the theory of 'quantics' (i.e. homogeneous polynomials) and partly by his own facility in algebraic manipulation.

The essence of Cayley's discovery can be stated as follows. In Euclidean space, if P,Q,R are three collinear points, if Q is between P and R

and if $\delta(P,Q)$, $\delta(Q,R)$, $\delta(P,R)$ are the Euclidean distances between P and Q, Q and R, P and R respectively, then

$$\delta(P,Q) + \delta(Q,R) = \delta(P,R).$$

The distances $\delta(P,Q)$ etc. are of course functions of the Cartesian coordinates of P,Q,R.

Cayley showed that there are other functions of these coordinates, which we may call 'Cayley distances', $\Delta(P,Q)$, $\Delta(Q,R)$, $\Delta(P,R)$, with precisely the same additive property, and that any such Cayley-distance function is related to a fixed conic in the plane, which he called the 'Absolute'.

To simplify the algebra, we restrict ourselves (following Cayley) to two-dimensional space S, and take *homogeneous* coordinates (x,y,z), in which the equation of the Absolute is

$$Q(x,y,z) \equiv ax^2 + by^2 + cz^2 + 2fyz + 2gzx + 2hxy = 0.$$

In the space S let the coordinates of the collinear points $P_\alpha (\alpha = 1, 2, 3)$ be $(x_\alpha, y_\alpha, z_\alpha)$, P_2 being between P_1 and P_3. Let P'_α be three points with the same numerical coordinates $(x_\alpha, y_\alpha, z_\alpha)$ in three-dimensional Euclidean space referred to oblique axes in which the Euclidean distance between the points P'_α and P'_β is

$$\delta(P'_\alpha, P'_\beta) = Q(x_\alpha - x_\beta, y_\alpha - y_\beta, z_\alpha - z_\beta).$$

The points P'_1, P'_2, P'_3 are collinear and P'_2 is between P'_1 and P'_3. Hence if O' is the origin, the angle $\angle P'_1 O'P'_3$ is the sum of the angles $\angle P'_1 O'P'_2$ and $\angle P'_2 O'P'_3$, i.e. if

$$(\alpha, \beta) = ax_\alpha x_\beta + by_\alpha y_\beta + cz_\alpha z_\beta + f(y_\alpha z_\beta + y_\beta z_\alpha) + g(z_\alpha z_\beta + z_\beta x_\alpha) + h(x_\alpha y_\beta + x_\beta y_\alpha)$$

then

$$\cos^{-1}\frac{(1,3)}{(1,1)^{\frac{1}{2}}(3,3)^{\frac{1}{2}}} = \cos^{-1}\frac{(1,2)}{(1,1)^{\frac{1}{2}}(2,2)^{\frac{1}{2}}} + \cos^{-1}\frac{(2,3)}{(2,2)^{\frac{1}{2}}(3,3)^{\frac{1}{2}}}$$

Thus the function

$$\Delta(P_\alpha, P_\beta) = \cos^{-1}\frac{(\alpha,\beta)}{(\alpha,\alpha)^{\frac{1}{2}}(\beta,\beta)^{\frac{1}{2}}}$$

(with suitable restrictions on the many-valued inverse cosine!) possesses the same additive property in two-dimensional space as the Euclidean distance function. This function is Cayley's distance function. When the Absolute consists of the pair of circular points at infinity, the Cayley distance reduces to the ordinary Euclidean distance.

7.11. *The Projective Theory of Measurement*

The whole of the metrical theory of measurement from Riemann and Helmholtz to Weyl and Eddington is based upon the assumption that it is possible to fix positions in space, or in space–time, by means of coordinates. But how can we introduce coordinates without employing some methods of measurement? And the object of all these theories was to justify the concept of measurement!

The argument therefore seems to revolve in a circle, but the difficulty is resolved by the researches of von Staudt (1847, 1856) who showed that it is possible to introduce coordinates without any appeal to quantitative measurement.

7.12. Von Staudt's Theory of 'Throws'

To introduce a system of numerical coordinates into a given straight line, von Staudt made use of the properties of the complete quadrangle, which had been known since the time of Möbius. Given any three points A,B,P on a straight line, a fourth point Q can be constructed (using only straight lines) so that $(A,B; P,Q)$ forms a harmonic range. With the metrical properties of Euclidean geometry this would imply that the ratio

$$(AP/PB) \div (AQ/QB) = -1.$$

But distances still have to be defined! A harmonic range is therefore defined projectively by the Möbius construction of the complete quadrilateral.

Again in Euclidean geometry a harmonic range would be formed by the points with coordinates

$$A, 0;\ B, 1;\ P, \tfrac{1}{2};\ Q, \infty.$$

Von Staudt therefore adopts the convention that a certain fixed point Q is to be given the coordinate ∞ and that if $(A,B; P,Q)$ is a harmonic range, and that if A,B have 'Staudtian' coordinates a,b then P has the Staudtian coordinates $\tfrac{1}{2}(a + b)$, and is the 'Staudtian bisector' of AB. Then, starting with two arbitrary points A_0 and A_1 with Staudtian coordinates 0 and 1, he proceeds by successive Staudtian bisectors ('throws') to assign Staudtian coordinates to an enumerable set of points on the given line. He proves that this process is self-consistent and that it can be extended to points in a given plane and then into points in space.

It is now possible to give a projective definition of a quadric using the definition due to Steiner (1881–2). The coplanar ranges of points $(P,Q, \ldots)$ and $(P',Q', \ldots)$ on two lines l and l' are projectively related if they can be obtained from a third range $(A,B, \ldots)$ by simple projection from a base point Ω. If O and O' are any other base points the intersections of the corresponding rays, $(OP,O'P')(OQ,O'Q') \ldots$, lie on a quadric passing through O and O'.

Von Staudt's investigations conclusively established the logical priority of projective over metrical geometry, but in extending his definition of distance from collinear ranges of points to coplanar sets of points he felt himself obliged to employ Euclid's parallel postulate.

7.13. Klein and Non-Euclidean Geometry

In fact it is possible to dispense with the parallel postulate altogether and this great achievement was made by Klein (1871, 1873).

The introduction of metrical coordinates by von Staudt and the possibility of alternative definitions of distance noticed by Cayley were fused together by Klein in a definitive manner. Klein starts with a special non-degenerate quadric surface, the 'Absolute', fixed once and for all, and defines the distance between any two points P,Q on a line l by considering the intersection A,B of l with the Absolute. The 'distance' between P and Q is then defined as $c \log r$ where r is the cross-ratio of the points (AB,PQ) and c is a constant chosen to make the distance a real positive number—a convention which is necessary because the Absolute may be an imaginary quadric and the intersections A,B may be imaginary points.

Klein classified the various species of 'non-Euclidean' geometry according to the character of the Absolute. In homogeneous coordinates the equation of the Absolute can always be chosen so that it has one of the forms

$$x^2 + y^2 = 0, \ z = 0 \quad \text{(Euclidean or parabolic),}$$
$$x^2 + y^2 - z^2 = 0 \quad \text{(Lobatchevsky or hyperbolic),}$$
$$x^2 + y^2 + z^2 = 0 \quad \text{(Riemannian or elliptic).}$$

Klein showed further that there are two species of Riemannian elliptic geometry accordingly as two straight lines intersect in one or two points and that the corresponding models of these geometries are provided by a complete spherical surface or a hemispherical surface (with the convention that antipodal points on the boundary are identified). The models of straight lines are the great circles or geodesics.

Finally Klein showed that the transformations in non-Euclidean geometry which correspond to the translations and rotations in Euclidean geometry are precisely the projective transformations which leave invariant the Absolute quadric.

There are a number of other interesting models of non-Euclidean geometry, which are most easily visualized in a non-Euclidean plane.

In his researches on automorphic functions, Poincaré (1882a) takes the Absolute to be a circle and the 'straight line' to be circles intersecting the Absolute at right angles.

7.14. *The Milne–Whitehead Criticism of Einstein*

In Einstein's general theory of relativity the interval ds between two events with coordinates x^i and $x^i + \mathrm{d}x^i$ is given by an expression of the form

$$\mathrm{d}s^2 = g_{\mu\nu}\mathrm{d}x^\mu \mathrm{d}x^\nu$$

where the coefficients $g_{\mu\nu}$ are themselves functions of the coordinates, determined by a certain set of partial differential equations and by the distribution of matter in the universe. But the distribution of matter cannot be specified without a knowledge of the metric, i.e. the coefficients $g_{\mu\nu}$. Hence both Milne (1935) and A. N. Whitehead (1922, p. v) conclude that Einstein's general theory is vitiated by a circularity in its fundamental definitions.

The problem which remained was to graft a theory of gravitation, or at least of the general distribution of matter, on to the special theory of relativity, accepted (with their own special interpretations!) by both Milne and A. N. Whitehead.

Milne solved this problem by positing a new general principle of relativity to the effect that the global distribution of matter (i.e. of galactic nebulae!) should appear the same to all observers in uniform relative motion, and he thus successfully predicted the recession of distant nebulae with velocities varying directly with their distance from the observer.

A. N. Whitehead adopted a much more radical approach by attempting to give *a priori* definitions of an 'event', a point, a rectilinear geodesic and a straight line. This ambitious theory was heavily indebted to the projective theory of measurement which we have described above (Sections 7.11 and 7.12).

7.15. The Topological Theory

It was undoubtedly a great achievement by von Staudt to show how metric relations could be defined in projective geometry, merely from the incidence properties of straight lines. But even when projective geometry had been formalized and given a satisfactory axiomatic basis by the Italian geometers, Pasch and Pieri, there remained the inevitable question—is it really necessary to adopt as undefined entities the 'point' and the 'straight line'?

The objects of sense perception are certainly not points nor straight lines. They are lumps of matter, and points are at best an abstraction or logical figment. It therefore seems desirable, and perhaps necessary, to investigate the possibility of basing geometry and measurement on lumps, or volumes, rather than points. Indeed it may be possible to go further, for the concept of a lump seems to imply a volume with a well-defined boundary—and this is something which transcends direct observation. Menger (1928, p. 15) certainly envisaged the possibility of a topology based on something 'still more vague' than lumps, something like a cloud of tobacco smoke, or the fuzzy clouds described by Perrin (1914) in his researches on Brownian movement. Such a theory did not appear until 1965, when 'Fuzzy systems' began to develop.

7.16. Huntingdon and 'Spheres'

The first attempts to construct a geometry of 'lumps' seems to have been made by Huntingdon (1913), who showed that Euclidean geometry could be based on the ideas of 'spheres' and a relation of 'inclusion'. In this scheme a point is defined as a sphere which does not include any other sphere.

If A and B are any two distinct points, and if X is a point such that every sphere which contains A and B also contains X, then X is said to belong to the segment $[AB]$ or $[BA]$, and there are similar definitions for a triangle and tetrahedron. Coplanar lines are said to be parallel if they have no point in common. Four points A,B,C,D form a parallelogram if AB is parallel to CD and if BC is parallel to DA. Huntingdon's postulates are sufficient to ensure that the diagonals AC and BD intersect in only one point, which is called the midpoint of AC and BD. Thus a Euclidean metric is introduced into the geometry.

Huntingdon's system is interesting for the use of a solid body (a 'sphere') instead of the point as an undefined concept, and for his examination of the independence, consistency and categoricity of his postulates.

7.17. Russell and the Elimination of Abstractions

The great contribution of Bertrand Russell to the study of the foundations of mathematics is a rigorous application of Ockham's razor—'entities are not to be multiplied without necessity'. He had before him the definitions of 'real' numbers offered by Dedekind and Weierstrass and he had the inspiration of close association with A. N. Whitehead. Close examination of the definitions of a real number reveal that the existence of real numbers is, in fact, postulated, and not established. Thus Dedekind constructs a 'cut' or species of dichotomy of the totality of rational numbers. Weierstrass in effect constructs a sequence of rational numbers with a certain internal property of convergence. And these logical constructions are then taken to be evidence of the existence of real numbers, said to be obtained by a process of 'abstraction'.

Russell claimed to sharpen these definitions by rephrasing them in the bald, drastic form that a real number *is* a Dedekind cut, or a real number *is* a convergent sequence of rational numbers. This completely eliminated real numbers as abstractions and replaced them by constructive definitions as cuts or sequences. Russell's criticisms of real number definitions are expounded in *The Principles of Mathematics* (Russell 1937), but they were also adumbrated in his earlier lectures on 'Our Knowledge of the External World . . .' (given in 1914). In Chapter IV of this work Russell explains how a similar reductive method had been invented by Whitehead to give a definition of 'points' in terms of such sense data as lumps of matter, by what we may call the 'Chinese Box' method of constructing a monotone decreasing and convergent sequence.

7.18. A. N. Whitehead and Extensive Abstraction and Connection

A. N. Whitehead had always been deeply concerned with the difficult and challenging cosmological problems of the origin and validity of our concepts of space and time. His two Cambridge tracts on projective and descriptive geometry (A. N. Whitehead 1906, 1907) give admirable accounts, in the Whiteheadian manner, of the theories summarized above of the foundations of metrical and projective geometry. His *Universal Algebra* (A. N. Whitehead 1897) envisages geometry, with mathematical logic and statics, as particular examples of a general system of algebra. But it was the advent of Einstein's special theory of relativity which impelled Whitehead to return to the cosmological problem, which now appeared as a pressing question of contemporary physics. Whether or not we are interested like S. Alexander in the philosophy of space time (S. Alexander 1916–18), or like Robb in the axiomatics of the Lorentz transformation (Robb 1936), A. N. Whitehead's investigations have a profound significance for the topologist interested in abstract spaces. The following account of his work may omit certain aspects which no longer seem to be vital issues, but it summarizes the results which are of perennial value.

In his *Enquiry concerning the Principles of Natural Knowledge* (A. N. Whitehead 1919) he began to develop the 'method of extensive abstraction' by which points in space in Newtonian cosmology or point-events in space–time in Einsteinian relativity can be defined in terms of a set of 'lumps' or extended events which are arranged in serial order by the undefined relation of extension (or its converse—inclusion). Such a set he calls an 'abstractive class'. This, as Menger has shown, is analogous to the definition of a real number as a nested sequence of rational intervals whose lengths approach zero. However, as soon as the ideas of measure and distance are discarded it is necessary to ensure that the sequence of 'lumps' converges in such a way that it can be used to define a point (or point-event).

A. N. Whitehead began to solve this problem by introducing the concepts of an abstractive class which is 'prime, or anti-prime, in respect of a formative condition σ'. Two abstractive classes are equivalent when each class contains an element which extends over some element (and therefore an infinity of elements) of the other. The concept of primes and anti-primes seems to be A. N. Whitehead's way of defining an 'abstractive element' as a set of equivalent abstractive classes. With this concept he is able to define those special abstractive elements which cover no other abstractive elements. In the modern nomenclature of Boolean algebra these would be called 'atoms' and they would correspond to Euclidean points or Einsteinian events.

Unfortunately the further development of Whitehead's theory (A. N. Whitehead 1919) was frustrated by the introduction of the concept of a 'duration', which may be evocatively described as a slab of space bounded by two planes, or the corresponding infinitely extended event in space–time. But, although this concept seems to reintroduce Newtonian simultaneity and to lack any topological significance, it is a curious anticipation of J. W. Alexander's theory of 'gratings' (J. W. Alexander 1938).

A. N. Whitehead's account of his definition of 'punctual elements' is by no means pellucidly clear and both Russell (1922) and Nicod (1924) have given illuminating commentaries on the method of extensive abstraction. It seems, however, that Whitehead's definition is quite adequate as a basis for the introduction of a Hausdorff space in which each pair of points have disjoint neighbourhoods.

7.19. The 'Points' of Wald and R. L. Moore

As early as 1928 Menger had stated the desirability of giving a definition of points as nested sequences of regions of space (Menger 1928, p. 15), and in 1940 he gave an account of a number of methods of ensuring that such a sequence has a unique limit, namely a single point (Menger 1940).

Thus Wald (1932) gave the following criterion for the 'convergence' of a strictly decreasing sequence of open sets $\{U_k\}$ to a single point: if the open set V does not contain any of the U_k completely, and if W is any open set completely contained in V, then W is disjoint from almost all U_k.

A rather more elaborate condition was given by R. L. Moore. He noted (R. L. Moore 1935) that Huntingdon's system 'makes much use of what may be termed the convexity of his undefined spatial elements'—a concept on which A. N. Whitehead (1929, pp. 427–38) relied for his definition of a straight line. Moore himself gave a 'system of axioms for plane analysis situs' based on the notion of a 'piece' or 'chunk' of space and the relation 'embedded in'.

7.20. A. N. Whitehead and 'Extensive Connection'

In A. N. Whitehead's Gifford Lectures delivered at Edinburgh in 1927–8 (Whitehead 1929, pp. 407, 416–38) he returned to the problem of an abstract characterization of points and now formulated the much more ambitious problem of a non-metrical definition of a straight line.

His theory of extensive abstraction had in fact already given a complete and satisfactory definition of a point as an 'atom' (although Whitehead seems not to have been explicitly aware of this achievement). To give an abstract definition of a straight line he needed to characterize 'convex' regions, and for this purpose he required some theory of the 'contact' of solids. He found the materials ready to hand in the investigations of de Laguna (see Section 7.21).

The theory of extensive *connection* provided a definition of inclusion and hence incorporated the theory of points as in the earlier theory of extensive *abstraction*, but it also provided a theory of external contact of regions and of 'connectedness', and hence a definition of a 'class of ovals'. At this point Whitehead introduces 'ovate abstractive sets', which are abstractive sets, serially ordered by the relation of inclusions, but consisting of ovals of the same class.

With considerable ingenuity Whitehead defines those complete ovate abstractive sets which cover two, three of four prescribed points.

If we picturesquely regard these ovate sets as gradually shrinking into the smallest convex shape which is inexorably attached to the prescribed points, we see that Whitehead has succeeded in defining straight lines, planes and three-dimensional regions. Moreover, the dimensionality, n, of the underlying space (or space–time) is defined by the condition that such ovate sets exist for n points but not for $(n + 1)$ points.

A. N. Whitehead left open the question of the existence of more than one ovate class for a given space (or space–time), but it is clear that from only one ovate class we can obtain another by a continuous transformation. Hence in the Euclidean plane the Whiteheadian 'straight line' joining points P and Q might be the arc of the circle passing through P and Q and a fixed base point O. It is implicit in Whitehead's topological approach that, although he has not achieved a unique and specific definition of straight lines and planes, he has obtained the best possible result.

7.21. De Laguna and 'Solidity'

Independently of A. N. Whitehead's researches, de Laguna (1922, p. 449–61) had already advanced a topological theory in which points were constructed from solids and which endeavoured to include in the axiomatic characterization of solids that quality of 'solidity' or impenetrability which appears as such an immediate sense experience. De Laguna introduces the undefined binary relation of 'extensive connection' existing between regions (i.e. lumps, solids or bounded, closed regions) which would be described informally as overlapping or as having at least one point in common.

Axiomatics in Geometry

7.22. The Axiomatic Foundations of Geometry

From the 'Elements of Euclid' in ca 300 BC to the *Eléments de Géometrie* of Legendre in 1794 it was agreed that the foundations of geometry were self-evident propositions, and the attempts of Clairaut in 1741 and L. Bertrand in 1778 to provide an observational and even experimental basis had little success. It was the rise of non-Euclidean geometry which really stimulated a more critical attitude, and among the first to emancipate themselves from what they stigmatized as the 'faux point de vue metaphysique' were Houel 1867 and Méray in 1874, together with a far greater mathematician—the incomparable Riemann, whose contributions to geometry we have described above (Sections 7.5–7.9).

According to A. N. Whitehead, a judicious and well-informed critic, 'the first satisfactory exposition of the subject', i.e. of the *axioms* of geometry, is due to Pasch (1882b), who explicitly recognized that the basic elements of any geometry must remain undefined and characterized only by the axioms which are asserted about those elements.

Pasch was the forerunner of an influential Italian school of geometers who developed the axiomatic method with great clarity and precision. The most important of these was undoubtedly Giuseppe Peano, in whose work (Peano 1894) the undefined ideas are a class of entities called 'points' and the class of points lying between any two given points, called a 'line'. He was followed by Pieri (1898) and Enriques (1898), and, surprisingly enough, by A. N. Whitehead (1906, 1907). A magisterial exposition of the axiomatic method for projective geometry was given by Veblen (1904) and Veblen and Young (1910, 1918). These researches are conveniently summarized by Whitehead (1906, 1907) and in the dissertation submitted by Russell for a Fellowship at Trinity College, Cambridge in 1895 (Russell 1897). This work is of immense value for its critical survey of the history of 'metageometry' and for its acute and critical account of previous philsophical theories.

It was Whitehead who interested Russell in the philosophical importance of projective geometry, and Russell's essay is invaluable also for the light it throws on the subsequent development of mathematical logic in which Whitehead and Russell collaborated with such remarkable success.

A. N. Whitehead (1906, 1907) gives a general exposition, with very full historical references, of the ideas of harmonic ranges, projectivity, order, cross-ratio and especially the introduction of numerical coordinates. A. N. Whitehead (1907) simplifies Lie's solution of the 'Riemann–Helmholtz problem' by taking the group of congruence transformations to be a subgroup of the general projective group.

7.23. Klein and the Erlanger Programme

The renewed interest in non-Euclidean geometries familiarized mathematicians with the idea that there are many geometries and raised the question, what is the formal relation between these disciplines? This was the question which Klein endeavoured to settle in the speech which he delivered in 1872 on the occasion of his admission to the faculty of the University of Erlanger (Klein 1893).

According to Klein, each geometry is specified by a group of transformations, and the theorems of a particular geometry concern the invariant structures under the relevant group. This scheme is admirably adapted to the so-called 'non-Euclidean geometries' which are specified by those subgroups of the three-dimensional projective group which leave invariant certain types of quadric, elliptic, hyperbolic or parabolic. Klein's classification can also be stretched to include differential geometry, regarded as the theory of the invariants and covariants of the continuous transformations of the quadratic differential form for the elementary distance function (Section 7.29).

However, later developments have broadened our concepts of geometry and Klein's programme no longer attracts an audience. Differential geometry is now developed from the concept of 'affine connection' as in Milnor (1963) which provides an axiomatic definition of a covariant derivative to replace the pioneering definition given by Christoffel (see Section 7.29). Even topology can be subsumed under the rubric of the group of all continuous transformations of a space into itself, but this is a motivation rather than a purposive definition.

7.24. Hilbert and the Foundations of Geometry

The other outstanding contribution to the study of the axiomatic basis of geometry is undoubtedly the great work of Hilbert, *Grundlagen der Geometrie*, first published in 1899, and subsequently in nine further editions, and in translations into many European languages, of which editions some are based on the French translation of Laugel in 1900. The best account is in the critical edition prepared by Rossier (Hilbert 1899).

Hilbert perfected the axiomatic treatment of the foundations of geometry. Where it was possible he reduced the number of axioms posited by previous workers, and when it was necessary he explicitly introduced a new axiom. He systematically classified the axioms into five classes which respectively clarified the concepts of incidence, of order, of congruence, of parallelism, and of continuity. He gave formal proofs that his system of axioms was consistent. He was, however, forced to admit the justice of the criticism made by Schur (1902) that his axioms did not form an independent system. This blemish was removed by Veblen and Whitehead. In the course of his researches Hilbert invented a non-Archimedean geometry, which allowed for infinitesimals (Chapter 3), and a non-Desarguesian geometry.

7.25. *The Problem of Congruence*

Riemann's researches contain in an informal, suggestive and enormously influential form not only a sketch of 'spherical geometry' but also the concepts of a manifold, of general coordinate systems, of the necessarily Pythagorean form for the infinitesimal line element, of geodetic coordinates and of the famous four-index symbols.

After Riemann the theory of measurement is the history of attempts to answer three questions of deepening abstraction and difficulty:

(1) What is the justification for the comparison of distances by the transport of rigid bodies?

(2) Is it possible to define distance without the use of rigid bodies, using only the projective geometry of straight lines?

(3) Is it possible to give a topological definition of a straight line?

The first question was attempted by Helmhotlz and answered by Lie and Poincaré. The special theory of relativity generalized the question to space–time measurements which required the transport not only of rigid scales but also of clocks. The generally accepted solution was given by Einstein in terms of the Lorentz transformation, but alternative and equally valid solutions were given by Robb, Milne and A. N. Whitehead.

These solutions of the first question were 'global' in form, but the question can also be posed in a 'local' form, where interest is concentrated on the infinitesimal expression for the line element. The necessary fundamental analysis was provided by Christoffel's theory of covariant differentiation, and by Ricci's general tensor calculus. Solutions of the 'transport' problem were given by Levi-Civita's theory of parallel displacement and by the theory of the mobile reference frame ('n-bein') due to Hessenberg and E. Cartan (1951). The last word seems to have been uttered by Weyl in his solution of the 'Raum-problem'.

The answer to the second question is provided by the projective theory of distance, implicit in the work of Cayley and explicit in the definitive researches of von Staudt. The necessary examination of the axiomatic foundation of projective geometry was given by Pasch and Pieri, and finalized in Hilbert's classical work.

The deepest and most difficult question is the third—what is the definition of a straight line? The first attack on this problem was made by Menger in an audacious construction of geometry without points. Independently Whitehead was led from his cosmological investigations applied to special relativity to formulate the method of extensive abstraction and to give definitions of 'point-events', of convex sets and of straight lines.

7.26. *Helmholtz's Theory of Congruence*

Helmholtz's interest in the foundations of geometry arose from his work on the physiology of vision and his researches exemplify Weyl's dictum that 'physics is too difficult for physicists', by which aphorism I presume he meant that the solution of physical problems may require very sophisticated mathematical techniques.

Lie's theory of continuous groups of transformations was not available when Helmholtz, stimulated by Riemann's theory of distance, undertook to deduce the expression for the line element from axioms of congruence. In fact, Helmholtz's proofs are far from rigid, but they provided the groundwork for Lie's investigations.

Helmholtz's researches (Helmholtz 1868) are based on four axioms which can be formulated as follows:

(1) The points of physical space form an n-dimensional manifold, i.e. they can be mapped into a Euclidean space of n dimensions.

(2) In any motion of a rigid body there is a numerical function of the $2n$ coordinates of any point-pair which remains invariant.

(3) A rigid body can be moved so as to transfer any point to any other assigned point.

(4) If $(n - 1)$ points of a rigid body are fixed, then in any motion of the body any point describes a closed curve.

From these axioms Helmholtz attempted to establish that if ds is the element of distance between neighbouring points with coordinates (x,y,z), $(x + \mathrm{d}x,\ y + \mathrm{d}y,\ z + \mathrm{d}z)$ then ds^2 is a homogeneous quadratic function of dx, dy, dz. It seems generally agreed that his argument is unconvincing, and Lie devoted 34 pages to a criticism, followed by his own two solutions of the 'Riemann–Helmholtz' problem (Lie 1886).

7.27. *Lie's Solution of the 'Riemann–Helmholtz' Problem*

A definitive solution of the problem of congruence, posed by the work of Riemann and Helmholtz, was finally given by Lie (1886); see also Lie (1890, 1893). In fact Lie gave two solutions starting from two different characterizations of the motion of 'rigid bodies' in a three-dimensional manifold, with a coordinate system (x,y,z).

Both solutions depend upon the expression of our physical experience in the form that the possible motions of a rigid body must form a continuous group of transformations. Once this fundamental fact had been explicitly stated, Lie was in a position to employ all the powerful machinery of his theory of continuous groups.

Lie's first solution depends upon his definition of 'free mobility in the infinitesimal' in the form that, if a point P and an arbitrary real line element passing through P are fixed, then continuous motion is still possible, but if in addition another different real line element passing through P is also fixed, then no continuous motion is possible. This is all that Lie requires in order to prove that the coordinates of the space can be chosen so that the group of rigid body motions is either

(1) the group of translations and rotations in Euclidean space, or
(2) the six-parameter group of all real projective transformations which leave invariant either the imaginary surface

$$x^2 + y^2 + z^2 + 1 = 0$$

or the real surface

$$x^2 + y^2 + z^2 - 1 = 0.$$

In other words the geometry of a space in which rigid bodies have free mobility in the infinitesimal must be either Euclidean or one of the species of non-Euclidean geometry.

Lie's second solution assumes that, when an arbitrary real point (a,b,c) is fixed, then the real points (ξ,η,ζ) into which it is possible to move an arbitrary point (x,y,z) satisfy an equation of the form

$$W(a,b,c;\ x,y,z;\ \xi,\eta,\zeta) = 0$$

and the inequalities

$$\xi \neq a,\ \eta \neq b,\ \zeta \neq c$$

the function W being continuous in ξ,η,ζ. Once again Lie shows that this assumption is sufficient to show that the space of rigid-body motion is either Euclidean or 'non-Euclidean'.

Lie's theory of continuous groups inspired Poincaré (1887a) to examine the problem of congruence for motion in a plane, regarded as a two-dimensional manifold in which the position of a planar rigid figure is completely specified by three conditions.

Poincaré shows that even this very general condition is sufficient to limit the possible geometries to

(1) Euclidean and 'non-Euclidean' geometries, and
(2) two somewhat bizarre geometries in which a uniplanar rigid figure can still move although two distinct points are fixed.

7.28. The Metric of Relativistic Space–Time

The classical or non-relativistic problem of congruence envisaged a rigid body *at rest* in two different positions, and the transformation from one such position to another. But the question was radically changed by the special and general theories of relativity.

In the special theory of relativity we envisage the same rigid body (or at least the same rigid measuring rod, or clock!) in two different frames or reference which may be in *relative motion*. As a result we have to deal with a manifold in which there are three space-like coordinates x,y,z and a fourth time-like coordinate t. The distance between two points in space is now to be replaced by the 'interval' between two 'events' with coordinates (x_1,y_1,z_1,t_1) and (x_2,y_2,z_2,t_2). In the special theory of relativity the square of this interval is given by the expression

$$c^2(t_1 - t_2)^2 - (x_1 - x_2)^2 - (y_1 - y_2)^2 - (z_1 - z_2)^2$$

where c is the velocity of propagation of light. In the general theory of relativity we have only an expression for the square of the interval ds^2 between two neighbouring events in the form of a quadratic in their coordinate differences.

To justify these generalizations of the classical space problem is the relativistic problem of congruence, i.e. of spatial and temporal measurement. The gradual appreciation of the nature of this problem from the pioneer researches of Fitzgerald, Poincaré (in 1900), Lorentz (in 1904), Einstein (in 1905) and Minkowski (in 1908) in the special theory of relativity, and in the researches of Einstein (in 1913–14), Weyl (in 1918) and Eddington (in 1921) in the general theory is admirably summarized in Whittaker's great history (Edmund Whittaker 1953).

The numerous axiomatic approaches to the special theory (Edmund Whittaker 1953, p. 43) attempt to characterize the group G of transformations which relate the space–time measurements made by rigid rods and clocks in uniform motion, and are usually based on the postulate of the Euclidean geometry of rigid bodies in relative rest. To this postulate is added the postulate that the group G leaves invariant the linear equations which purport to describe the motion of a particle free from external influence.

The congruence problem has then three possible solutions.

(1) The Newtonian solution, in which G leaves invariant the spatial distance

$$(x_1 - x_2)^2 + (y_1 - y_2)^2 + (z_1 - z_2)^2$$

and the temporal duration

$$t_1 - t_2$$

between any pair of events with coordinates (x_1,y_1,z_1,t_1) and (x_2,y_2,z_2,t_2).

(2) The Minkowskian solution in which G leaves invariant only the square of the 'interval'

$$c^2(t_1 - t_2)^2 - (x_1 - x_2)^2 - (y_1 - y_2)^2 - (z_1 - z_2)^2.$$

(3) The bizarre solution in which G leaves invariant the expression

$$c^2(t_1 - t_2)^2 + (x_1 - x_2)^2 + (y_1 - y_2)^2 + (z_1 - z_2)^2.$$

In the Minkowskian world, the past and future relative to the event 0 $(x = y = z = t = 0)$ lie in the regions

$$t < - \sqrt{(x^2 + y^2 + z^2)}$$

and

$$t > \sqrt{(x^2 + y^2 + z^2)}$$

respectively, while rays of light converging to, or diverging from, this event are the generators through 0 of the light cones $t = \pm \sqrt{(x^2 + y^2 + z^2)}$. The events 'outside' the light cones are 'copresent' with 0 and inaccessible by any signals.

In the world corresponding to the third solution there is no distinction of past and future.

Whereas the special theory of relativity is essentially a 'global' theory, the general theory is a 'local' theory and represents a return to Riemann's conception of the characterization of a manifold by its infinitesimal properties. Again in the special theory, there was a privileged set of frames of reference which were in uniform motion of translation with respect to one another. In Einstein's general theory, there are no such priviliged systems of reference. However, the Einstein world is locally Minkowskian in the immediate neighbourhood of any event, and *local* coordinates X,Y,Z,T can be determined such that the general expression for the square of the interval, i.e.

$$ds^2 = \sum_{\mu,\nu=1}^{4} g_{\mu\nu} dx^\mu dx^\nu$$

where the coefficients $g_{\mu\nu}$ are functions of x^1, x^2, x^3, x^4, reduces to

$$ds^2 = c^2 dT^2 - dX^2 - dY^2 - dZ^2$$

near any chosen event.

The Absolute Differential Calculus

7.29. Christoffel and Covariant Differentiation

The familiar spherical coordinates r,θ,φ had been used by Euler and Laplace, but it was Lamé (1833, 1859) who introduced general curvilinear coordinates, in order to obtain new solutions of the equation for the conduction of heat.

Lamé's work is mainly of interest now because of his discovery of new functions required in these problems and because he familiarized geometers with the idea that the intrinsic properties of a surface are implicit in the expression for the distance ds between two points on the surface with curvilinear coordinates (u^1, u^2, u^3) and $(u^1 + du^1, u^2 + du^2, u^3 + du^3)$

$$ds^2 = \sum_{\mu,\nu} g_{\mu\nu} du^\mu du^\nu$$

where the coefficients $g_{\mu\nu}$ are functions of u^1, u^2, u^3.

Riemann's researches (Riemann 1861) on curvilinear coordinates were undertaken to solve a problem proposed by the Académie des Sciences of Paris in 1858—to determine a system of surfaces $f(x,y,z) = $ constant, such that the equation of heat conduction possesses a solution in which the temperature at each instant is uniform over each such surface.

These investigations were naturally restricted to two-dimensional surfaces but they led Christoffel (1869) to consider the geometry of an n-dimensional manifold which is prescribed only by the quadratic form

$$ds^2 = \sum_{\mu,\nu=1}^{n} g_{\mu\nu}dx^\mu dx^\nu$$

which determines the distance ds between two neighbouring points.

In the first phase of differential geometry, the starting point is the expression for the element of distance ds between two neighbouring points with coordinates $(x^1, x^2, \ldots, x^n)$ and $(x^1 + dx^1, x^2 + dx^2, \ldots, x^n + dx^n)$ in an n-dimensional manifold V_n, in the form

$$ds^2 = g_{\mu\nu}dx^\mu dx^\nu$$

where $g_{\mu\nu}$ is a function of the variables $(x^1, x^2, \ldots, x^n)$ and where we use the most useful convention, due to Einstein (1916) that any expression which contains the same index (such as μ) twice, once as a subscript and once as a superscript, is to be summed over all values of this index, $\mu = 1$, $2, \ldots, n$.

In another system of coordinates $(y^1, y^2, \ldots, y^n)$ the line element would have a similar form,

$$ds^2 = h_{\mu\nu}dy^\mu dy^\nu$$

with different coefficients $h_{\mu\nu}$, which are now functions of $(y^1, y^2, \ldots, y^n)$ and the two sets of coordinates would be related by equations of the form

$$x^\mu = f^\mu(y) \equiv f^\mu(y^1, y^2, \ldots y^n).$$

The intrinsic properties of the manifold V_n must, of course, be independent of the system of coordinates which is employed, and must therefore be expressed in a form which remains invariant under all transformations of the form

$$x^\mu = f^\mu(y).$$

Two closely connected questions arise at once:

(1) What are these invariants and what form do they take?

(2) What are the conditions that two quadratic forms, $G_2 = g_{\mu\nu}dx^\mu dx^\nu$ and $H_2 \equiv h_{\mu\nu}dg^\mu dy^\nu$ are equivalent, in the sense that they are transformed into one another by coordinate transformations of the form $x^\mu = f^\mu(y)$?

In the case of a two-dimensional manifold the simplest invariant is the Gaussian curvature K (Gauss 1827). To define the Gaussian curvature K of a surface S at a point P we consider the normal $n(Q)$ to the surface S at a point Q near P and map Q into the point Q' on a unit sphere with fixed centre O, such that OQ' is parallel to $n(Q)$. Then as Q describes a small area A on the surface S, its image Q' describes a small area A' on the unit sphere. The Gaussian curvature K is then defined as the limit of the ratio A'/A as A shrinks to the point P.

The second problem for n-dimensional manifolds was considered by Riemann (1861) when the functions have the special values $h_{\mu\nu} = 1$ if $\mu = \nu$, $h_{\mu\nu} = 0$ if $\mu \neq \nu$, appropriate to a Euclidean manifold.

The first systematic examination of the general problem was made by Christoffel (1869) whose results may be summarized as follows: the equivalence of the two quadratic forms G_2 and H_2 is expressed by the $\frac{1}{2}n(n+1)$ differential equations

$$g_{\mu\nu}\frac{\partial x^\mu}{\partial y^\varrho}\frac{\partial x^\nu}{\partial y^\sigma} = h_{\varrho\sigma}$$

By differentiating these equations twice and eliminating all derivatives of the second and third orders, Christoffel obtained relations of the form

$$(\alpha\beta\gamma\delta) = \sum (ghik)\frac{\partial x^g \partial x^h \partial x^i \partial x^k}{\partial y^\alpha \partial y^\beta \partial y^\gamma \partial y^\delta}$$

where the four index symbols are the functions of the coefficients g_{ik} and their first derivatives, or of $h_{\beta\delta}$ and their first derivatives, originally introduced by Riemann. Thus the equivalence of the quadratic forms G_2 and H_2 implies the equivalence of the quartic forms

$$G_4 = (ghik)d^{(1)}x^g d^{(2)}x^k d^{(3)}x^h d^{(4)}x^i$$

$$H_4 = (\alpha\beta\gamma\delta)d^{(1)}y^\alpha d^{(2)}y^\delta d^{(3)}y^\beta d^{(4)}y^\gamma$$

where there are four sets of differentials in each form.

Christoffel also considered the generalization of these results to the problem of the equivalence of two homogeneous polynomial forms of order m in the infinitesimals dx^μ and dy^μ, i.e. the consequences of a relationship of the form

$$(l_1 l_2 \ldots l_\mu)dx^{l_1}\partial x^{l_2} \ldots \partial x^{l_\mu} = (\alpha_1\alpha_2 \ldots \alpha_\mu)\partial y^{\alpha_1}\partial y^{\alpha_2} \ldots \partial y^{\alpha_\mu}$$

the number μ of indices $l_1, l_2, \ldots, l_\mu$, being greater than 2 and less than n, the dimension of the space. He showed that this relation implies a further similar relation between two similar polynomial forms of order $\mu + 1$,

$$(ll_1l_2 \ldots l_\mu)\partial x^l \partial x^{l_1}\partial x^{l_2} \ldots \partial x^{l_\mu} = (\alpha\alpha_1\alpha_2 \ldots \alpha_\mu)\partial y^\alpha \partial y^{\alpha_1}\partial y^{\alpha_2} \ldots \partial y^{\alpha_\mu}$$

where

$$(ll_1l_2 \ldots l_\mu) = \frac{\partial(l_1l_2 \ldots l_\mu)}{\partial x^l}$$

$$- \sum_\lambda \left[\left\{ \begin{matrix} ll_1 \\ \lambda \end{matrix} \right\}(\lambda l_2 \ldots l_\mu) + \left\{ \begin{matrix} ll_2 \\ \lambda \end{matrix} \right\}(\lambda l_1 \ldots l_\mu) + \ldots \left\{ \begin{matrix} ll_\mu \\ \lambda \end{matrix} \right\}(\lambda l_1 \ldots l_{\mu-1}) \right]$$

and there is a corresponding expression for $(\alpha\alpha_1\alpha_2 \ldots \alpha_\mu)$. The 'three-index symbols' are defined by

$$\left\{ \begin{matrix} ll_m \\ \lambda \end{matrix} \right\} = \frac{1}{2}\left\{ \frac{\partial g_{l_mk}}{\partial x^l} + \frac{\partial g_{lk}}{\partial x^{l_m}} - \frac{\partial g_{ll_m}}{\partial x^k} \right\}g^{k\lambda}$$

and better written as $T^\lambda_{ll_m}$ to indicate the quasi-covariant or contravariant character of the suffixes λ or l and l_m.

7.30. Ricci and Covariant Differentiation

Christoffel seems to have been almost exclusively occupied with the problem of the equivalence of two polynomials in the differentials of the coordinates, but his results were utilized by Ricci and Levi-Civita (1901) in their formulation of the 'absolute differential calculus' (ADC).

This is the natural generalization of vector analysis in Euclidean space to tensor analysis in any metric space with n coordinates. Vectors in the classical sense play a very subordinate role and the main subject of the ADC is 'tensors'.

The simplest examples of these are the differentials $(dx^1, \ldots, dx^n)$ which transform according to the law

$$dy^\mu = (\partial y^\mu / \partial x^\nu) dx^\nu$$

(when we pass from coordinates x^ν to y^μ) and the derivatives $(\partial \varphi / \partial x^1, \ldots, \partial \varphi / \partial x^n)$ which transform according to the law

$$\partial \varphi / \partial y^\mu = (\partial \varphi / \partial x^\nu)(\partial x^\nu / \partial y^\mu).$$

Tensors like dx^μ are called contravariant and tensors like $\partial \varphi / \partial x^\mu$ are called covariant.

The most general tensors considered by Ricci and Levi-Civita have both covariant and contravariant characters and transform according to the law

$$T^{\alpha\beta}_{\mu\nu}\cdots = S^{ab}_{mn}\cdots \frac{\partial y^\alpha}{\partial x^a} \frac{\partial y^\beta}{\partial x^b} \cdots \frac{\partial x^m}{\partial y^\mu} \frac{\partial x^n}{\partial y^\nu} \cdots$$

It must be admitted that the simplicity and elegance of the ADC is somewhat marred by the proliferation of suffixes. Schouten has devoted much energy to the devising of a better notation—although with little success. For tensors which are anti-symmetric in each pair of suffixes the exterior differential forms of H. Cartan provide a much simpler notation.

The addition and multiplication of tensors are defined in the natural manner, but it is the definition of covariant differentiation which is the central achievement of the ADC. It is fundamental in the ADC that if $S_{mn}\ldots$ is any convariant tensor, then

$$S_{mn}\ldots \ dx^m d\xi^n \ldots$$

is an invariant for all coordinate transformations. It therefore follows from Christoffel's work that so also is the expression

$$S_{mn}\ldots_{/k}\partial x^m \partial \xi^n \ldots \partial \omega^k$$

where
$$S_{mn}\ldots_{/k} = \frac{\partial S_{mn}\ldots}{\partial x^k} - \Gamma^\lambda_{km} S_{\lambda n}\ldots.$$

These functions form the components of another covariant tensor, called the covariant derivative of $S_{mn}\ldots$. The additional terms in the three index symbols compensate for the curvilinear character of the manifold (Ricci 1886).

7.31. A. N. Whitehead's Tensors of the Third Rank

In giving an account of the absolute differential calculus as a preparation for one of his three relativistic theories, A. N. Whitehead (1922; pp. 173, 176) proved a curious property of the 'three-index' symbols, which has not been noticed by any other writers. As a preliminary he remarked that a Christoffel three-index symbol, $H\{\mu\nu,\lambda\}$, can be derived from *any* second-rank tensor $H_{\mu\nu}$ by the formula

$$2H\{\mu\nu,\varrho\} = H^{\varrho\lambda}\{\partial H_{\mu\lambda}/\partial x^\nu + \partial H_{\nu\lambda}/\partial x^\mu - \partial H_{\mu\nu}/\partial x^\lambda\}.$$

He then proved that if $K\{\mu\nu,\lambda\}$ is the corresponding three-index symbol derived from another tensor $K_{\mu\nu}$ then

$$H\{\mu\nu,\varrho\} - K\{\mu\nu,\varrho\}$$

is a tensor of the third rank.

Tensors of the third rank are so scarce that this is a noteworthy result.

7.32. Levi-Civita and Parallel Displacement

In the special theory of relativity there is an obvious sense in which the 'world lines'

$$x_\mu = a_\mu + \lambda_\mu \sigma \qquad (\mu = 1, 2, 3, 4)$$

(where the a_μ and λ_μ are constants and σ is a parameter) are *parallel*, but in the general theory of relativity there is in general no possibility of defining parallelism as a symmetrical and transitive relation. However, it is still possible to construct a local theory of parallelism.

Stimulated by Einstein's general theory of relativity, Levi-Civita (1917) gave the first solution of this problem by considering any n-dimensional manifold with a distance element,

$$ds = \{g_{\mu\nu}dx^\mu dx^\nu\}^{\frac{1}{2}} \qquad (\mu,\nu = 1, 2, \ldots, n).$$

In Euclidean space, or in a space of constant curvature, a rigid body can be freely moved to any prescribed position and orientation, and two rigid structures are 'congruent' if one of them can be displaced so as to be in coincidence with the other, the path along which the first is moved having no influence on the result. But in a general manifold the concepts of a rigid structure, freely movable, and of the congruence of rigid structures have no significance.

Nevertheless in Einstein's general theory of relativity the possibility of measurement must depend on the transport of rigid measuring rods and of clocks from one observer in space–time to another. The theory of 'parallel displacement' was invented by Levi-Civita (1917) to take the place of the theory of congruence in Euclidean space. The main qualitative features of the theory are as follows.

Instead of considering finite displacements of finite structures (as in Euclidean geometry) we consider infinitesimal displacements of infinitesimal structures, and thus return to Riemann's principle that a manifold is characterized by its infinitesimal properties.

An infinitesimal structure, be it a small distance determined by a measuring rod, or a short time determined by a clock, is represented by an infinitesimal vector with components δx^i ($i = 1,\ 2,\ 3,\ 4$) or more rigorously by a tangent vector ξ^i to the space–time manifold at the event $A(a^1, a^2, a^3, a^4)$ where it is located. We propose to define what is meant by the parallel transport of this vector ξ^i along a world line $x^i = \gamma^i(s)$ passing through the event A. Here s is a parameter running along the world line. To determine the vector $\xi^i(s)$ at the event s, Levi-Civita expresses the derivative $d\xi^i(s)/ds$ in terms of s and the functions $\gamma^i(s)$. It is to be expected that an infinitesimal change $\delta \xi^i$ in ξ^i will be a linear function of the displacement $\delta \gamma^i$ and of the vector ξ^i itself so that the required relation is of the form

$$\frac{d\xi^i}{ds} + \Gamma^i_{jk}\frac{d\gamma^i}{ds}\xi^k = 0$$

The problem is to determine the coefficients Γ^i_{jk} which determine the 'parallel transport'.

Although this problem arose from the physical theory of special relativity it was clearly a problem which could be considered in relation to any n-dimensional manifold and whose solution might rightly be expected to have a great influence on differential geometry in general. There are in fact many ways of solving this problem. The method employed by Levi-Civita was to consider the curved n-dimensional manifold V_n to be immersed in a Euclidean manifold S_N of N dimensions. [The possibility of such immersion was subsequently established by Whitney (1935) when $N = \frac{1}{2}n(n + 1)$]. The vector $\xi^i(s)$ in V_n then becomes a vector $\alpha^\nu(s)$ ($\nu = 1$, $2, \ldots, N$) in S_N.

Now let t^ν be any fixed vector in S_N lying in the tangent plane to V_n at a point A on the curve C, $x^i = \gamma^i(s)$, ($i = 1,\ 2,\ \ldots,\ n$). Then in general the projection of the vector $\alpha^\nu(s)$ on the fixed vector t^ν, i.e. the scalar product,

$$\sum_{\nu=1}^{N} \alpha^\nu(s) t^\nu$$

will vary as the point with parameter s describes the curve C. But we can choose the rate of change of α^ν at the point A, so that the rate of change of the scalar product

$$\sum_{\nu=1}^{N} \alpha^\nu t^\nu$$

is zero. Levi-Civita showed that, for this condition to be realized, the necessary and sufficient condition is that

$$\frac{d\xi^i(s)}{ds} + \sum \left\{\begin{matrix} jk \\ i \end{matrix}\right\} \frac{d\gamma^j}{ds}\xi^k = 0$$

at the point A, where the symbol $\begin{Bmatrix} jk \\ i \end{Bmatrix}$ denotes Christoffel's three-index symbol.

The parallel displacement of the vector $\xi^i(s)$ along the curve C is then defined by the condition that the preceding relation should hold at each point s (in relation to the tangent plane at s). Thus the components of $\xi^i(s)$ must satisfy the differential equation

$$\frac{d\xi^i}{ds} + \Gamma^i_{jk} \frac{dy^j}{ds} \xi^R = 0$$

where

$$\Gamma^i_{jk} = \begin{Bmatrix} jk \\ i \end{Bmatrix}.$$

Since the three-index symbol depends only on the metric of V_n this condition is intrinsic to V_n, i.e. independent of the mode of immersion in S_N.

The significance of the concept of parallel displacement is illustrated by the following two theorems.

(1) A geodesic is characterized by the property that its tangent vector is transported along the geodesic by parallel displacement.

(2) If a vector ξ^i is transported by parallel displacement around an infinitesimal 'parallelogram' or surface element with two adjacent edges specified by the vectors dx^j and δx^j then ξ^i does not return to congruence with its original vector but suffers a change given by

$$\Delta \xi^i = R^i_{jkl} \xi^j dx^k \delta x^l$$

where the coefficients R^i_{jkl} are the components of the Riemann Christoffel tensor.

There are other solutions of the problem of parallel displacement due to Fermi (1922) and Walker (1955, 1959).

7.33. *Weyl's Solution of the Space Problem*

According to Einstein's general theory of relativity, the transport of measuring rods and of clocks is subject to the condition of parallel displacement, and the changes in the components of a vector which is so transported from A to B depends upon the path C from A along which the vector is carried. Hence Lie's solution of the 'Riemann–Helmholtz' problem, which depends upon the group properties of congruence displacements, is no longer relevant. This problem—that of finding an axiomatic basis for the differential geometry of space–time—is one of great difficulty and therefore naturally attracted the attention of Weyl (1923).

Weyl's solution depends upon two principles:

(1) The definition of congruence at neighbouring events by means of parallel transport, the conditions for a vector ξ^i being of the form

$$d\xi^i = \Lambda^{rs}_i \xi^r dx^s$$

where the coefficients Λ_i^{rs} cannot, as yet, be identified with the three-index symbols, since, as yet, no metric has been introduced.

(2) The definition of congruence for vectors at the same event by means of a group G of transformations corresponding to rotations and to translations.

The second condition allows a limited use of Lie's theory of continuous groups. It implies that there are N infinitesimal transformations

$$\mathrm{d}\xi^i = \mathrm{A}_{kr}^i\xi^k \qquad (r = 1, 2, \ldots, N).$$

From the consistency of these two conditions Weyl inferred (albeit with some diffidence) that the group G is isomorphic with the general Lorentz group and therefore leaves invariant a quadratic form

$$\mathrm{d}s^2 = g_{\mu\nu}\mathrm{d}x^\mu\mathrm{d}x^\nu.$$

7.34. *Schouten and 'Geometric Objects'*

According to Schouten, the definition of a 'geometric object' is one of the most fundamental concepts of differential geometry, and it is undoubtedly the first problem to arise as soon as we realize that there is no priority among the infinity of allowable coordinates which determine the points of a differential manifold. The transformation from one such system S of allowable coordinates to another such system S' must carry with it the description of a geometric object in the system S into its description in the system S'. So much is clear, but the precise expression of this idea in its most general form is a difficult and delicate matter.

The problem of defining a geometric object (Nuenhius 1952) has been considered by Klein (1909), Oswald Veblen and Thomas (1926) and Veblen and J. H. C. Whitehead (1932), but according to Schouten it was Wundheiler (1937) who was 'the first writer who really tried to establish a theory of objects'. Schouten and van Dantzig (1935) attempted to give a more exact definition and Schouten's definitive views were published in a paper written in collaboration with Haantjes (Schouten and Haantjes 1936). All Schouten's papers were notorious for the wealth and complexity of the notations with which he experimented and this paper is no exception.

The complex general theory can be illustrated from the simplest example of the geometry of the Euclidean plane, in which the familiar features of straight lines and circles furnish the simplest examples of 'geometric objects' whose equations are transformed cogrediently with the translations and rotations of the plane. The reader will appreciate that difficulties arise as soon as we begin to consider inversions.

Perhaps the most interesting feature of Schouten's paper is the restriction which he places upon the transformations from one 'allowable' coordinate system to another, i.e. if S and T are two such allowable transformations then the domain of T must coincide with the codomain of S (using the modern terminology in which a transformation T carries a set of points X, its domain, into another set, Y, its codomain). It is remarkable that this relation of 'exactness' in the composition TS of S and T should appear here for the first time.

8. The Algebraic Origins of Modern Algebraic Geometry

8.1. Introduction

When he contemplates the enormous subject of algebraic geometry, the historian can find some consolation and encouragement in the fact that even such a master among mathematicians as Lefschetz could describe the relevant literature as being 'as indigestible as it is vast', and as 'undergoing an extensive process of recasting and reorganization'. The one unifying description of algebraic geometry in all its manifold forms, from Serret (1854), who wrote as an algebraist, to Grothendieck who writes in terms of functors, is the deceptively simple statement of Serge Lang that 'algebraic geometry is the study of systems of algebraic equations'.

Following the same expositor we may distinguish four great drives which have been cut through the vast and tangled forest of algebraic geometry, viz. the analytic, the topological, the algebraico-geometric and the arithmetic. Moreover, these highways tend to converge in the functorial theory of schemes.

More fortunate than Dante, we have three magisterial guides through this world of abstractions, for Dieudonné (1966, 1972) Zariski (1971) and van der Waerden (1970) have each given us illuminating perspectives and panoramas. In this chapter we attempt only a description of the algebraic origins of algebraic geometry.

8.2. The Algebraic Theory

The algebraic theory may be naïvely described as the study of 'solutions' of a system of polynomial equations in many variables—a description which immediately suggests the geometric interpretation as the intersection of a set of $(n - 1)$-dimensional surfaces in n-dimensional Euclidean space in a set of manifolds, such as curves and points, of lower dimensions than $(n - 1)$. This introductory view of the algebraic theory is formally developed in the 'elimination' theory due to Kronecker, and his expositors Molk, J. König and Netto. This line of advance naturally leads to the 'intersection' theory of the Italian school of Castelnuovo, Enriques, Severi, and to the 'enumerative' geometry of Schubert.

What gives the especially algebraic flavour to these investigations is the restriction of the coefficients and variables in the polynomials considered to certain specific and restricted classes, such as rational numbers or algebraic numbers, and the systematic use of the theory of 'ideals' and 'modules'.

This algebraic theory approaches and, some would say, includes the whole of diophantine analysis, or the solution of equations in terms of integers.

The main purpose of Molk's lengthy and pleasantly discursive paper (Molk 1885) is to develop a generalization of the concept of a 'divisor'. In a given polynomial ring P a function $f(x_1, x_2, \ldots, x_n)$ has divisors $g(x_1, x_2, \ldots, x_n)$ and $h(x_1, x_2, \ldots, x_n)$ in the ring P if $f = gh$. According to Molk a function f contains a system of divisors $f_1, f_2, \ldots, f_k$ if

$$f = \varphi_1 f_1 + \varphi_2 f_2 + \ldots + \varphi_k f_k$$

and each of the functions φ_k belongs to P. In Kronecker's notation

$$f = 0 \pmod{f_1, f_2, \ldots, f_k}.$$

A system of functions $(F_1, F_2, \ldots, F_n)$ 'contains' another system $(f_1, f_2, \ldots, f_k)$ if each function F_g contains $(f_1, f_2, \ldots, f_k)$ and Kronecker writes

$$(F_1, F_2, \ldots, F_n) = 0 \qquad \pmod{f_1, f_2 \ldots, f_k}.$$

Two systems of functions (F_i) and (f_j) are 'equivalent' if

$$F_i \equiv 0 \qquad \pmod{f_i, \ldots, f_k}$$
and
$$f_j \equiv 0 \qquad \pmod{F_1, \ldots, F_n}$$

for all i and j.

The 'composition' of two systems (F_i) and (f_j) is defined to be the system $(F_i f_j)$ containing all the products of the elements of the first and second systems.

8.3. The Concept of Elimination

The subject which has dominated algebraic geometry throughout its history from Kummer to Grothendieck is the solution of systems of polynomial equations in many variables

$$f_i(x_1, x_2, \ldots, x_n) = 0 \qquad (i = 1, 2, \ldots m).$$

This is essentially a purely algebraic problem, but in fact it has been the geometric interpretation which has provided the stimulus and inspiration for its elucidation.

Thus in elementary coordinate geometry, a single polynomial equation in three variables, x, y, z, represents a surface in Euclidean space, and a system of two (or three) such equations represents a set of curves (or points). The most elementary and naïve process for the determination of these intersections of surfaces is by the systematic elimination of the variables x,y,z, one by one. Thus it is evident to the geometer that the two cones,

$$-ax^2 + y^2 + z^2 = 0,$$
$$x^2 - \beta y^2 + z^2 = 0, \qquad (a,\beta > 0)$$

intersect in the set of straight lines, given by the parametric equations,

$$x = \pm(1+\beta)^{\frac{1}{2}}t, \; y = \pm(1+a)^{\frac{1}{2}}t, \; z = \pm(-1+a\beta)^{\frac{1}{2}}t.$$

The actual process of elimination is trivial, but, when it is accomplished, there remains the really serious problem of deciding why the intersection is represented more satisfactorily by the parametric equations than by the original pair of equations of the cones.

The answer to this question is that the parametric equations exhibit the intersection of the two cones as the union of a number of algebraic loci, each of which cannot be further resolved by algebraic means into simpler loci. This is the origin of the concepts of the 'reducibility' and 'irreducibility' of systems of algebraic loci.

Again the very process of elimination, which yields the equations of two cylinders,

$$(1+a)z^2 + (1-a\beta)y^2 = 0,$$
$$(1+\beta)z^2 + (1-a\beta)x^2 = 0,$$

or of a pair of planes and a paraboloid,

$$(1+a)x^2 = (1+\beta)y^2,$$
$$2z^2 = (1-a)x^2 + (1-\beta)y^2,$$

poses the question, what is the relation between these different expressions for the intersection of the two cones? The answer to such a question gives rise to the theory of ideals and modules.

8.4. Sylvester and the Dialytic Method

The simplest problem in elimination is to determine the common zeros (if any) of two polynomials in a single variable x

$$F = \sum_{i=0}^{m} f_i x^i, \quad G = \sum_{i=0}^{n} g_i x^i.$$

The interesting, but futile, question of priority can be closed by describing the standard solution of this problem as Sylvester's 'dialytic method', as a tribute to Sylvester's facility in coining appropriate names for various algebraic operations (Sylvester 1839, 1841).

In this method the equations

$$x^a F = 0 \qquad (a = 0, 1, 2, \ldots, n-1)$$
$$x^b G = 0 \qquad (b = 0, 1, 2, \ldots, m-1)$$

are regarded as $(m + n)$ equations in the variables x^c ($c = 0, 1, 2, \ldots, m + n - 1$).

The condition for the existence of common zeros of F and G is then given by the vanishing of the determinant of the coefficients of these equations, and the common zeros can be found by evaluating the appropriate cofactors.

8.5. Algebraic Numbers

The publication of Serret's *Cours d'Algèbre Supérieure* (Serret 1854) marks the end of the period in which the central problem of algebra was the study of the zeros of a polynomial

$$f(z) = z^n + a_1 z^{n-1} + \ldots + a_n,$$

i.e. the complex numbers $z = \xi + i\eta$, ($i = \sqrt{-1}$, ξ and η real) such that $f(z) = 0$. Most, if not all, of subsequent algebraic research has grown from this narrowly restricted problem by abstraction from the properties of complex numbers and of polynomials.

Gauss had already generalized the classical concept of a real integer to a complex integer of the form $a + bi$ where a and b are integers. Just as in the Euclidean theory of integers, a Gaussian complex integer, $a + bi$, is either composite, i.e. the product of two other complex integers (neither of which is unity), or is a prime, such as $1 + i$. And exactly as in the Euclidean theory a Gaussian integer has a unique representation as the product of prime Gaussian integers.

Again Gauss had proved that the polynomial $f(z)$ in the complex number z can be uniquely represented as a product

$$(z - z_1)(z - z_2) \ldots (z - z_n)$$

where $z_1, z_2, \ldots, z_n$ are its zeros.

When, however, we turn to the practical question of actually computing the exact zeros of a given polynomial, all limiting processes must be excluded, and we are restricted to a finite number of operations on integers. This of course implies that the coefficients of the polynomial, a_1, $a_2, \ldots, a_n$, must themselves be integers (or rational fractions such as p/q where p and q are integral).

Only Kronecker has been bold enough to accept the full consequences of this severe limitation of algebraical armaments. If this devastating restriction is not accepted, then we are driven to admit that the domain of the 'known quantities', in terms of which the solutions of our equation $f(z) = 0$ are to be expressed, must be extended by the 'adjunction' of such irrationals as $a + b\sqrt{\mu}$ where a, b and μ are rational numbers.

These irrationals, like the complex numbers of Gauss, form a field, in the sense that they are the elements of an algebra in which addition and multiplication are commutative, associative and distributive. By a bold extension of the irrationals, $a + b\sqrt{\mu}$, which are the solution of a quadratic equation with integral coefficients, Kummer (1847) introduced the 'cyclomatic' field which contains the complex zeros of $z^p - 1$.

Kummer believed that by means of this new set of algebraic numbers he could establish Fermat's 'last theorem', viz. that the equation

$$x^p + y^p = z^p$$

has no solutions if x,y,z are integers and p is an odd prime. The algebraic numbers introduced by Kummer were of the form

$$a_0\xi^p + a_1\xi^{p-1} + \ldots + a_p$$

where ξ^1, ξ^2, ..., ξ^p are distinct zeros of $z^p - 1$, and the coefficients a_0, a_1, ..., a_p are rational numbers. Kummer gave the appropriate definitions of integers and primes in this field, and assumed that the Euclidean law of unique factorization into primes was still valid.

Dirichlet pointed out to Kummer that this assumption was not true for all odd prime numbers p. This fortunate error led Kummer to make an extensive study of the 'cyclomatic' field, which presents almost all the properties of the general case.

8.6. Dedekind and Ideals

The difficulties encountered by Kummer in his theory of cyclomatic fields can be illustrated by examples given by Dedekind. Thus in the field obtained by adjoining to the integers the irrational $\sqrt{-5}$ the number 9 can be resolved into prime factors in two different ways, viz.

$$9 = 3^2 = (2 + \sqrt{-5})(2 - \sqrt{-5}).$$

To overcome these difficulties Kummer (1847) invented a theory of ideal numbers, which was described by H. J. S. Smith (1894). This complex theory was completely superceded by a totally different theory of ideals due to Dedekind, but in fact Kummer had discovered the theory of 'valuations' (Bourbaki 1969; pp. 122–5).

The complex numbers, $a + bi$, introduced by Gauss, are the zeros of quadratic polynomials, $(x - a)^2 + b^2$, and they form a closed system under the operations of addition, multiplication and division (if we exclude the zero 0). We owe to Dedekind (1877) the concept of algebraic numbers which are the zeros of a polynomial

$$x^r + a_1 x^{n-1} + \ldots + a_r,$$

with rational numbers a_1, a_2, ..., a_r as coefficients.

Dedekind showed that these algebraic numbers, like the Gaussian numbers, $a + bi$, form a system which is closed under addition and multiplication and in which the ordinary algebraic laws of complex numbers are still valid.

Algebraic numbers include all rational numbers and many, but by no means all, irrational numbers. The irrational numbers which are not algebraic are known as 'transcendental' numbers, but it requires great expertise to demonstrate that even the best known irrationals are transcendental.

Hermite (1873) proved that e, the basis of natural logarithms, is transcendental, and Lindemann (1882) proved that π is transcendental, but it is still unknown whether Euler's constant

$$C = \lim_{n \to \infty} \left(1 + \frac{1}{2} + \frac{1}{3} + \ldots + \frac{1}{n} - \log n \right)$$

is algebraic or transcendental

The study of algebraic numbers may seem to form a very restricted domain of mathematics, but it is precisely the limitations of this study which constituted part of its attraction as a challenge to mathematicians. The other part of the attraction is that algebraic numbers have so many properties in common with rational numbers and certain unexpected eccentricities, the elucidation of which has led to the creation of some of the most important structures in modern algebra, viz. fields, modules and ideals.

These concepts were first introduced by Dedekind, who published his discoveries in Supplement 10 to the second edition of Dirichlet's *Zahlentheorie* (Dirichlet 1871).

The concept of a field was only gradually elaborated as a subset of the complete system of all algebraic numbers. In the first place Dirichlet defines a 'Körper' to be the system Ω of all algebraic numbers ω of the form

$$\omega = x_0 + x_1 \theta + \ldots + x_{n-1} \theta^{n-1},$$

where θ is an algebraic number of degree n, i.e. the zero of a polynomial with rational coefficients of degree n. But the name of Körper or 'field' was soon extended by Dedekind to mean any set F of (non-zero) algebraic numbers such that if α and β belong to F, so also do $\alpha \pm \beta$, $\alpha\beta$ and α/β. In particular, if θ is any algebraic number, the field $\Omega(\theta)$ consists of all the algebraic numbers of the form

$$\alpha = f_1(\theta)/f_2(\theta)$$

where f_1 and f_2 are rational polynomials and f_2 does not vanish identically. It is a remarkable fact that any number α of the field $\Omega(\theta)$ can be uniquely expressed in the form

$$\alpha = c_0 + c_1 \theta + \ldots + c_n \theta^n$$

where the coefficients $c_0, c_1, \ldots, c_n$ are rational.

It was, perhaps, the existence of such a representation of any algebraic number α in the field $\Omega(\theta)$ which suggested to Dedekind the study of the structure which he called a 'module', namely the aggregate of all algebraic numbers of the form

$$a_1 \alpha_1 + a_2 \alpha_2 + \ldots$$

where $\alpha_1, \alpha_2, \ldots$ are *fixed* elements of $\Omega(\theta)$ and $a_1, a_2, \ldots$ are rational numbers.

But it was in the construction of 'ideals' that Dedekind made the greatest advance in the theory of algebraic integers. Ideals were defined as systems of algebraic integers, which are defined as the zeros of polynomials

$$x^r + c_1 x^{r+1} + \ldots + c_r,$$

with ordinary integer numbers c_1, c_2 ..., c_r as coefficients, and the coefficient of the leading term x^r being unity. The precise definition of a Dedekindian ideal I is that I is an infinite set of algebraic integers, such that, if α and β belong to I, so also do $\alpha + \beta$ and $\mu\alpha$, where μ is any algebraic integer. In particular a 'principal' ideal in a field Ω, say J, is an infinite set of algebraic integers of the form $\mu\alpha$, where α is a fixed algebraic integer of Ω and μ any algebraic integer in Ω. Each ideal I in a field K has a 'basis' of algebraic integers α_1, α_2, ... such that each element of I has the form $\mu_1\alpha_1 + \mu_2\alpha_2 + \ldots$ where $\mu_1, \mu_2, \ldots$ are elements of K, and the ideal I is then written as $[\alpha_1, \alpha_2, \ldots]$.

A similar concept of an ideal had occurred independently to Kronecker, in whose writings an ideal is not an aggregate of algebraic numbers, but is represented by a single expression of the form

$$\alpha_1 u_1 + \alpha_2 u_2 + \ldots + \alpha_k u_k$$

where α_1, α_2, ..., α_k are fixed algebraic integers and u_1, u_2, ..., u_k are 'indeterminates' used mainly to facilitate the construction of the product of two ideals in the form of the product of the representative polynomials.

In Dedekind's theory the product K of two ideals I and J in a field Ω is the ideal which contains the product $\alpha\beta$ of each element α in I and each element β in J. Among the elements of this product are each of the elements of I and each of the elements of J. Thus we can define the ideals I and J to be factors of the ideal K if K is the product of I and J, and a prime ideal K is such that

$$K = IJ \quad \text{only if} \quad K = I \quad \text{or} \quad K = J.$$

It is, of course, important to show that the Dedekindian system of ideals and prime ideals is a genuine generalization of the system of rational integers and prime numbers. A simple example will illustrate this point. In Dedekind's theory, for example, the numbers 3 and 7 are replaced by the ideals

$$[3] = (3, 6, 9, \ldots) \quad \text{and} \quad [7] = (7, 14, 21, \ldots)$$

while the number 21 is replaced by the ideal

$$[21] = (21, 42, 63, \ldots)$$

Clearly the ideal [21] contains all the products of the elements 3m . 7n which are in [3] and [7], whence

$$[21] = [3] \cdot [7].$$

With this definition Dedekind was able to establish the central theorem of ideal theory, that if I is not a prime ideal, then it has a unique representation as the product of prime ideals.

Thus in the field of the numbers of the form $a + b\theta$ where $\theta = \sqrt{-5}$ the prime ideals are

$$p_1 = [3, \, 1 + 2\theta], \; p_2 = [3, \, 1 - 2\theta], \; p_3 = [7, \, 1 + 2\theta], \; p_4 = [7, \, 1 - 2\theta]$$

and the ideal [3] has the prime factors p_1 and p_2.

The concept of an ideal admits of very wide generalization and can be developed in the most general form of set theory in which the only undefined notions are

(1) an element (or point), say x, y, ...,
(2) a set, say α, β, ...,
(3) a relation of inclusion, $x \in \alpha$, i.e. the point x belongs to the set α.

Then the union $\alpha \cup \beta$ of two sets α and β is defined by

$$x \in \alpha \cup \beta \quad \text{if} \quad x \in \alpha \quad \text{or} \quad x \in \beta,$$

and the intersection $\alpha\beta$ of two sets α and β is defined by

$$x \in \alpha\beta \quad \text{if} \quad x \in \alpha \quad \text{and} \quad x \in \beta.$$

Then an ideal is a class of sets I such that, if $\alpha \in I$, and $\beta \in I$, then $\alpha \cup \beta \in I$ and $\mu\alpha \in I$ if μ is any set.

There is clearly complete duality in the logical connectives 'and' and 'or', and therefore in the concept of union and intersection. Thus the dual of an ideal is a 'filter' F, i.e. a class of sets such that, if $\alpha \in F$ and $\beta \in F$, then $\alpha\beta \in F$ and $\mu \cup \alpha \in F$, if μ is any set.

8.7. *Gaussian Complex Integers*

The theory of algebraic numbers has its origins in Gauss' arithmetic theory of complex integers and in the geometric theory of polynomials.

Gauss' invention of complex integers was a by-product of his endeavours to generalize Legendre's law of 'quadratic reciprocity', by considering the existence of rational integers x such that $x^n - q$ is exactly divisible by p, where p,q,n are given positive integers. If the symbol (p/q) takes the values $+1$ or -1 accordingly as such an integer x exists or not, then Legendre's law takes the elegant form

$$(p/q)(q/p) = (-1)^r, \quad r = \tfrac{1}{2}(p - 1)(q - 1),$$

when p,q are positive odd primes and $n = 2$. In searching for the corresponding relation when $n = 4$, Gauss was led to introduce the Gaussian complex integers, i.e. the complex numbers of the form $a + bi$, where a and b are rational integers and $i = \sqrt{-1}$

8.8. *The Geometric Theory of Polynomials*

The need for an historical survey of the development of algebra is clearly if unconsciously exposed in the opening paragraphs of Hilbert's magisterial survey of the theory of algebraic number fields (Hilbert 1897). It is not merely the absence of any historical information to which we refer, but the disarming way in which all the researches of Kronecker and Dedekind on polynomial algebra are so taken for granted that the main theorems are assumed without any formal enunciation. In order to uncover the roots of the theory of algebraic numbers it is therefore necessary to summarize the work of the earlier nineteenth-century algebraists.

By common consent the title of the 'fundamental theorem of algebra' is given to Gauss' theorem that a polynomial in z,

$$A(z) = a_0 z^m + a_1 z^{m-1} + \ldots + a_m,$$

with real or complex coefficients $a_0, a_1, \ldots, a_m$ has at least one root ζ, which is also a complex number.

After this discovery there naturally arises the problem of investigating the common zeros of two polynomials $A(z)$ and $B(z)$ of degrees m and n respectively. The simplest account seems to be due to Sylvester already mentioned in 8.4 consists in forming $m + n$ equations,

$$A(z) = 0, \quad zA(z) = 0, \ldots, \quad z^{n-1}A(z) = 0,$$

$$B(z) = 0, \quad zB(z) = 0, \ldots, \quad z^{m-1}B(z) = 0,$$

and considering these as $m + n$ linear equations in the unknowns, z, z^2, $\ldots$, z^{m+n-1}. The determinant R of the coefficients of these equations is called their *resultant*, and the vanishing of R is the condition that the polynomials $A(z)$ and $B(z)$ should have a common zero. This result was known to Euler. Sylvester's contribution is in the simplicity of the dialytic method which can easily be generalized.

It is clear that Sylvester's dialytic method is equally applicable to a pair of homogenous polynomials $A(x,y)$ and $B(x,y)$ of the same or different degrees, for we have only to make the substitution $z = (y/x)$. But the significance of this restatement of the method is that it immediately suggests a geometric interpretation in terms of the points of intersection of the algebraic curves $A(x,y) = 0$ and $B(x,y) = 0$.

8.9. Kronecker and the Resultant

The problem of elimination for n homogeneous polynomials $F_1, F_2, \ldots,$ F_n in n variables presents a number of difficulties which are absent in the special case when $n = 2$, but Kronecker (1882) has given a general theory for the construction of the 'resultant', R, i.e. a determinant constructed from the coefficients of the polynomials and such that the equation $R = 0$ is the necessary and sufficient condition that the given polynomials should have a common zero [i.e. a set of values $\xi_1, \xi_2, \ldots, \xi_n$ such that

$$F_i(\xi_1, \xi_2, \ldots, \xi_n) = 0 \quad \text{for} \quad i = 1, 2, \ldots n].$$

Kronecker's theory is expounded and generalized in J. König (1903).

The main properties of the resultant R are (1) that it is irreducible, i.e. it is not the product of two factors, each of which is an integral function of the coefficients of the polynomials, and (2) that by means of the Liouville substitution,

$$x = u_1 x + \ldots + u_n x_n,$$

we can find the solutions of the equations

$$F_1 = 0, \; F_2 = 0, \; \ldots, \; F_r = 0$$

in n variables $x_1, x_2, \ldots, x_n$ if $r \leqslant n$.

8.10. Modules of Polynomials

The next step is clearly to consider the general problem of 'elimination' when we are given n homogeneous polynomials $F_1, F_2, \ldots, F_n$ in n variables $x_1, x_2, \ldots, x_n$ of degrees $l_1, l_2, \ldots, l_n$. All the algebraic methods of eliminating the variables from the equations $F_i = 0$ depend essentially upon the construction of a system M of polynomials, such that if the polynomials G and H belong to M so also do $G + H$ and AG where A is any homogeneous polynomial in $x_1, x_2, \ldots, x_n$. Such a collection of polynomials is called a 'module' or a modular system. Such a construction clearly leaves invariant the common solutions of n independent members of the module. In the simplest form of the algebraic theory the coefficients of the polynomials are restricted to rational integers.

A modular system of polynomials can be considered, independently of the mode by which it is constructed, as an enumerable set forming a 'ring' M in the algebraic sense of being a complete set of homogeneous polynomials in the same n variables such that if G and H belong to M so also do $G + H$ and AG when A is any polynomial in the same variables.

In 1888 Hilbert proved the astonishing result that any such enumerable modular system has a finite basis, so that any polynomial F of the module M has the form

$$F = A_1 F_1 = \ldots + A_m F_m$$

where $F_1, F_2, \ldots, F_m$ are a finite set of 'basic' polynomials in M, and $A_1, A_2, \ldots, A_m$ are suitable polynomials not necessarily in M. This discovery arose in Hilbert's researches on invariants (see Section 8.16).

8.11. Zolotareff's Theory of Ideals

In order to determine under what conditions the integral

$$\int \frac{(x + A)}{(x^4 + \gamma x^3 + \delta x^2 + \varepsilon x + \xi)^{\frac{1}{2}}} \, dx$$

can be expressed by means of logarithmic functions, Zolotareff (1876, 1880) devised a new theory of ideal complex numbers.

If x_0 is a zero of the polynomial

$$F(x) = x^n + a_1 x^{n-1} + \ldots + a_n$$

(with integral coefficients), we associate a number of the form

$$\xi_0 + \xi_1 x_0 + \ldots + \xi_{n-1} x_0^{n-1}$$

where ξ_0, ξ_1, ..., ξ_{n-1} are integral. A number of this type, say p, is a 'prime number' if

$$f(x) = V_0^{m_0} V_1^{m_1} \dots V_s^{m_s} + pF_1(x)$$

where $F_1(x)$ is a polynomial with integral coefficients and V_0, V_1, V_2, ... are irreducible functions, and m_0, m_1, m_2, ... are positive integers (if necessary, a preliminary transformation is made to ensure that F_1 is of lower degree than F).

It must be confessed the Zolotareff's work has had little influence.

8.12. Abstract Ideals

Although ideals were first introduced by Dedekind in the context of algebraic number fields, they were speedily emancipated from this restriction and given a more abstract definition in terms of 'rings'. The concept of a ring is due to Kronecker (who called them 'orders') and Dedekind, but the name of 'ring' is due to Hilbert.

An abstract ring is any set of elements in which addition and multiplication obey some of the usual laws of classical arithmetic (commutativity etc.), and which is *closed* under these operations. Division, however, is not introduced as a generally admissible operation.

The introduction of subrings of a given ring R does not by itself greatly advance the search for structural properties of rings, but a decisive step is the introduction of ideals. An ideal I of a ring R is a subring of R with the crucial property that if $a \in I$ and $r \in R$ then $ra \in I$, i.e. the ideal I is invariant under multiplication by any element of the ring R.

One of the most familiar examples of a ring is the collection of all polynomials with real (or complex) coefficients, but the theory of polynomial rings developed by Kronecker is much more sophisticated than the name may at first suggest. In the first place the coefficients of the polynomial are restricted to be rational numbers, or, more abstractly, to be elements of a field K. Secondly the polynomial, say

$$a_0 + a_1 x + \dots + a_n x^n$$

is not to be regarded as a function of the variable x. In fact x is expressly described as an 'indeterminate', certainly not an element of K, and, in fact, merely an 'umbral' symbol introduced to give a concise account of the addition and multiplication of the coefficients.

Kronecker used the term 'modular system' in a sense analogous to Dedekind's 'ideal', i.e. a modular system is any subset M of a ring R of polynomials (in n variables) such that if f_1 and f_2 belong to M and g belongs to R, then $f_1 - f_2$ and gf_1 also belong to M. A modular system M has a 'basis' $(b_1, b_2, \dots)$ if any polynomial in M can be expressed in the form

$$f_1 b_1 + f_2 b_2 + \dots$$

where b_1, b_2, ... are elements of M and f_1, f_2, ... are elements of R.

8.13. Kronecker's Theory of Elimination

The first problem concerns the polynomials

$$f(x) = a_0 x^n + \ldots + a_n,$$

with integral coefficients. Such polynomials form a ring P; $f(x)$ is reducible if there exist two polynomials $g(x)$, $h(x)$ in the same ring P such that

$$f(x) = g(x)h(x).$$

$g(x)$ and $h(x)$ are then divisors of $f(x)$.

By an ingenious use of Lagrange's interpolation formula, Kronecker and Molk constructed an algorithm which determined, in a finite number of rational operations on integers, whether or not $f(x)$ was reducible, and which also determined all the divisors of $f(x)$ if $f(x)$ was reducible. And all this was achieved without any appeal to the 'fundamental theorem of algebra', which analysts would cite as proving the existence of zeros of the polynomial $f(x)$!

The next problem is to establish the unicity of the resolution of a polynomial $f(x)$ in P into irreducible divisors $f_1(x)$, $f_2(x)$, ... also in P, such that

$$f(x) = f_1(x)f_2(x) \ldots.$$

Similar considerations apply to polynomials in n variables x_1, x_2, ..., x_n. Such polynomials form a ring P_n and Kronecker and Molk established the unique decomposition of such a polynomial into irreducible divisors by the device of making the transformation

$$x_k \to x^{g^{k-1}}.$$

This transforms a polynomial F in x_1, x_2, ..., x_n into a polynomial $f(x)$ in the one variable x, and each term in the first polynomial into a different term in $f(x)$ if the integer g is made sufficiently large. The unique decomposition of $f(x)$ then establishes the unique decomposition of F.

8.14. Kronecker's Theory of Algebraic Functions

By a strict analogy with Dedekind's theory of algebraic numbers (Section 8.6), Kronecker (1882) developed a theory of algebraic functions, of which the motivation was the consideration of polynomials of the form

$$F(z) = a_0 z^n + a_1 z^{n-1} + \ldots + a_n$$

where the coefficients $a_0, a_1, \ldots, a_n$ are each rational integral functions of a variable x, i.e.

$$a_k = \sum_{l=1}^{p} c_{k,l} x^l.$$

In Dedekind's theory x is a real or complex variable, and an algebraic function $\theta(x)$ is defined to be any zero of the polynomial $F(\theta) = a_0 \theta^n + $

$a_1 \theta^{n-1} + \ldots + a_n$, where a_0, a_1, ... are rational functions of x (Dedekind and Weber 1882).

In Kronecker's theory the domain of the variable x is restricted to the rational numbers and the coefficients $c_{k,l}$ are also rational numbers.

A mathematician whose researches have been in the field of analysis may find this concept an intolerable restriction on the immense liberty of the theory of functions of a real or complex variable, but the genius of Kronecker is especially shown in the masterly way in which he exploits this very limitation in the construction of a theory in which every definition and theorem is expressed in terms of positive integers. Negative integers, zero and even rational numbers are only allowed as a concession to human weakness, and as useful abbreviations, since any expression involving such numbers can easily be replaced by an equivalent expression involving positive integers only.

8.15. Kronecker's Method of Elimination

A common zero $(\xi_1, \xi_2, \ldots, \xi_n)$ of m polynomials $f_1, f_2, \ldots, f_m$ in the variables $(x_1, x_2, \ldots, x_n)$ represents a point on the intersection of the manifolds $f_i = 0$. The general problem of elimination as it was posed by Kronecker (1882) is to classify the points of intersection according to the irreducible varieties of different dimensions to which they belong.

It is convenient to make all the polynomials homogeneous by introducing a variable x_0. By means of a preliminary linear transformation of the variables it is possible to ensure that each of the functions f_i is regular. Then any common divisor of $f_1, f_2, \ldots, f_n$ will be a rational function of *all* the variables.

The condition for the existence of a common zero of a set of polynomials is found by constructing a sufficient number of 'elementary members' of the module $(f_1, f_2, \ldots, f_n)$, i.e. polynomials of the form ωf_i where ω is of the form $x_0^{a^0} x_1^{a^1} \ldots x_n^{a^n}$. If f_i has degree l_i and $l = l_0 + l_1 + \ldots + l_n - n - 1$ the elementary members are to be of degree l. Then the highest common factor of the determinants in the matrix of the coefficients of all these elementary members is called the 'resultant' R of $f_1, f_2, \ldots, f_m$.

The vanishing of the resultant is the necessary and sufficient condition for the existence of varieties contained in the intersections of the surfaces $f_i = 0$, but to determine these varieties explicitly by Kronecker's method it is necessary to construct the 'resolvent' of the equations.

By a painstaking and elaborate process (restricted to non-homogeneous equations) Kronecker showed that by purely algebraic methods it is possible to construct a function of the form $S = \Sigma g_i f_i$, where $g_1, g_2, \ldots$ are polynomials such that

$$S = D D^{(1)} D^{(2)} \ldots D^{(n-1)}$$

where D is the highest common factor of $f_1, f_2, \ldots, f_m$ (or unity if no such H.C.F. exists), regarded as a polynomial in the single variable x_1. If $f_i = D \varphi_i$ the resultant of the two polynomials $\Sigma \lambda_i \varphi_i$ and $\Sigma \mu_i \varphi_i$ regarded as polynomials in x_1 is constructed in the form $\Sigma \varrho_j F_j^{(1)}$ where the coefficients

ϱ_j are different power products of the λ's and μ's, while the functions $F_j^{(1)}$ are polynomials in x_2, x_3, $\ldots$, x_n only. If $D^{(i)}$ is the H.C.F. of the functions $F_j^{(i)}$ regarded as polynomials in x_2, x_3, $\ldots$, x_n, we write $F_j^{(i)} = D^{(i)}\varphi_j^{(i)}$ and repeat the process as often as is necessary.

A full and detailed description of the process is given by J. König (1903) and Molk (1885). A concise and critical account is given by Macaulay (1916), the importance of which was first noticed by Emmy Noether.

8.16. Hilbert's Basis Theorem

What has been described by Weyl as 'one of the most fundamental theorems of algebra' was discovered by Hilbert when he undertook to reorganize the theory of invariants, which had slowly ground to a standstill in the prodigious efforts of Cayley, Sylvester and Gordon to enumerate all the invariants of algebraic forms.

Gordon himself is justly famous, not so much for his work on binary forms (i.e. polynomials in two variables) (Gordon 1868), but for formulating his great problem—the examination of the structure of all the invariants of a collection of polynomials in any number of variables x_1, x_2, $\ldots$, x_n. Any linear transformation of these variables induces a linear transformation of the coefficients, and certain functions of the coefficients remain invariant under all linear transformations.

Thus in the simple case of two quadratics

$$u = \lambda(x + \alpha y)(x + \beta y)$$
$$v = \mu(x + \gamma y)(x + \delta g)$$

there are four invariants, viz.

$$I_1 = \lambda\mu(\alpha - \beta)^2,$$
$$I_2 = \lambda\mu(\gamma - \delta)^2,$$
$$I_3 = \lambda\mu(\alpha\gamma + \alpha\delta + \beta\gamma + \beta\delta - 2\gamma\delta - 2\alpha\beta),$$
$$I_4 = \lambda\mu(\alpha + \beta - \gamma - \delta)^2,$$

which are connected by the 'syzygy',

$$I_4 = I_1 + I_2 - 2I_3.$$

The word 'syzygy' has almost disappeared from the literature of invariants and is only remembered by astronomers to refer to a conjunction of the Earth, Moon and Sun. But the concept of a syzygy still dominates the algebra of invariants and signifies an identical relation between a set of invariants of the form $F(I_1, I_2, \ldots, I_n) = 0$.

In England the study of these invariants was vigorously pursued under the title of the 'theory of quantics' by Cayley, Salmon, Elliott and many other lesser workers. The lengthy and elaborate researches of Gordon and the special results obtained for binary forms led him to conjecture the truth of two great theorems:

(1) The invariants of a given system of polynomials have a finite basis, i.e. there is a finite number of basic invariants, such that any invariant is a polynomial in these basic invariants.

(2) If $F_i = 0$ $(i = 1, 2, \ldots)$ is the set of polynomials in the basic invariants which vanish when the variables x_j take values ξ_j, then any of these, say F, can be expressed in terms of a finite 'basis', F_1, $F_2, \ldots, F_k$, such that

$$F = \sum Q_i F_i$$

where the coefficients Q_i are polynomials in the basic invariants.

These results were sufficiently startling in invariant theory, but Hilbert (1889, 1890) proceeded to give two generalizations which still dominate polynomial algebra. For simplicity we restrict ourselves to polynomials with complex coefficients.

(1) *The Basis Theorem:* In any polynomial ideal I (where the polynomials have real coefficients) there exists a finite basis consisting of polynomials $f_i \in I$ $(i = 1, 2, \ldots, n)$ such that any polynomial f in I can be expressed in the form

$$f = \sum_{i=1}^{n} c_i f_i$$

where the coefficients are polynomials (not necessarily in I).

(2) *The Zero Theorem:* If the polynomials $f_i(x_1, \ldots, x_n)$ $(1 \leqslant i \leqslant h)$, and $g(x_1, \ldots, x_n)$ are such that $g = 0$ at all the common zeros of the f_i, then there exists an integer s and polynomials $c_i(x_1, x_2, \ldots, x_n)$ such that

$$g^s = \sum_{i=1}^{h} c_i f_i.$$

8.17. *Lasker's Resolvent*

The theory of the resolvent of a set of polynomial equations was given a definitive form by Lasker (1905). The method given by Kronecker and expounded by Molk and König for the solution of equations by the construction of a 'resolvent' appears to depend upon the successive calculation of solutions of the form

(1) $x_1 = f(x_2, x_3, \ldots, x_n)$,

(2) $x_1 = f_1(x_3, x_4, \ldots, x_n)$

$x_2 = f_2(x_3, x_4, \ldots, x_n)$,

(3) $x_1 = g_1(x_4, x_5, \ldots x_n)$

$x_2 = g_2(x_4, x_5, \ldots x_n)$

$x_3 = g_3(x_4, x_5, \ldots x_n)$.

Lasker gave the theory a more symmetric form in which all the variables are treated alike.

But the fundamental theorem established by Lasker is couched in terms of the theory of modules, and states that any given module is the L.C.M. of a finite number of primary modules.

The actual resolution of a given module into primary modules may, however, be an arduous matter. A process for this resolution has been given by Macaulay (1913), who has also given a critical account of the subject as it existed in 1916. Macaulay's work has been strangely neglected, but was warmly praised by Emmy Noether.

8.18. *M. Noether's Fundamental Theorem*

If $U = 0$ and $V = 0$ are any two curves then it is clear that the curve

$$F \equiv AU + BV = 0$$

must pass through the intersection of the given pair of curves. The converse theorem is that any curve passing through the intersections of the given pair must have the form

$$F \equiv AU + BV = 0.$$

Here U,V,A,B are each polynomial functions of x and y.

This converse is of fundamental importance in algebraic geometry and the difficulty in establishing it arises from the necessity of allowing for the points of intersection to be not simple, but multiple.

The proof given by Brill and M. Noether (1894) has been simplified by Scott (1899).

8.19. *Emmy Noether and Abstract Ring Theory*

A new era in algebraic theory was inaugurated by Emmy Noether in her epoch-making paper 'Idealtheorie in Ringbereichen' (E. Noether 1921), which exhibits her remarkable talent for abstraction, generalization and simplification. Lasker (1905) had given a representation of ideals as intersections of prime ideals for polynomial rings, and his work had been extended by Macaulay (1913), but Emmy Noether's researches were so much more fundamental and influential that Weyl regarded them as 'the unmovable foundation of General Ideal Theory'.

In Emmy Noether's work the problem of 'elimination' for simultaneous polynomial equations is raised to its most abstract level as the problem of the factorization of ideals in an abstract ring; and this problem is itself reduced to its most useful and manageable form by the restriction to rings satisfying any one of the three equivalent conditions. In such 'Noetherian' rings (1) every ideal has a finite basis, (2) every strictly ascending sequence of ideals $A_1 \subset A_2 \subset \ldots$ is finite, and (3) in any non-empty family of ideals there exists a maximum element.

The celebrated Hilbert basis theorem (Section 8.16) shows that Noetherian rings form an extensive and important class of rings. The name of Emmy Noether may fittingly close this sketch of the algebraic origin of modern algebraic geometry.

9. The Primitive Notions of Topology

9.1. The Origins of Topology

Topology may be briefly, if unintelligibly, described as the study of those properties of general and abstract spaces which are invariant under continuous transformations, but it is more easily and significantly appreciated in its historical perspective as the confluence of three streams of mathematical thought arising from the theory of convergence initiated by Cantor, the theory of functional analysis developed from the calculus of variations by Volterra, and the theory of the connectivity of manifolds due to Riemann and Poincaré.

The Theory of Convergence

9.2. The Metric Theory of Convergence

The history of the theory of convergence is essentially the record of the gradual emancipation of the concept of a limit from its early metrical restrictions. The development of the theory is expressed in the sequence of technical terms now familiar to undergraduates, but which only slowly took form as well-defined concepts: convergent sequence of numbers, a convergent sequence of points, limit points, points of accumulation, neighbourhoods of points in space, metric function space, écart or deviation, closure of a set of points, derived sets, open and closed sets, directed sets and filters. The concepts become more and more abstract, more and more general, and exhibit continually increasing utility and simplicity.

Cantor's great contributions to the theory of real numbers are described elsewhere (Chapter 4). The essence of his contribution to the theory of convergence may be compactly described here by a definition and a theorem.

A sequence of real numbers $\{a_n\}$, $(n = 1, 2, \ldots)$ is convergent if, for any assigned 'tolerance' $\varepsilon > 0$ the 'scatters' $|a_n - a_n|$ are less than ε for 'almost all' m and n, i.e. there is only a finite number of scatters such that $|a_m - a_n| \geqslant \varepsilon$. The theorem is that if $\{a_n\}$ is a convergent sequence of real numbers then there exists a real number l such that $|a_n - l|$ is less than ε for 'almost all' n. This elegant version of the usual definition was suggested by Kowalewski.

The definition is easily extended from real numbers representing points on a straight line to vectors representing points in space. A sequence of points $\{p_n\}$ in Euclidean space, of any finite number of dimensions, is convergent if $|p_m - p_n|$ is less than an arbitrary tolerance $\varepsilon > 0$ for almost all m and n, $|p|$ being the absolute value of the vector drawn from the origin to the point p. Euclidean space is 'complete' in the sense that a convergent sequence of points $\{p_n\}$ possesses a limit p such that $|p_n - p| < \varepsilon$ for almost all n.

To the despair of the historian, the next step in the development of the concept of convergence had been taken long before, by Bolzano in 1817, and again by Weierstrass in his unpublished lectures (Section 2.4). This decisive step was the extension of the concept of convergence from a *sequence* of numbers (or points) to any bounded infinite collection of numbers, such as the real numbers between 0 and 1.

For a bounded infinite collection C of real numbers there is, in general, no question of a unique limit, but only of the existence of one or more 'points of accumulation' p, such that an enumerably infinite subset of the collection C exists in any 'neighbourhood' of p, i.e. there is a sequence $\{p_n\}$ belonging to C and $|p - p_n| < \varepsilon$ for any assigned tolerance ε. The famous Bolzano–Weierstrass theorem declares that a bounded infinite collection C of real numbers possesses at least one point of accumulation. This theorem is valid for points in a Euclidean space of any finite number of dimensions.

Two points need to be stressed.

(1) The concept of the 'neighbourhood' of a point, defined as the set of points in the space at a distance less than ε from p:
(2) The possibility of generalizing the Bolzano–Weierstrass property of Euclidean space to more general spaces.

This property of a space X, called 'compactness' by Fréchet (1906) (to whom we owe this concept), is that any bounded subset of X which contains an infinite number of points must contain at least one point of accumulation. The main property of 'compact spaces' is expressed by the famous covering theorem associated with the names of Heine, Borel and Lebesgue.

Heine (1872) proved that any numerical function point-wise continuous in a closed interval is necessarily uniformly continuous in that interval, but he does not appear to have recognized that his method of proof could be easily generalized to establish the covering theorem. This powerful and influential theorem states that if a compact space is covered by a collection of open sets $\{E_n\}$ then it is also covered by a finite subset of the collection $\{E_n\}$. The proof was given by Borel (1895) (for points on a straight line) and by H. Lebesgue (1904).

9.3. *The Metric Spaces of Fréchet*

The metric theory of convergence was primarily restricted to sequences (or sets) of points on a straight line or in a Euclidean space endowed with a system of coordinates, but it was Fréchet (1906) who first liberated topology from the prison of Cartesian coordinates by introducing the concept of a 'metric' space. Such a space is characterized by the existence of a 'metric', i.e. a numerical function $\varrho(x,y)$ of any pair of points, such that

(1) $\varrho(x,y) = \varrho(y,x) > 0$, unless $x \equiv y$,

(2) $\varrho(x,x) = 0$,

(3) $\varrho(x,y) + \varrho(y,z) \geqslant \varrho(x, z)$,

i.e. the same properties as those which characterize distance specified by Cartesian coordinates, or given by Euclid's axioms.

The notion of convergence can be generalized to a metric space with some reservations. A sequence of points $\{x_n\}$ in a space with a metric function ϱ converges to a limit x if $\varrho(x_n, x) \to 0$ as $n \to \infty$, *but a sequence of points* $\{x_n\}$ which satisfies the generalized Cauchy criterion that $\varrho(x_m, x_n) \to 0$ as $m, n \to \infty$, does not necessarily possess a limit.

Cantor had introduced the concepts of 'closed' and 'open' sets in n-dimensional Euclidean space. Fréchet showed that these ideas could be readily extended to any metric space.

If x is the limit point of a sequence which belongs to a set of points E, then x may be called a 'point of accumulation' of the set E, using the term introduced by Jules Tannery. Fréchet proposed to consider the totality E' of all the points of accumulation of a set E as the 'derived set' of E. A 'closed set' K in a metric space is then defined by the condition that $K' \subseteq K$ and the 'closure' of K as $\bar{K} = K \cup K'$.

Fréchet (1926, p. 154) examined, but without complete success, the problem of metricizing the space of continuous curves, which requires the definition of the 'distance' between two oriented curve segments g and h. If H is *any* one–one relation connecting points p on g and q on h, we can define δ_H as the maximum of the distances between p and q, and then define the distance between g and h as the lower bound δ of δ_H for all H. But for applications to the calculus of variations Morse (1934, p. 209) found it better to take H to be the special relation which connects points p and q dividing g and h in the same ratio with respect to an appropriate definition of length.

9.4. Fréchet's Theory of Neighbourhoods

The definition of a derived set, given by Fréchet in terms of his sequential definition of a limit, does not entail Cantor's theorem that the nth derived set contains the $(n + 1)$th derived set. To remedy this defect, Fréchet made a fresh start with the spaces (V) with an 'écart' $\varrho(A,B)$ such that, if A and B are two elements of V, then

(1) $\varrho(A,B) = \varrho(B,A) > 0$ if $A = B$,

(2) $\varrho(A,A) = 0$,

(3) if $\varrho(A,B) < \varepsilon$ and $\varrho(B,C) < \varepsilon$,

then there exists a function f such that

$$\varrho(A,C) < f(\varepsilon),$$

and $$f(\varepsilon) \to 0 \quad \text{as} \quad \varepsilon \to 0.$$

Clearly it is possible to define convergence (and neighbourhoods) in terms of the écart, but Chittendon (1917) has shown that if a space possesses an écart, then it also possesses a metric, so that no effective generalization has been obtained.

9.5. Hilbert and Neighbourhoods

One of the earliest contributions to the modern theory of the topology of sets of points is to be found, suprisingly enough, in Hilbert's solution of one of the problems which he himself had proposed in Paris in 1900. Helmholtz (Section 7.26) had sought to characterize the congruence group of rigid body motions and Lie (Section 7.27) had given what seemed a final form to the solution of this problem by employing his own theory of continuous groups. But this theory required the transformation functions of the group to be differentiable. Hilbert (1900) raised the question—is a continuous transformation group necessarily differentiable?—and settled this question in the affirmative by a proof which even Hermann Weyl describes as 'difficult and laborious'.

Hilbert (1902) defines a two-dimensional manifold by means of neighbourhoods and introduces a class of 'admissible' neighbourhoods defined in terms of their mappings upon Jordan domains. When Weyl lectured on Riemann surfaces at Gottingen in 1912 he noticed that the neighbourhoods themselves (apart from any mapping) could be used to characterize the 'admissible' class.

Finally Hausdorff (1914, p. 209) achieved the final and perfect definition of a neighbourhood in the axiomatic form:

(1) Every point x has at least one neighbourhood U_x, and U_x always contains x.
(2) For any two neighbourhoods U_x and V_x of the same point, there exists a third, $W_x \subseteq U_x V_x$.
(3) Every point $y \in U_x$ has a neighbourhood $U_y \subseteq U_x$.
(4) For two different points x,y there exist two disjoint neighbourhoods U_x, U_y.

However, to introduce the special and important case of a 'differentiable' manifold we still have to return to Hilbert's method.

Since the purpose of Hilbert's investigation was to characterize the group of rigid body displacements in a plane by means of the concept of continuous groups, he began by defining the plane as a collection of (geometric) points which is in one–one correspondence with the (arithmetical) points, i.e. the ordered pairs of finite real numbers (x,y). The main tool of the theory is Jordan's theorem, according to which a closed continuous curve without double points divides the plane into two regions—the interior and the exterior of the curve. The interior is called a 'Jordan domain'.

The fundamental assumption is that to any geometric point A there corresponds in the arithmetic plane a Jordan domain containing the image A^x of A. Any Jordan domain $J(A^x)$ containing A^x is called a 'neighbourhood' of A, and is also a neighbourhood of any point B in $J(A^x)$. (There is here some confusion which the reader can easily penetrate!) Finally if A and B are any geometric points there exists a neighbourhood of A which contains B.

9.6. The 'Abstract Set Theory' of F. Riesz

The great ingenuity and fertile imagination of Fréchet did not absolve him from a severe stricture by Weil who declared that the mathematicians who worked in general topology in the period 1900–35 introduced a completely disordered collection of notions and axioms such as can be found in the table of contents of the book which the pioneer in this subject, Fréchet, published on 'Les Espaces Abstraits' (Fréchet 1926). Happily, said Weil, the majority of these notions are devoid of interest, and the only spaces worth considering are those now called topological spaces. But if Fréchet sometimes appears to be verbose, by contrast F. Riesz is tantalizingly terse. In a short communication to the Fourth International Congress of Mathematics at Rome he created the modern abstract theory of topology (F. Riesz 1909a).

Riesz entirely dispensed with any metrical considerations and seized upon the concept of the 'points of accumulation' (or limit points) E' of a given set of points E, as a fundamental and undefined element, characterized only by the three axioms:

(1) If $A \subset B$ then $A' \subset B'$,
(2) $(A \cup B)' = A' \cup B'$,
(3) $A' = \phi$ the null set, if A consists of a single point.

Riesz therefore proposed to add as a fourth axiom that if p is a point of accumulation of a set E, and if $\mathcal{F}$ is any subset of E such that $p \in \mathcal{F}'$, then p is the unique point of intersection of all the sets $\mathcal{F}$.

Riesz did little more than propose this new axiomatic basis for topology, but, almost casually, he introduced the concept of an 'ideal' $\mathcal{F}$ with the following properties.

(1) If A is any set belonging to $\mathcal{F}$ and $A \subset B$, then $B \in \mathcal{F}$.
(2) If $A = X \cup Y$ then either $X \in \mathcal{F}$ or $Y \in \mathcal{F}$.
(3) If $A \in \mathcal{F}$ and $B \in \mathcal{F}$, then A and B are 'verkettet'.
(4) $\mathcal{F}$ is not contained in any other collection of sets which satisfies axioms (1), (2) and (3).

These axioms are a most remarkable anticipation of the concept of an 'ultrafilter' introduced in 1937 by Cartan (Section 9.12).

9.7. Hausdorff's Theory of Neighbourhoods

The great achievement of Hausdorff was to give a definitive theory of 'neighbourhoods' in a form which was sufficiently abstract to transcend the limitations of metric space and to furnish a satisfactory basis for general topology. This theory is expounded in the second half of his treatise *Grundzuge der Mengenlehre*, published in 1914, but, strangely enough, was omitted in the third edition (1935) and in its English translation (Hausdorff 1957). Hausdorff's axioms are as follows:

(i)   If $N(x)$ is a neighbourhood of x, then $x \in N(x)$,
(ii) If $U(x), V(x)$ are two neighbourhoods of x, then there is also a neighbourhood $N(x)$ such that $N(x) \subset U(x) \cap V(x)$,
(iii) If y is any point in $N(x)$, then there is a neighbourhood of y such that $N(y) \subset N(x)$,
(iv) If x and y are distinct points, then there exist two neighbourhoods $N(x)$ and $N(y)$ which have no common point.

These axioms are clearly satisfied in a metric space in which for each point x a neighbourhood $N_\sigma(x)$ can be taken as the 'ball' consisting of the points y for which $\varrho(x,y) < \sigma$.

9.8. Kuratowski and Closure Topology

F. Riesz and Fréchet had opened the path towards a characterization of abstract topological spaces. In 1914, Hausdorff had given a systematic exposition of the topology of sets of points in terms of the concept of 'neighbourhood'. A further great advance was made by Kuratowski (1922b).

In an n-dimensional Euclidean space it is easy to define the 'closure' of a given set of points A as the intersection of all closed sets containing A. If the closure of A is denoted by $\bar{A}$ then the following results can be obtained at once:

(1) $\overline{A \cup B} = \bar{A} \cup \bar{B}$,
(2) $A \subset \bar{A}$,
(3) $\bar{\varphi} = \phi$,
(4) $\bar{\bar{A}} = \bar{A}$.

Kuratowski proposed to adopt these relations as axioms characteristic of the 'closure' operation which leads from a set A to the set $\bar{A}$, and to define an abstract topological space X by the existence of a closure operation for each subset of X. This definition is more general (i.e. more abstract) than the definition of the abstract spaces of Fréchet (Fréchet 1926), and lends itself more easily to further definitions of important topological concepts.

9.9. The Extended Topology of Hammer

The closure topology of Kuratowski can be regarded as based on a mapping of the subsets of a space onto the closures $\bar{E}$ of the subset E. Hammer has studied a generalization of this concept by considering mappings f of all the subsets M of a space M into M (Hammer 1961, 1962). The simplest such functions are

(1) 'isotonic' if $X \subseteq Y \Rightarrow fX \subseteq fY$,
(2) the complement function c, if $cX = M - X$,
(3) the identity e, if $eX = X$,
(4) the limit function, l, if $l \subseteq c$.

9.10. *Abstract Spaces and Sierpinski*

In the abstract theory of topology, the basic notions are necessarily undefined and are characterized solely by systems of axioms. The self-consistency of these axioms is easily demonstrated by a reference to the simple example of Euclidean space of n dimensions.

Following the work of Fréchet, F. Riesz, Hausdorff and Kuratowski, Sierpinski (1927) showed that it was possible to take the concept of the 'derived set' E' of a set E as a primitive notion, characterized by the following axiom—$E \subset F \Rightarrow E' \subset F'$.

But a much more influential suggestion was mentioned almost casually in the same paper, viz. that topology can be based on the concept of a 'closed' set characterized by the properties that

(1) the whole space is closed,
(2) the intersection of any number of closed sets is closed.

With this definition as a basis, the derived set of a set E' is given by the condition that p is a point of E' if each closed set containing $E - \{p\}$ also contains p.

Later writers have preferred to work with the complement of a closed set, i.e. an 'open' set. Such sets are exemplified in a metric space by any set of points S such that a neighbourhood of any point p in S is also in S. In an abstract space open sets are characterized solely by the conditions that

(1) the union of any collection of open sets is open,
(2) the intersection of any *finite* number of open sets is open.
(3) ϕ is open,
(4) the whole space is open.

9.11. *E. H. Moore and Directed Sets*

In a lengthy series of memoirs E. H. Moore (1910) constructed a very general abstract definition of convergence, starting with the elementary concept of the convergence of a sequence of real numbers. Instead of concentrating on the concept of a neighbourhood he focused attention on the idea of a sequence.

The need for this investigation arose partly from problems in the calculus of variations, where it is required to minimize an integral of the form

$$\int_a^b F(x, y, \, \mathrm{d}y/\mathrm{d}x) \, \mathrm{d}x$$

for a given class of functions $y(x)$. To define this minimum as a limit we must either introduce neighbourhoods given in terms of some metric or norm, or generalize the concept of a sequential limit.

Pursuing this latter idea, E. H. Moore and H. L. Smith (1922) gradually elaborated the theory of 'directed sets' based on the concept of a partially ordered set.

A space E is said to be partially ordered if there is a relation denoted by $x < y$ (where $x, y \in E$) such that

(1) $x < x$,
(2) $x < y$ and $y < x \Rightarrow x = y$,
(3) $x < y$ and $y < z \Rightarrow x < z$.

If either $x < y$ or $y < x$ for all x, y in E then E is said to be totally ordered.

Now instead of thinking about a sequence of numbers $a_1, a_2 \ldots, a_n, \ldots$ indexed by the natural numbers $1, 2, \ldots, n, \ldots$ Moore and Smith invite us to consider, in a space X, a directed family of elements x_α indexed by points α in a partially ordered space E, in which the concept of 'neighbourhood' has been defined. Then the directed family $\{x_\alpha\}$ is said to converge to the limit l if all the elements of the directed family 'ultimately' lie in any prescribed neighbourhood N of l, i.e. if there exists an index ν such that $x_\alpha \in N$ when $x_\nu < x_\alpha$.

This general theory applies to nearly all known examples of limits of numerically valued functions, but there is one important limit not covered by the theory, viz. the so-called approximate limit used in the advanced theory of integration. To include the approximate limit H. L. Smith (1938) presented a new general theory which is, in effect, an independent discovery of the theory of 'filters', announced by H. Cartan (1937a, b).

According to Kelley (1955, p. 83) it is part of the folklore, current among American mathematicians, that the theory of directed sets is equivalent to the theory of filters. Presumably it is this mistake, rather than national pride, which accounts for the popularity of directed sets in the USA and the unpopularity of filters.

It is one of the ironies of mathematics that the theory of filters was discovered independently by H. L. Smith (1938) at almost the same time as H. Cartan published his own decisive and final definition.

Smith's definition is transparent and explicit, but A. N. Whitehead must be given full credit for the definition of 'extensive connection' which a benign interpretation will recognize to be a translucent and implicit description of the process of filtration.

Whitehead's definition is given in terms of the relation of 'connection', introduced by de Laguna (1922; 393–407, 421–40, 449–59) which is an abstraction from the physical concept of the connection of solid bodies by inclusion or by touching. This concept is, however, really irrelevant, and Whitehead's relation of inclusion can be accepted as fundamental in the definition of a 'geometric element', which is effectively a filter.

9.12. H. Cartan and Filters

A new epoch in topology was opened by two papers by H. Cartan in which he introduced the concept of a 'filter', which was destined to revolutionize the axiomatic basis of general topology (H. Cartan 1937a, b). The idea of a filter appears to provide the ultimate generalization of a sequence. It applies to any space E which is partially ordered by a relation of inclusion.

A filter $\mathcal{F}$ is a collection of sets $A, B, C, \ldots$ of E such that

(1) $A \in \mathcal{F}$, *and* $A \in B \Rightarrow B \in \mathcal{F}$,
(2) $A \in \mathcal{F}$ and $B \in \mathcal{F} \Rightarrow A \cap B \in \mathcal{F}$,
(3) ϕ does not belong to $\mathcal{F}$

It will be obvious that a filter $\mathcal{F}$ is the dual of an ideal $\mathcal{I}$ when the latter is given an abstract definition in the form

(1) $A \in \mathcal{I}$ and $B \in A \Rightarrow B \in \mathcal{I}$,
(2) $A \in \mathcal{I}$ and $B \in \mathcal{I} \Rightarrow A \cup B \in \mathcal{I}$,
(3) E does not belong to $\mathcal{I}$.

A critical comparison of the theories of filters and of directed sets is given by Tukey (1940).

9.13. *Weil and Uniform Spaces*

The existence of a system of 'open sets' U in a topological space X, in the sense introduced by Sierpinski (Section 9.10), allows us to define a continuous function f from a topological space X to another topological space Y by the non-metrical condition that the inverse of each open set is open, i.e. that $T = f(S)$, $S \subset X$, $T \subset Y$, and T is open, then S is open.

Weil (1937) discovered a topological method of characterizing the *uniform* continuity of a function f. This depends upon the existence of a certain system of subsets of the cartesian product space, $X \times X$. These subsets are called a 'uniformity' for the set X and their properties generalize the properties of a distance function $d(x,y)$ for pairs of points in a metric space.

A subset U of $X \times X$ is a uniformity of X if it possesses the following properties:

(1) Each member of U contains the 'diagonal' set, i.e. the set of all pairs (x,x), $x \in X$.
(2) If M, with elements (x,y), is a subset of U so also is its 'inverse' M^{-1}, with elements (y,x).
(3) If $M \subset U$, then $V \circ V \subset U$, for some V in U, the circle denoting a 'composition', such that $(x,z) \in M \circ N$ if, for some y, $(x,y) \in M$ and $(y,z) \in N$.
(4) If $M \subset U$ and $N \subset U$, then $M \cap N \subset U$.
(5) If $M \subset U$ and $U \subset V \subset X \times X$ then $V \subset U$.

The simplest example of a uniform space is the set of real numbers $\{x\}$, with the uniformity defined by

$$(x,y) \in U \text{ if } |x - y| < r,$$

for some positive number r.

If f is a function on a space X (with a uniformity U) with values in a space Y (with a uniformity V), then f is 'uniformly continuous', relative to U and V, if for each element N of V, the set (x,y) such that $(f(x), f(y)) \in N$ is an element of U.

9.14. J. W. Alexander and Gratings

The complete elimination of points from topology was achieved by J. W. Alexander (1938) by boldly defining a 'space' to be an element x of a Boolean algebra A, in which there is prescribed an arbitrary preferred class Ω of elements which play the part of open sets.

To discuss the connectivity of the space x, Alexander introduces a 'covering pair' (a,c) of x as an ordered pair of subspaces a and c of x such that $a \cup c = x$ with a 'barrier' $b = ac$. A 'grating' T is an arbitrary set of indexed covering pairs (a_s,c_s). For example, in n-dimensional Euclidean space R^n, with coordinates ξ_i, a_s, b_s and c_s may be given by the conditions $\xi_i \leqslant 0$, $\xi_i = 0$, $\xi_i \geqslant 0$ respectively.

9.15. Wallman and Lattices

Stone (1936, 1937) approached the problems of general topology from a detailed and exhaustive algebraic analysis of representations of Boolean rings, but Wallmann (1938) employs a strictly 'geometrical' approach based on a distributive lattice L with elements 0 and 1. In this lattice L a divisorless, additive ideal, not containing zero, is called a 'point', and the elements of L in this ideal are called its 'coordinates'. Wallman then proceeds to give constructive definitions of closed sets and of bicompact spaces. This algebraic characterization of a point independently of numerical coordinates is an unexpected and noteworthy anticipation of the theory of 'schemes'.

In a further generalization Wallmann starts with an abstract multiplicative complex K which is the 'nerve', in Alexandroff's sense of an infinitive covering of a space, and which generates a bicompact space S.

These theories, together with those of Huntingdon, Nicod, Wald, E. H. Moore and Milgram are given a critical discussion by Menger (1940).

The Beginnings of Functional Analysis

9.16. Volterra and Functions of Lines

We owe to Volterra (1887, 1913, 1936) a synthesis of the many diverse ideas which he combined into a new mathematical discipline—now called 'functional analysis'.

For Volterra the fundamental concept of the classical analysis, as it had been developed, for example, by Newton and Cauchy, was the passage to a limit, e.g. the passage from the discrete to the continuous, from a sum to an integral, issuing in the theory of functions of one or of a finite number of variables. He noted a number of problems of mathematical physics which demanded a further development of this procedure.

In the theory of the electromagnetic field the phenomena depend on the distribution of electric and magnetic intensities throughout space. In the theory of elastic bodies with a 'memory' the present depends on the forces to which they have been subjected throughout their past history. Even in ancient Greece we find isoperimetric problems such as the determination of which plane curve of given length encloses the greatest area—and such a problem leads us to consider the area as a function of the curve. Similar considerations arise in the calculus of variations where the value of an integral depends on the functions which are being integrated, and also in the theory of integral equations.

In order to provide a systematic method of dealing with these problems Volterra proceeded to construct a theory of functions which depend on other functions—a theory which he published under the modest title of *Leçons sur les Fonctions de Lignes*.

The simplest functionals, such as

$$F[f(x)] = \int_a^b f(x)\phi(x)\,\mathrm{d}x$$

where $\phi(x)$ is fixed, depend linearly on the function $f(x)$ which corresponds to the 'independent variables' of classical analysis. Hadamard (1903a) has shown that the most general expression for a linear functional of $f(x)$ has the form

$$F[f] = \lim_{n \to \infty} \int_a^b f(x)\phi_n(x)\,\mathrm{d}x.$$

9.17. *Fréchet and Functionals*

In 1904 Fréchet attempted to extend to functionals the theorem by which Weierstrass had established that a function, which is continuous in a closed interval, necessarily attains its maximum in this interval. The arguments of the functionals considered were elements of a collection C such as lines or surfaces, and the limit A of a convergent sequence of elements $A_1, A_2, \ldots$ of C was characterized by the properties

(1) if $\lim_{n \to \infty} A_n = A$, then $\lim_{n \to \infty} A_{p_n} = A$,

    where $\{A_{p_n}\}$ is any subsequence of $\{A_n\}$ for which $p_{n+1} > p_n$;

(2) if $A_n = A$ for all n, then $\lim_{n \to \infty} A_n = A$.

A set E of elements of C was said to be closed if E contained the limit of each convergent sequence of distinct elements of E.

With these definitions Fréchet defined a continuous functional $f(X)$, $(X \in C)$, by the condition that the sequence $\{f(A_n)\}$ converges to the limit $f(A)$ whenever the sequence $\{A_n\}$ converges to a limit A. Finally he defined a subset E of the collection C to be compact, if any sequence of closed sets such that

$$E_{n+1} \subset E_n$$

possessed at least one common element.

Fréchet's most significant contribution to the theory of functionals was his definition of the derivative of a functional.

9.18. Schmidt and Sequence Spaces

In Hilbert's researches on integral equations, a function is determined by
its 'Fourier coefficients' with respect to an orthonormal complete system
of functions. These coefficients $(e_1, e_2 \ldots)$ are such that the infinite series
of real numbers,

$$e_1^2 + e_2^2 + \ldots$$

is convergent. Schmidt (1907a, b) takes any such series $\Sigma|z_p|^2$, con-
structed with *complex* numbers z_p, to specify a point z in an infinite
dimensional space, and the distance between two such points z and w to
be (in modern notation)

$$(z,w) = \Sigma z_p w_p$$

and the distance of z from the origin to be

$$\|z\| = \sqrt{(z,z)}.$$

A sequence of points $z^{(n)}$ is said to converge strongly to a limit point z if
$\|z^{(n)} - z\| \to 0$ as $n \to \infty$; and the criterion for strong convergence is that
$\|z^{(p)} - z^{(q)}\| \to 0$ as $p,q \to \infty$.

These concepts clearly have their analogues in the theory of functions
which are 'square summable' as was pointed out by Schmidt (1907a, b)
and Fréchet (1907), and a little later clarified by F. Riesz (1910).

9.19. Banach and Normed Spaces

A further development of the abstract theory of function space was taken
by Banach (1932), who abandoned the concept of the 'inner product'
(z,w), but retained the concept of the 'norm' $\|z\|$. The Banach space of the
points, or elements, z, is a linear vector space with a norm $\| . \|$, such that

$$\|z + w\| \leq \|z\| + \|w\|,$$

and such that a Cauchy sequence $\{z^{(p)}\}$ with strong convergence, namely

$$\|z^{(p)} - z^{(q)}\| \to 0 \quad \text{as} \quad p,q \to \infty,$$

has a strong limit z such that

$$\|z^{(p)} - z\| \to 0 \quad \text{as} \quad p \to \infty.$$

A 'functional' as defined by Fréchet (1926) was a mapping of the points
of a Banach space into the space of real numbers. Banach extended this
concept by introducing the idea of a 'linear operator' U by which one
passed from a point x of a Banach space X to a point $y = U(x)$ of another
(or the same) Banach space Y, with the restriction that

$$U(x_1 + x_2) = U(x_1) + U(x_2).$$

Perhaps the most important class of operators introduced by Banach
are not necessarily linear but enjoy the more significant property of
'contraction'. Such an operator U maps a space X into itself in such a way

that

$$\|U(x) - U(y)\| \leq \alpha \|x - y\|$$

where $0 < \alpha \leq 1$. For such a contraction operator there is a unique fixed point z such that $U(z) = z$, and this point is the limit of the sequence $\{z_n\}$, where $z_{n+1} = U(z_n)$ and $z_0 \in X$.

According to Dunford and J. T. Schwartz (1957) one of the three most fundamental principles of modern functional analysis is the Hahn–Banach theorem which establishes the existence of the 'extension' of a linear functional $f(x)$ defined on a linear subspace E of a Banach space X. By an appeal to Zorn's lemma (Section 5.9) Hahn (1927) and Banach (1929) showed that there is a linear functional $g(x)$ defined in X such that $f(x) = g(x)$ in E. Moreover if we define the 'norm' of $f(x)$ in E to be the least constant $N_E(f)$ such that

$$\|f(x)\| \leq N_E(f) \cdot \|x\|,$$

then

$$N_E(f) = N_X(g).$$

Fixed-point Theorems

9.20. *Kronecker's Integral and Index*

One of the most fruitful studies in topology has considered the mapping T of a set of points S into S, and the existence of fixed points x such that

$$Tx = x.$$

The importance of these studies is largely due to their application to ordinary and partial differential equations which can often be transformed into a functional equation $Fx = 0$ with $F = T - I$ where $Ix = x$.

The origin of these investigations can be precisely located in Kronecker (1869) and an elegant account is given by Picard (1891a), (see also Picard 1891b, c; Kronecker 1891), who illustrates the general theory of Kronecker for functions of n real variables by the simplest case of two functions, $f(x,y)$ and $g(x,y)$.

One of the simplest formulae of two-dimensional potential theory is

$$\oint \text{grad} \log 1/R \cdot dC = -2\pi \quad \text{or} \quad 0,$$

accordingly as the simple curve C does or does not encircle the origin. If we expand this integral as

$$I = -\oint (X dY - Y dX)/R^2,$$

and make the transformation

$$X = f(x,y), \quad Y = g(x,y),$$

we find that

$$I = -\oint (P dx + Q dy) = -\oint d \text{ arc tan } (g/f),$$

where

$$P = (fg_x - gf_x)/(f^2 + g^2),$$

$$Q = -(fg_y - gf_y)(f^2 + g^2).$$

Each critical point, i.e. each solution (ξ, η) of the equations $f = 0$, $g = 0$ within the contour of integration, makes a contribution to the integral of 2π if the determinant

$$D = \begin{vmatrix} f_x & f_y \\ g_x & g_y \end{vmatrix}$$

is positive and -2π if D is negative, at the point (ξ, η). Thus with each point of intersection of the curves $f = 0$, $g = 0$ we can associated its 'Kronecker index', i.e. the value of $D/|D|$ evaluated at the point of intersection.

Kronecker's integral does not give the number of points of intersection within a prescribed contour, but only the sum of the indices of these points. Picard, however, showed how to modify Kronecker's method so as to obtain the actual number of intersections.

9.21. The Poincaré Index

Poincaré, who spent little, if any, time in searching the literature of mathematics, rediscovered the Kronecker index in the course of his researches on the 'three-body problem' in celestial mechanics. He defined Kronecker's integral in terms of the vector V with components (f, g) attached to each point M of the contour C. As M describes the curve C, the vector V makes n complete revolutions, and clearly n is an integer or zero.

By purely geometrical arguments Poincaré identified n with the sum of the indices of the critical points of the vector V which lie inside the contour C. If the contour is chosen to surround only one critical point O then the Kronecker–Poincaré integral gives the 'index' of this point.

Stimulated by his astronomical problems, Poincaré developed powerful topological methods for studying ordinary differential equations of the form $\dot{x} = f, \dot{y} = g$) where f and g are functions of x and y only. Previous investigators had usually restricted themselves to the determination of particular solutions, represented by trajectories, $x = x(t; a, b)$, $y = y(t; a, b)$ issuing from a prescribed point (a, b). Poincaré made the decisive advance of considering the totality of all the trajectories of the given differential equation. In the dynamical applications x is usually a coordinate and y the corresponding velocity $\dot{x}$. The totality of the trajectories then furnishes the 'phase portrait' of the system.

The trajectories can also be regarded as the contour lines (or 'level lines') of the surface $z = \varphi(x, y)$ determined by the equation $f\varphi_x + g\varphi_y = 0$. The critical points can then be classified geometrically as centres (i.e. maxima or minima), nodes or foci, with index $+1$, or cols (i.e. saddle points, or minimaxes), with index -1.

9.22. The Brouwer Degree of a Mapping

The decisive step of incorporating these fragments into a complete topological theory was taken by Brouwer (1912, 1913a) who defined the 'degree' of a mapping of closed orientable manifolds into orientable manifolds, and proved the crucial theorem that if an n-cell is mapped into itself then there exists at least one fixed point.

Apparently in ignorance of the earlier work, J. W. Alexander (1922) gave Kronecker's index-integral and hence obtained Brouwer's index for a continuous transformation of an n-sphere into itself.

Extensions of Brouwer's theorem by G. D. Birkhoff and Kellogg (1922), Schauder (1930) and Tychynoff (1935) are all included in Lefschetz's fixed-point theorem or its extensions (Lefschetz 1926, Brown 1971).

These theorems are *global* in character, i.e. they give the algebraic number of fixed points of a continuous mapping f of a space X into itself, but, for applications to partial differential equations, what is needed is the number of fixed points on open subsets of X. A second new era in the theory was opened by Leray and Schauder (1934) who provided the necessary *local* theorems for continuous mappings of certain 'convexoid' spaces. The essential feature of this theory was the definition of the *local* topological, degree of a mapping, relative to some open set G of X.

The work of Leray and Schauder is based on the use of simplicial approximations to the given mapping f. A simpler approach has been given by Nagumo (1950, 1951a, b) and Browder (1960) has extended the Leray–Schauder theorem to more general spaces.

Connectivity

9.23. The Problem of Multiform Functions

The study of the theory of connectivity of surfaces was initiated by Riemann's researches on multiform functions of a complex variable (Riemann 1851, 1857a). If $f(w,z)$ is a polynomial in the complex variables w and z of degree n in w, then the roots $w_1, w_2, \ldots w_n$ of the equation in w,

$$f(w,z) = 0,$$

vary continuously with z as z pursues a path in the z-plane, and also pass continuously into one another at the branch-points of $f(w,z)$, i.e. at the points where two or more roots w_k become equal. This leads to an inevitable confusion unless some method is devised to reduce the multiformity of w as a function of z. Two such methods have been proposed, one, said by Forsyth to be substantially due to Cauchy, and the other due to Riemann.

In Cauchy's method the domain of the complex argument z is restricted by the removal of the points at which w becomes infinite and by the removal of certain 'branch-lines' or curves joining branch-points of w. These deleted lines act as 'barriers' obstructing the free movement of the variable z. The equation in w, $f(w,z) = 0$, has n distinct roots, each of which is now single-valued in the punctured and barriered plane. The disadvantage of this method is that there is a certain degree of arbitrariness in the introduction of the barriers.

Thus for the polynomial

$$f(w,z) = w^2 - (1 - z^2)(1 - k^2z^2) \qquad (0 < k < 1),$$

associated with the elliptic integrals, we can introduce as barriers either

(1) the lines

$$-1 \leqslant \mathrm{Re}(z) \leqslant 1, \ -\infty < \mathrm{Re}(z) \leqslant -k^{-1}, \ k^{-1} < \mathrm{Re}(z) < \infty$$

or

(2) the lines

$$-k^{-1} \leqslant z \leqslant -1, \ 1 \leqslant z \leqslant k^{-1}.$$

In Riemann's method the domain of definition of the complex variable z in the polynomial $f(w,z)$ is replaced by a 'Riemann surface', custom-tailored to the function $f(w,z)$ in such a way that the function $w(z)$ defined by the equation $f(w,z) = 0$ is single-valued on the Riemann surface.

This remarkable surface was constructed by Riemann by taking n replicas of the barriered z-plane, placing these n sheets one above the other so that the corresponding barriers coincide, and then making interpenetrating connections across the barriers, in such a way that a single connected manifold is obtained.

To depict or even to imagine such a mysterious surface is most difficult, and it is a great help to map a Riemann surface on a suitable two-dimensional curved manifold lying in ordinary Euclidean space.

The credit for the idea of mapping the complex plane of z on to a sphere is usually given to C. Neumann, who employed the stereographic projection from the south pole of the sphere onto the z-plane placed so as to touch the sphere at its north pole (C. Neumann 1884). The equations

$$X/x = Y/y = \tfrac{1}{2}(Z + 1)/(X^2 + Y^2 + Z^2) = 1$$

connect the point $z = x + iy$ with the point X,Y,Z on the sphere and the neighbourhoods of the origin, $z = 0$, and of the point of infinity, $|z| = \infty$ are mapped on neighbourhoods of the north pole ($Z = +1$) and the south pole ($Z = -1$) respectively. But it was Prym who, in an opportune conversation in 1874, gave Klein the idea of mapping a Riemann surface on a suitable curved surface in three-dimensional space (Klein 1893).

The great advantages of this representation are threefold. It presents a vivid pictorial image of a Riemann surface. It allows physical intuition to formulate, if not to demonstrate, the existence theorems for algebraic functions on the Riemann surface. And, above all, it enables the full resources of topology to characterize and classify the possible Riemann surfaces.

9.24. *The Dissection of Surfaces by Cross-cuts*

Riemann (1851) laid the foundations of the topology of what we should now call differentiable manifolds. The simplest illustration of such surfaces is provided by attaching a number p of 'handles' to a spherical surface. This yields the 'normal surfaces' introduced by Klein (1882). If $p = 0$ we have a sphere, if $p = 1$ an anchor-ring or torus. In the general case we have a surface of 'genus' p.

To define and discuss the connectivity of such a surface two methods are available. We can dissect the surfaces into a number of simply connected pieces, or we can draw on the surface a number of closed curves which cannot be continuously deformed into one another or continuously shrunk into points.

The distinction between surfaces which are bounded by one or more closed curves and surfaces (like the sphere) which are finite in extent but unbounded or closed in the sense that they have no boundary is not of prime importance, for a boundary can always be introduced into a closed surface by punctures, i.e. the removal of a point or points. What is of the first importance is the distinction, between two-sided surfaces and one-sided surfaces such as the famous 'Möbius strip'.

Riemann employed the method of dissection, and divided the complete surface by 'cross-cuts' which run from one point of the boundary (if any) to another, or by 'loop-cuts' which are closed curves on the surface, none or which intersects itself. A surface is then said to be 'simply connected' if it is resolved into two distinct pieces by every cross-cut or loop-cut. Since a loop-cut can be regarded as a cross-cut from an artificial point boundary, we need speak only of cross-cuts. In general m cross-cuts may divide the surface into n simply connected pieces. Riemann proved that, however this dissection is effected, the number $m - n$ is the same for each method of dissection.

Thus for a spherical surface with h holes bored through it, $h - 1$ cross-cuts dissect it into a single simply connected domain, and for Klein's normal surface with p handles, p cross-cuts are necessary to cut through each handle and reduce it to a spherical surface with p holes.

The number $n - m$ is the first topological invariant of a surface to be discovered, and the number $N = n - m + 2$ is called the 'connectivity' of the surface. For a simply connected surface $N = 1$ and for Klein's normal surface with p handles $N = 2p + 1$.

Perhaps the most important theorem regarding the dissection of surfaces is the famous proposition that if a closed surface of connectivity $2p + 1$ is dissected into any number of simply connected pieces each in the form of a curvilinear polygon, then if F is the number of faces, E is the number of edges and V the number of vertices

$$F - E + V = 2 - 2p.$$

The special case of this formula for a sphere was discovered by Euler in 1752. The general result was given by L'Huilier (1812), but the first proof for the general result was due to Poincaré in 1894. Curiously enough the only reference given by Poincaré to the early history of this famous formula is to 'M. l'amiral de Jonquières' (Jonquières 1890).

9.25. *Irreducible Circuits*

The second method of studying the connectivity of a surface is the method of 'irreducible circuits' due to Jordan (1866a).

Any simple closed curve on the surface which does not intersect itself is called a 'circuit'. If a circuit can be reduced to a point by continuous deformation, without crossing the boundary of the circuit, it is called 'reducible'. Thus the circles of latitude on a sphere are reducible to a north pole and a south pole. When an irreducible circuit is subjected to a continuous transformation without crossing the boundary two cases may present themselves:

(1) its transform may always be a single circuit, or
(2) its transform may consist of repetitions of one or more circuits.

Thus an endless string wound n times around an anchor-ring can be continuously deformed into a meridian curve and n repetitions of a circle of latitude.

The operations of describing a circuit, of repeating this process or of reversing it by describing it in the opposite sense, and of transforming a circuit into one or more circuits and their repetitions can be represented algebraically as follows. Any simple circuit, together with all the simple circuits into which it can be continuously transformed, is represented by a symbol such as a, and the same circuit is denoted by $-a$ when described in the reverse direction. The multiple circuit consisting of m repetitions of the simple circuit a, all in the same sense as a, is denoted by ma. If a compound circuit c can be continuously transformed into a system of simple irreducible circuits, a_1, a_2, ..., a_n, which are respectively described m_1, m_2, ..., m_n times, then c is denoted by

$$c = m_1a_1 + m_2a_2 + \ldots + m_na_n.$$

A point, a point circle, or a simple reducible circuit is denoted by 0 (zero). The circuit c is reducible if

$$m_1a_1 + m_2a_2 + \ldots + m_na_n = 0.$$

A system of irreducible simple circuits, a_1, a_2, ..., a_n, is independent if they are not connected by any relation of the form

$$p_1a_1 + p_2a_2 + \ldots + p_na_n = 0$$

with integral coefficients p_k, unless all the coefficients are zero. Such a system is complete if any circuit c on the surface can be expressed in the form

$$c = m_1a_1 + m_2a_2 + \ldots + m_na_n.$$

Such a representation is not necessarily unique, but if

$$c = m_1a_1 + \ldots + m_na_n = \mu_1a_\nu' + \ldots + \mu_\nu a_\nu'$$

are two representations, each in terms of complete independent sets of irreducible circuits, then $n = \nu$.

The number of circuits in a complete set of independent circuits is therefore a topological invariant of the surface, and it is in fact equal to $N - 1$, where N is the connectivity of the surface. (In the special case of a simply connected surface, $N = 1$, and all circuits are reducible.) Jordan showed that the necessary and sufficient condition that two surfaces should be applicable to one another (if we supposed them to be extensible and deformable), is that they should have the same connectivity or genus.

9.26. Connectivity in Higher-dimensional Space

The generalization of the theory of connectivity is briefly considered in the famous 'Fragment aus der Analysis Situs' found among Riemann's papers after his death in 1866, and more fully developed in a paper by Betti (1871), with whom Riemann had been in correspondence, although we have no letters bearing on the theme of connectivity.

Listing (1861) had considered these problems for regions of three-dimensional Euclidean space, but Betti studied the general problem for a region R of n dimensions which is bounded by one or more hypersurfaces each of $n-1$ dimensions.

Even the simple passage from two to three dimensions considerably increases the complexity of the problem. Consider, for example, the region R bounded externally by a large sphere and internally by a small anchor-ring. Then in the region R there are simple curves reducible to a point and simple curves, threaded through the anchor-ring and therefore irreducible. There are also closed surfaces in R which can be collapsed like bubbles to a point and closed surfaces which embrace the anchor-ring and are irreducible. Thus there are different species of connection according as we consider closed curves or closed surfaces in the region R.

In the general case of a region R of n dimensions Betti proceeds by considering a system of closed manifolds each of dimension $m(m < n)$, say

$$V_1, V_2, \ldots, V_{q_m}.$$

No one of these manifolds taken by itself is the boundary of an $m+1$-dimensional space, and no two of these manifolds can be continuously deformed into one another, i.e. they are irreducible and independent. Moreover, the totality of these manifolds does not form the boundary of $m+1$-dimensional space by itself, but only when we adjoin any other independent and irreducible m-dimensional manifold. Then we can say that the order of connection of the n manifold R is $p_m = q_m + 1$, with respect to the submanifolds of m dimensions.

To justify this definition Betti showed that the number p_m is independent of the choice of the $q_m = p_m - 1$ manifolds $V_1, V_2, \ldots, V_{q_m}$. A more formal version of this theorem is given by Picard and Simart (1897). Thus if R is the region between two concentric spheres S_1 and S_2, then $p_1 = 1$ and $p_2 = 2$, for any closed curve in R bounds a surface, and there cannot be more than two independent surfaces, one of which embraces the inner sphere S_2 while the other is external to S_2.

This great advance in the theory of connectivity appears to have attracted little attention until Poincaré started to publish his great series of papers on 'Analysis situs' (Poincaré 1895a), which inaugurated the modern theory of connectivity. In fact the definition given by Betti is open to criticisms made by Heegaard (1898) to which Poincaré replied in his second paper (Poincaré 1899).

9.27. Poincaré and Homology

It is one of the main tasks of the historian of mathematics to discover the persons and the ideas which influenced a great mathematician and prepared the way for his own researches. This task is peculiarly difficult in the case of Poincaré, especially in relation to his vast contributions to topology. The fact is that although Poincaré does mention Riemann and Betti by name it is almost certain, according to Freudenthal (1955) speaking at the centennial celebrations of the birth of Poincaré, that the celebrated mathematician had never read Riemann and the authors who continued his theory.

However this may be, Poincaré (1895a) presents his theory of connectivity with startling brevity.

A set of points whose coordinates $(x_1, x_2, \ldots, x_n)$ satisfy p independent equations

$$F_\alpha(x_1, x_2, \ldots, x_n) = 0 \qquad (\alpha = 1, 2, \ldots, p),$$

and q inequalities

$$\varphi_\beta(x_1, x_2, \ldots, x_n) > 0 \qquad (\beta = 1, 2, \ldots, q),$$

is said to form a 'variety' V of $n - p$ dimensions. The complete frontier of V consists of the union of q sets $v_1, v_2, \ldots, v_q$, of which a typical one is given by a relation of the form

$$F_\alpha = 0, \; \varphi_\beta > 0, \quad \text{if} \quad \beta \neq \gamma \quad \text{and} \quad \varphi_\gamma = 0$$

$$(\alpha = 1, 2, \ldots, p; \beta = 1, 2, \ldots, \gamma - 1, \gamma + 1, \ldots, q).$$

Now let the equations $F_\alpha = 0$ be replaced by the p equations $\Phi_\alpha = 0$, where

$$\Phi_\alpha = \sum_{\beta=1}^{q} A_{\alpha\beta} F_\alpha$$

where the coefficients $A_{\alpha\beta}$ are functions of $x_1, x_2, \ldots, x_n$. If Δ, the determinant of these coefficients does not vanish in the variety V, then the equations $\Phi_\alpha = 0$ (together with the inequalities $\Phi_\beta > 0$) represent the same set of points as the original set $F_\alpha = 0$. If Δ is positive then this set of points is identical with V, but if Δ is negative, this set of points is said by Poincaré to represent the 'opposed variety'. This is the germ of all subsequent definitions of 'orientation'.

In order to give symbolical expression to the fact that the q-varieties, $v_1, v_2, \ldots, v_q$ together form the complete frontier of V, Poincaré writes

$$v_1 + v_2 + \ldots, v_q \sim 0$$

or, 'more generally'

$$k_1 v_1 + k_2 v_2 + \ldots + k_q v_q \sim 0$$

to show that the frontier is composed of k_1 varieties 'a little different' from v_1, of k_2 varieties, 'a little different' from v_2, etc., the sign of a coefficient indicating whether the variety $k_\lambda v_\lambda$ is opposed ($k_\lambda < 0$) or not ($k_\lambda > 0$) to the variety v_λ.

The reader may find the explanation a little brief and may notice the absence of any proof of its internal consistency, such as Betti (1871) and

Jordan (1866b) give in the two-dimensional case. But Poincaré merely asserts that relations such as

$$k_1 v_1 + k_2 v_2 + \ldots + k_\lambda v_\lambda \sim 0$$

can be combined with one another like ordinary equations and gives to these relations the name of 'homologies'. Thus he created what we now know as algebraic topology.

9.28. Betti Numbers

The definition of the Betti numbers given by Poincaré is undoubtedly clearer and more concise than the definitions given by Betti, Riemann and Picard, once the key concept of homology is accepted.

The varieties

$$v_1, v_2, \ldots, v_\lambda$$

are said to be linearly independent if they are not connected by any homology $k_1 v_1 + k_2 v_2 + \ldots + k_\lambda v_\lambda \sim 0$, except the trivial relation in which each coefficient is zero. If there exist $P_m - 1$ linearly independent closed m-dimensional varieties in a space V, and no more than $P_m - 1$, then the order of connection of V with respect to the m-dimensional varieties in V is equal to P_m. The numbers $P_1, P_2, \ldots, P_{m-1}$ are the Betti numbers of V.

These numbers are not all independent, for Poincaré proved that

$$P_m = P_{n-m} \quad (m = 1, 2, \ldots, n - 1),$$

a result which had been previously conjectured but not demonstrated.

Moreover Poincaré[1] succeeded in establishing the complete generalization of Euler's theorem that, if a convex polyhedron in three-dimensional space has V vertices, F faces and F edges,

$$V - E + F = 2.$$

For any polyhedron, not necessarily convex, Jonquières (1890) had obtained the result

$$V - E + F = 3 - p_1.$$

Poincaré considered a closed p-dimensional variety V subdivided into α_p varieties v_p each of p dimensions, while the varieties v_p have in all α_{p-1} $(p-1)$-dimensional frontiers v_{p-1}, and these have in turn altogether α_{p-2} $(p-2)$-dimensional frontiers, etc., all the varieties $v_p, v_{p-1}, \ldots, v_1$ being simply connected. Then Poincaré's result is that

$$\alpha_p - \alpha_{p-1} + \alpha_{p-2} - \ldots + (-1)^p \alpha_0 = 3 - p_1 + p_2 - \ldots + p_{p-1} \quad (p \text{ even})$$

$$= 0 \qquad (p \text{ odd}).$$

9.29. Poincaré and Homotopy

A second method of studying the connectivity of a variety, also invented by Poincaré, depends on the concepts of 'equivalences' and the 'fundamental group' of a variety (Poincaré 1895a).

[1] See Section 9.24.

If we consider the closed paths which start and finish at a prescribed point M of an m-dimensional variety V and denote by Γ_p the operation of describing a closed path γ_p then $\Gamma_q\Gamma_p$ corresponds to the description of γ_p followed by the description of γ_q. Further, Γ_0 will be the identical operation if γ_0 is the complete frontier of a two-dimensional variety forming part of V, for, says Poincaré, in this case γ_0 'can be decomposed into a very great number of loops', each of which can collapse into the point M.

From these brief considerations Poincaré inferred that the operations Γ_p formed a group, which he called the 'fundamental group' of V, and he noted that the operations of this group do not necessarily commute, that the group is generated by a finite number of operations, $S_1, S_2, \ldots, S_p$ which may well be connected by certain 'fundamental equivalences' of the form

$$S_1^{k_1}S_2^{k_2} \ldots = \text{the identical operation.}$$

This is the origin of the theory of homotopy.

9.30. Homology Theories of Topological Spaces

Homology theory was developed first for spaces which admitted a triangulation, i.e. spaces which are homeomorphic to a simplicial complex in Euclidean space, but the theory has gradually been freed from these shackles, first imposed by Poincaré.

The first step was taken by Alexandroff (1928a) for compact, metric spaces R. The famous covering theorem of Borel (Section 11.12) asserts that such a space can be covered by a finite set of (overlapping) open sets, $U_1, U_2, \ldots$, each of which has a diameter less than some prescribed positive number ε (the norm of the covering). The 'nerve' N of this covering is defined as consisting of

(1) the set of abstract 'vertices' $U_1, U_2, \ldots, U_k$,
(2) the abstract n-simplexes (or 'skeletons') consisting of sets of $n + 1$ vertices, $U_{\alpha_0}, U_{\alpha_1}, \ldots, U_{\alpha_n}$, with a non-empty intersection.

Thus the given covering of the space R is mapped into a simplicial complex N.

By considering successive refinements $N_1, N_2, \ldots$ of the covering with a succession of norms $\varepsilon_1, \varepsilon_2, \ldots$ which tend to zero Alexandroff establishes a homology theory in which each cycle C_{v+1}^n in N_{n+1} determines a homologous cycle C_v^n in N_n and in which such a sequence of cycles $\{C_1^n, C_2^n, \ldots\}$ is taken to be a cycle in R.

This abstraction was taken a stage further by Čech (1932) by removing the restrictions that the space R should be metric and compact. If the space R is a topological space it can still be covered by a finite collection of open sets, and the corresponding 'nerve' can still be constructed, but there is now no definition of 'norm'. Čech is driven to consider, not a sequence of nerves but the family of *all* nerves, following the example of Hurewicz (1927) in his studies of the theory of dimension.

Further developments of the nerve theory have been given by Vietoris (1927).

The Theory of Dimensions

9.31. *Introduction*

Since the work of Bolzano[1] the theory of dimensions has developed from a concept which was merely intuitive and almost trivial into a profound topological investigation.

In elementary geometry to say that the physical space of everyday experience is a space of three dimensions means that the position of a point in this space can be completely specified by the numerical values of *three* coordinates, the cartesian coordinates (x,y,z) or the spherical polar coordinates (r,θ,φ), or any equivalent set of independent coordinates, and that it cannot be specified by only two coordinates. Similarly, to say that a surface has two dimensions would mean that the position of a point on the surface can be completely specified by the numerical values of *two* coordinates. Thus on the sphere $x^2 + y^2 + z^2 = a^2$, the position of a point is specified by its latitude $(\frac{1}{2}\pi - \theta)$ and its longitude φ where

$$x = a \sin \theta \cos \varphi$$

$$y = a \sin \theta \sin \varphi$$

$$z = a \cos \theta.$$

Again, to say that a curve has one dimension would mean that the position of a point of the line can be completely specified by the numerical value of *one* coordinate. Thus on the equator

$$z = 0, \ x^2 + y^2 = a^2$$

the position of a point is specified by its longitude φ, where $x = a \cos \varphi$, $y = a \sin \varphi$.

This concept of dimensions is merely a way of saying that a surface (or a curve) is a mapping of the plane with coordinates (θ,φ) (or the straight line with coordinate φ) into three-dimensional Euclidean space. It is therefore restricted to *manifolds* which are, by definition, obtained by mapping portions of Euclidean space, of say, m dimensions, into a topological space.

The object of the theory of dimension is to generalize the concept of dimension so that it shall be applicable if possible to any space, or, as a more modest aim, to characterize those spaces which do possess a dimension. The spaces to be considered are not necessarily manifolds, as described above. They may be metric spaces or non-metrizable spaces.

The simplest spaces would seem to be curves in a Euclidean plane. Now Jordan (1887) had given what seemed to be a satisfactory definition

---

[1] Bolzano's work lies outside our period and had little influence on subsequent researches. It was admirably described by Dale M. Johnson (York University) in a lecture on 'Problems in the history of dimension theory' in 1974.

of a plane curve as a mapping of the unit interval, $0 \leqslant t \leqslant 1$, into the plane by means of two functions $x = f(t)$, $y = g(t)$.

The naïve opinion that the dimension of a space was equal to the number of numerical coordinates required to specify the position of a point received two rude shocks from the discoveries of Cantor and of Peano.

Cantor (1878, 1879) showed that it is possible to establish a one–one correspondence between the points of a line and the points of a plane, while Peano (1890b) constructed a space-filling curve which mapped an interval, $0 \leqslant t \leqslant 1$, on the whole of the square, $0 \leqslant x,y \leqslant 1$. However, Cantor's mapping is not continuous, and Peano's mapping is neither single-valued nor continuous. The impossibility of a one–one continuous mapping from an interval of a straight line to the points of a square was proved later by Netto (1879).

The importance of dimension theory is that the dimension of a space is its simplest and most fundamental topological invariant, and that many of the topological properties of a space are crucially dependent on its dimension.

There are at least half a dozen different theories of dimensions, and dimension theory consists essentially of the construction of the various different definitions of dimension and the investigation of their equivalence. In each case it is of crucial importance to examine the topological invariance of dimension, i.e. to prove that two spaces which are respectively of dimensions m and n cannot be 'homeomorphic' unless n equals m.

According to Fréchet (1910) it is to Hadamard that we owe this term which denotes a one–one transformation of a set S onto a set T such that, if the elements $\{p_n\}$ of S converge to a point p of S then the corresponding elements $\{q_n\}$ of T converge to the corresponding point q of T, and vice versa.

9.32. *Preparatory Studies*

The preparatory studies which, unwittingly, led to the modern theory of dimension were concerned with the simplest version of this result, viz. that it is impossible to establish a homeomorphism between two Euclidean spaces of different dimensions n and $n + p$, $(p \geqslant 1)$. This result was established by Lüroth (1907) for the cases $n < 3$ and for the general case by Baire (1907) and Brouwer (1911).

The possibility of extending the concept of dimension to more general spaces was adumbrated by Baire (1907) and Fréchet (1910). Baire (1899a) also introduced the idea of the *zero*-dimensional space in which each point A is a sequence $\{a_n\}$ of positive integers, and in which a set of points $A_\nu = \{a_{n,\nu}\}$ is said to converge to a limit point $A_0 = \{a_n\}$ if for all sufficiently large ν

$$a_{n,\nu} = a_n.$$

We note that, in the non-standard analysis employed to define infinitesimals, Luxembourg (Section 3.7) uses the zero-dimensional space in which each point A is a sequence of real numbers.

Fréchet's paper explicitly considers the possibility of defining the dimension of an abstract space and of the existence of spaces of fractional dimension, but his actual achievement seems to fall short of this ambitious project. In fact he does little more than give a descriptive definition of the 'dimension-type' dG of a space G, such that $dG_1 \leq dG_2$ if G_1 is the image of G_2 or a part of G_2 under a homeomorphism.

9.33. 'Covering' Dimension

Lebesgue (1911) not only established the theorem of Baire and Brouwer to which we have just referred but prepared the way for a rigorous definition of the 'covering' dimension of a space.

It is well known that a plane area can be covered by regular hexagons (viz. six-sided polygons together with their edges and vertices) in such a way that no point of the area is covered by more than three hexagons—as in the Giant's Causeway in Co. Antrim, Northern Ireland. And it is clear that the set of hexagons is topologically equivalent to a set of staggered squares. Lebesgue proved that a domain D in n-dimensional Euclidean space could be covered by arbitrarily small 'cubes' in such a way that there must be points covered by at least $n + 1$ 'cubes'—a 'cube' being a *closed* n-dimensional rectangular parallelipiped, and he rightly stated that the covering can be constructed so that no point is covered by more than $n + 1$ 'cubes'. A formal proof of this theorem was given by Brouwer (1913a). The ideas of Lebesgue (1911) and Brouwer (1913) lead to a very general and precise definition of the dimension of a space R in terms of the concept of the 'covering' of a space.

An (open) covering of a space R is a collection of open sets whose union is R. If R is a metric space the cover is called an 'ε-cover' if each of these open sets has a diameter less than ε. In any space R the *order* of the covering is the greatest integer n such that there are $n + 1$ open sets of the covering which have a non-empty intersection. A covering β is a *refinement* of a covering α if each open set of β is covered by some open set of α.

We can now define the covering dimension of a space R. If each open covering α of R has a refinement β of order not exceeding $n + 1$ then R has a covering dimension, dim R, not greater than n. If dim $R \leq n$, and if it is false that dim $R \leq n - 1$, then dim $R = n$. For the empty set it is convenient to set dim ϕ equal to -1.

Alexandroff (1928b) further developed these concepts by introducing the idea of the 'nerve' of a covering. The nerve of any finite collection γ of sets of points $A_1, \ldots, A_s$ is the complex with vertices $a_1, \ldots a_s$ in one–one correspondence with the sets $A_1, \ldots, A_s$, and simplexes $|a_p, a_q \ldots|$ corresponding to those subcollections $A_p, A_q, \ldots$ of γ which have a non-empty intersection. The nerve of a covering is thus a combinatorial configuration approximating to the original space, and the dimension of the nerve γ is the order of the covering.

9.34. 'Inductive' Dimension

Another approach to the theorem of dimension was initiated by Poincaré (1912a) in an informal and popular article which describes (two-dimensional) surfaces as the boundaries of three-dimensional solids or as cuts which divide three-dimensional space, curves as boundaries of surfaces or as cuts dividing surfaces into two, and points as boundaries of curves or as cuts dividing curves.

This vague and familiar concept was transformed into a precise definition of dimension by Urysohn (1926), and independently by Menger[1]. According to Alexandroff (1925), Urysohn constructed his theory of dimension during the years 1921–2, and during this period communicated his results to the Mathematical Society of Moscow. A preliminary notice was published in 1922 (Urysohn 1922) and a full account in 1925 (Urysohn 1925). Urysohn's definition of the dimension of a space C at a point x applies to metric spaces only and depends upon a decomposition of the space C into three disjoint sets, A,B,D, such that

(1) x is a point of A,
(2) $A\bar{B} \cup \bar{A}B = \phi$ ($\bar{A}$ and $\bar{B}$ the closures of A and B), and
(3) $A \cup B$ is contained in a ball of centre x and radius ε.

If this decomposition is possible for any ε, and with B an empty set (!) then C has dimension zero at the point x. Higher dimensions are defined recursively.

Similarly Menger's dimension, which is not restricted to metric spaces, proceeds by successively specifying the spaces which have dimensions -1, 0, $+1$, $+2$, ... and it is therefore distinguished as the 'inductive' definition. The empty set ϕ has (strong) inductive dimension, ind ϕ, equal to -1.

If for any disjoint closed sets F and G in a topological space R there exists an open set U such that

$$F \subset U \subset R - G,$$

and the boundary of U, $B(U)$ has inductive dimension

$$\text{ind } B(U) \leqslant n - 1,$$

then ind $R \leqslant n$. If ind $R \leqslant n$ and ind $R \nleqslant n - 1$, then we define ind R to be n.

Menger's definition is undoubtedly simpler and more general than Urysohn's, and the question of priority is of minor importance.

---

[1] A sealed envelope containing his definitive definition of a curve and a suggestion of his concept of dimension was deposited by Menger with the Vienna Academy of Sciences in the autumn of 1921, while he was in the throes of a serious disease. The envelope was opened in 1926 and its contents published in the following periodicals: *Koninklijke Akademie von Wetenschappen te Amsterdam*, **29**, 1122–4 (1926); *Monatsch. Math.*, **36**, 411–32 (1926). Meanwhile Menger had also published the following papers: 'Über die Dimensionalität von Punktmengen', *Monatsch. Math.*, **33**, 148–60 (1923); **34**, 137–61 (1924); 'Zur Begrundung einer axiomatische Theorie der Dimension', *Monatsch. Math.*, **36**, 193–218 (1926). Further papers followed in 1926, 1927 and 1928 (Menger 1926d; 1927, 1928).

PART III. ANALYSIS

10. The Concept of Functionality

10.1. Introduction

We owe to the Arabs the name of 'Algebra' which is derived from a word meaning the surgical treatment of fractures by bone-setting, or, more generally, the reunion of broken parts. This name is admirably adapted to modern abstract algebra, each variety of which is characterized by functions which coalesce a set of elements of the algebra into a single element. Thus in the most familiar example, a pair of elements, say x and y, are united by addition into a single element $p(x,y) = x + y$ or by multiplication into a single element $q(x,y) = xy$.

But the name of 'Analysis', which still means literally the resolution of a complex entity into its constituent simple elements, and therefore the inverse concept to algebra, is far from suggesting the nature of those mathematical topics which are usually treated in those magnificent 'Cours d'Analyse' for which the French mathematicians are now so justly famous. If we are to attach any significance to the opinions of that famous detective M. C. Auguste Dupin (Poe 1845), the mathematicians 'with an art worthy of a better cause . . . have insinuated the term "analysis" into applications to algebra. The French are the originators of this particular deception'.

Unfortunately for this historian, Dupin did not develop his strictures nor give specific references. Indeed it now appears that there is only one venial fault chargeable against the French analysts—they did not provide any effective definition of their subject. To remedy this defect I suggest that 'Analysis' may be taken to mean the study of the local and global properties of functions regarded as mappings of one space on another, and especially the study of the relations between these two species of property.

From this point of view the subject divides naturally into four parts:

(1) the study of local properties such as continuity and differentiability;
(2) the study of global properties such as integrability;
(3) the deduction of global properties from local properties, for example the construction of a power series for a function from the values of its derivatives at a single point;
(4) the deduction of local properties from global properties, for example the numerical evaluation of a function at the points of a domain from the property that the function minimizes some integral taken over the whole domain.

Each of these topics can be further divided accordingly as the functions studied are mappings from one real variable to another, or from one complex variable to another, or from many variables, real or complex, to one or more variables, real or complex. However, we shall be almost exclusively concerned with functions of real variables.

This section of the history will therefore discuss

(1) the development of the concept of a function (Chapter 10);
(2) the theories of differentiation and integration from Cauchy to Laurent Schwartz (Chapters 11 & 12);
(3) ordinary differential equations (Chapter 13);
(4) the calculus of variations (Chapter 14) and the grand theory of potential and harmonic functions (Chapter 15).

10.2. Continuity and Arbitrary Functions

The early history of the development of the concept of a 'function' from Newton to Cauchy was dominated by two ideas—that of algebraic construction and that of kinematic continuity.

Starting from the unanalysed idea of a 'real variable' capable of taking all real values in some interval, the algebraic processes of addition, subtraction, multiplication and division, together with the extraction of square roots and the solution of algebraic equations yielded familiar and well-known constructions for a wide variety of functions. On the other hand, the concept of the motion of a particle along a straight line introduced an idea of geometric or kinematic continuity which persisted until Darboux (1875) proved conclusively that it was inadequate for the purposes of analysis.

If $x = f(t)$ is the distance x of a moving point from a fixed origin on a fixed straight line at a time t, the function $f(t)$ possesses 'kinematic continuity' if the point in moving from the position $x_1 = f(t_1)$ at time t_1 to the position $x_2 = f(t_2)$ at a subsequent time t_2 passes through all the positions intermediate between x_1 and x_2. This definition is so simple and natural that it is hard to accept that it is inadequate.

A counter example is provided by the function

$$f(t) = \begin{cases} \sin{(1/t)}, & t \neq 0, \\ 0, & t = 0. \end{cases}$$

Darboux proved that if $f(t)$ is differentiable in a closed interval, then its derivative possesses 'kinematic continuity' in this interval, even if it is discontinuous!

The accepted definition of continuity is due to Cauchy (see Section 11.2) and proved so useful that for some time it seemed that 'analysis' meant no more than the study of continuous functions, which were presumed to be differentiable and integrable.

H. Lebesgue (1928) has remarked that the definition of continuity criticized by Darboux in 1875 was still taught in 'un lycée de Paris' in 1903!

In England, Frost's presentation of the earlier sections of Newton's

Principia (Frost 1863) also taught the kinematic definition of continuity, and was still a set book when Forsyth was an undergraduate.

Indeed the same distinguished mathematician confessed in a nostalgic lecture (Forsyth 1935), 'I can remember a college question, set years after my student days, "Define a function, and prove that every function has a differential coefficient"'.

A third tributary to the main stream of function theory was the subject of the trigonometric functions—the sine, cosine, tangent etc.—which possessed such easy geometric definitions, enjoyed so many interesting algebraic properties and could be proved to possess an infinite number of derivatives.

All three tributaries converged in the investigations of the problems of vibrating chords (see Section 12.1) which compelled the consideration of 'arbitrary' functions which were neither algebraic, nor kinematic, nor trigonometric, but which could be represented by infinite, convergent series of trigonometric functions. The researches of Fourier on the conduction of heat, which began in 1807, carried these studies a step further by providing an informal but suggestive proof that some not unduly restrictive conditions were sufficient to ensure that an arbitrary function could be expressed in a trigonometric series[1].

The formal and rigorous examination of this problem begins with Dirichlet in 1829, the importance of whose researches is not merely that he obtained sufficient conditions for the convergence of Fourier's trigonometric series, but that he definitely codified the concept of a function.

10.3. Set Theory versus Function Theory

The formal definition of a function of a real variable contemplates two sets of real numbers, X and Y, and their 'cartesian product' $X \times Y$ which consists of the ordered pairs (x,y), such that $x \in X$ and $y \in Y$. A function is then defined merely as a subset F of the cartesian product space such that, if

$$(x,y_1) \in F \quad \text{and} \quad (x,y_2) \in F$$

then $$y_1 = y_2.$$

i.e. the function is 'single-valued', and only one number y 'corresponds' to one number x.

There is no doubt that this definition tells us everything we need to know about the function as a basis for a rigorous theory, but generations of students have found it unacceptably arid after the vague but evocative descriptions given by Newton, D'Alembert and Euler. This uneasy feeling is exemplified in a notorious passage from Forsyth's *Theory of Functions of a Complex Variable* (Forsyth 1893), pilloried by Littlewood in *A Mathematician's Miscellany* (Littlewood 1953).

Forsyth was an eloquent lecturer, never at a loss for a word, or even for a whole paragraph, and in this passage he gives full rein to his

---

[1] Detailed references will be found in the writings of Jourdain, especially his *Contributions to the Founding of the Theory of Transfinite Numbers* (Jourdain 1915).

oratundical verbosity in order to give the beginner some idea of what it means to be a function without ever repelling him by a formal definition.

Forsyth was not alone in voicing in public his disquiet with the accepted definition of a function. A much greater mathematician, namely Hermann Weyl, actually went so far as to say, 'We do not really know what a function is.'

But the most striking expression of the revolt against the 'set-theoretic' definition of a function is due to J. von Neumann (1928) who showed that the concepts of a set and of a function are equivalent and that analysis can be based on the undefined idea of a function just as well as, if not better than, on the undefined idea of a set.

10.4. *Category Theory*

The homology theory, whose origins we have briefly sketched in Section 9.26, developed in numerous variants, of different degrees of validity and complexity. The two main difficulties appear to have been the covering of a given space by triangulations, or by the more abstract complexes of Vietoris, Lefschetz and Čech, and then the consistent and unambiguous assignment of an orientation to the cells of a complex and their boundaries. Eilenberg and Steenrod (1952) invented an axiomatic treatment of homology theory which eliminated the concepts of complex, orientation and chain from the foundations of the subject and produced a great simplification in the proofs of the theorems.

These authors did very much more than offer a radically new approach to homology theory. In fact they created not merely a new abstract algebra, but what can be taken to be the most abstract and fundamental formulation of mathematics—at least for those who are prepared to waive the rigorous canons of constructibility enforced by Brouwer.

As our purpose is to exhibit this new theory as the latest development of the concept of functionality we shall adopt the formulation given by MacLane (1971).

In the first axiomatic presentation of function theory, a function f was defined in terms of the two sets, say X and Y, of which X is transformed into Y by f, i.e. if x is an element of X, $f(x)$ is an element of Y. The motivation of category theory may be described as the elimination of elements and the presentation of functions as entities which are basic or primary concepts. The function f, in the earlier presentation, was denoted by $f(X) = Y$ or by $f:X \to Y$, but in category theory the essential feature is the 'arrow', $\to$.

This theory then falls into three parts, a metacategory theory of arrows, an interpretation of this theory in terms of sets (this is the 'category' theory) and 'functor' theory, described below.

A metacategory C consists of 'arrows'[1], denoted here by $f, g, h \ldots$, and the operation $g \circ f$ which assigns to certain ordered pairs f and g another arrow denoted by $g \circ f$, and called the composite of f and g. This is an

---

[1] According to S. MacLane the idea of representing a function by an arrow is probably due to Hurewicz (1941).

abstraction from the elementary concept of forming the function $g[f(x)]$ from the functions f and g—a function which exists only if the domain of g is exactly the same as the 'co-domain' or 'range' of f. The triple composite $(k \circ g) \circ f$ is defined if and only if $k \circ (g \circ f)$ is defined, and then is identical with $k \circ (g \circ f)$. To each arrow g there correspond 'identity arrows', u and v, such that

$$g \circ u = g \quad \text{and} \quad v \circ g = g.$$

This last axiom enables us to define the domain of g as u and the codomain of g as v, and thus establish the primacy of arrows (or functions) over their domains and codomains (or the argument and values of a function).

J. von Neumann's theory (Section 10.3) is thus an unconscious anticipation of metacategory theory, but falls short of the category theory in which the arrows form a 'set' C and the identity arrows form another set O, and there is also a certain morphism, or mapping, of categories called a functor.

A 'functor' F is a function which carries arrows f of a category C into arrows $g = Ff$ of a category K, in such a way that each identity arrow u of C is carried into an identity arrow fu of K and each 'product' $g \circ f$ of C into the product $Fg \circ Ff$ of K.

There is an extensive and rapidly expanding literature on the subject of category theory and its applications to topology, algebraic geometry and mathematical logic. The preceding notes give no more than an indication of the genesis of the subject.

10.5. *Functions of a Complex Variable.*

Although the history of the theory of functions of a complex variable is not included in this volume, it is only fitting to notice the unique position which this theory occupies in the general topic of functionality.

In an n-dimensional space R^n with coordinates $(x_1, x_2, \ldots, x_n)$ we can define a function $f(x_1, x_2, \ldots, x_n)$ as a mapping from the points of R^n into the points y of a real straight line, and we can define the partial derivatives

$$\partial f/\partial x_k = \lim \, [f(x_1, \ldots, x_k + h, \ldots, x_n) - \\ f(x_1, \ldots, x_k, \ldots, x_n)]/h \text{ as } h \to 0.$$

In the space of the complex variable $z = x + iy$ we can do more than this by constructing the incrementory ratio

$$[f(z + \eta) - f(z)]/\eta, \quad f(z) = f(x + iy)$$

where $\eta = h + ik$. Cauchy must be credited with the discovery that there is a whole new world of functions $f(z)$ for which the incrementory ratio tends to a unique limit $f'(z)$ as $|\eta| \to 0$, independently of the path by which the point $z + \eta$ tends to z; $f'(z)$ is then appropriately called *the* derivative of $f(z)$.

If $f(z)$ possesses a unique derivative $f'(z)$ at all points of a connected domain D, then $f(z)$ is an 'analytic' function of z in D, and $f'(z)$ itself has a unique derivative $f''(z)$ at each point of D. Indeed $f(z)$ possesses derivatives $f^{(n)}(z)$ of all orders at each point of D and a power series expansion

$$f(z + \eta) = \sum_{n=0}^{\infty} \eta^n f(z),$$

in some disc $|\eta| < \varrho$ surrounding each point z of D.

The whole of the great theory of functions of a complex variable depends on this property of the existence of iterated derivatives.

The necessary and sufficient condition that a function $f(x,y)$ should be an analytic function of the complex variable $z = x + iy$ is expressed by the justly famous 'Cauchy–Riemann' equations

$$\partial u/\partial x = \partial v/\partial y \quad \text{and} \quad \partial v/\partial x + \partial u/\partial y = 0,$$

where $u(x,y)$ and $v(x,y)$ are the real and imaginary parts of $f(z)$.

The basis of this theory is clearly the existence of a mapping of the points (x,y) of the complex plane into a field of elements $z = x + iy$ which form an algebra with addition, multiplication, subtraction and division. There is no such mapping of the points (x,y,z) of three-dimensional space, although there is a mapping of four-dimensional space (x,y,z,w) into the Hamiltonian algebra of quaternions (Section 6.2).

11. DERIVATIVES AND INTEGRALS

11.1. Introduction

A survey of the development of 'analysis' from Cauchy to Laurent Schwartz suggests that its formal object may be conveniently described as the relation between the local and global properties of mappings, i.e. of functional relations. Its material object has been throughout the study of the processes of differentiation and of integration. The methods of investigation, which began with the use of infinitesimals, were given a more rigorous and powerful technique by the theory of convergence, and were transformed by the application of point-set topology.

11.2. The Legacy of Cauchy

We owe to Cauchy the liberation of analysis from its domination by the concept of infinitesimals by the dawning light of the theory of limits.

The concept of a function $f(x)$ of a real variable was still vaguely described in terms of the 'dependence' of the dependent variable $y = f(x)$ on the independent variable x, but a continuous function was correctly defined by the condition that, at each point ξ in its domain of definition,

$$\lim_{\eta \to 0} f(\xi + \eta) = f(\xi).$$

The limiting process was thought of as the result of allowing the variable to 'tend to ξ' through positive or negative values, but was explicitly defined by the condition that the differences,

$$|f(\xi + \eta) - f(\xi)|$$

between the values of $f(x)$ at $x = \xi$ and at points $x = \xi + \eta$ in some neighbourhood of ξ, could be made 'smaller than any assigned positive number ε', by taking a sufficiently small neighbourhood

$$-\delta < \eta < \delta.$$

It was erroneously thought that the continuity of the function $f(x)$ guaranteed the existence of its derivative $f'(\xi)$ defined by the limiting process

$$f'(\xi) = \lim_{\eta \to 0} \frac{f(\xi + \eta) - f(\xi)}{\eta}$$

as η tended to zero through positive or negative values, or, at least, that the derivative only failed to exist at a few points, such as the origin $x = 0$ for the function $f(x) = |x|$.

141

To define the integral of a function $f(x)$ over an interval $a \leqslant x \leqslant b$, Cauchy divided the interval by a finite number of points $x_0, x_1, \ldots, x_n$, such that

$$a = x_0 < x_1 < x_2 < \ldots < x_n = b,$$

and $$x_{k+1} - x_k < \varepsilon \quad \text{for all } k,$$

and then calculated the sum

$$I_n = \sum_{k=0}^{n-1} (x_{k+1} - x_k) f(x_k).$$

If this sum 'tended to a unique limit I' as ε tended to zero, and the number of points $\{x_n\}$ increased indefinitely, then I was said to be the value of the integral

$$\int_a^b f(x) \, dx.$$

The meaning of this definition is perfectly clear, but generations of schoolboys have been puzzled by the fact that they are required to consider, not the covergence of a function of one or of a fixed number of variables, but the convergence of the numbers I_n which depend on an ever-increasing number of variables as ε tends to zero. This characteristic of Cauchy's definition disappeared in the theory given later by Darboux, but received a satisfactory treatment only in terms of H. Cartan's theory of 'filters' (Section 9.12).

11.3. Non-differentiable Functions

It was commonly believed for many decades after Cauchy that any continuous function is differentiable everywhere, except for a few isolated points of negligible importance. The first and the simplest counter-example was invented by Bolzano at some time not later than 1830 (Bolzano 1834, 1950).

Bolzano's function (which remained hidden in his manuscripts until 1921) has no finite derivative at *any* point of its interval and no definitely positive (or negative) infinite derivative at *any* internal point.

But it was Weierstrass who first gave to the mathematical world a conclusive counter-example, which circulated orally from 1861 and was originally published in 1872 (Weierstrass 1876). Weierstrass' function is

$$f(x) = \sum_{n=0}^{\infty} a^n \cos (b^n \pi x),$$

where $0 < a < 1$ and b is an odd integer such that $ab > 1 + \tfrac{3}{2}\pi$. This function is continuous for all values of x but does not possess a finite derivative anywhere.

The construction of non-differentiable continuous functions now has a considerable literature. The outstanding contributions are due to Knopp (1918) and to van der Waerden (1930).

Closely related to the non-differentiable but continuous functions are the still 'wilder' functions which possess singularities at an infinitely numerous set of points in any interval. These pathological examples illustrate the way in which the limit of a sequence of continuous and differentiable functions may behave.

To construct these functions Cantor (1882) and Hankel (1882) invented the 'Principle of the condensation of singularities'. In Cantor's method, a function $\varphi(x)$ which is continuous at all points in the interval $[-1,1]$ except at the origin, where it vanishes, is used to generate the infinite series

$$f(x) = \sum_{n=0}^{\infty} c_n \varphi(x - w_n)$$

where the coefficients c_n are positive numbers, chosen so that the series

$$\sum_{n=0}^{\infty} c_n$$

is convergent, and where the points w_n form any enumerable aggregate. The function $f(x)$ then exhibits at each such point w_n the same type of discontinuity as $\varphi(x)$ does at the origin.

In Hankel's method the function $\varphi(x)$ generates the infinite series

$$f(x) = \sum_{n=0}^{\infty} \frac{\varphi(\sin n\pi x)}{n^s} \qquad (s > 1)$$

and the discontinuities of $f(x)$ are found only for rational values of x.

11.4. *The Riemann Integral and Fourier Series*

The theory of the definite integral given by Cauchy suffered from the defect that it did not give any indication of the conditions which were necessary or sufficient for its existence. Fortunately the mathematicians of the early nineteenth century were mainly concerned with 'tame' functions which were amenable to Cauchy's treatment.

However, when Riemann[1] began to give a rigorous theory of Fourier series the first problem which he encountered was how to discover

---

[1] The famous paper by Riemann, 'Über die Darstellbarkeit einer Funktion durch eine trigonometrische Reite' (Riemann 1876), was read as his Habilitation to a small group in Göttingen in 1854, but was not published, until after his death, by Dedekind.

sufficient conditions for the existence of the integrals which gave the Fourier coefficients of a function $f(x)$, viz.

$$a_n = \int_0^{2\pi} f(x) \cos nx\, dx, \quad b_n = \int_0^{2\pi} f(x) \sin nx \quad dx, \; n = 0, 1, 2, \ldots.$$

In the earlier researches on Fourier series by Dirichlet (1829) the function $f(x)$ was restricted to be 'piece-wise monotonic' i.e. to possess only a finite number of turning values, either maxima or minima, in the interval $[0, 2\pi]$ and to be monotomic between any consecutive pair of turning values. Such a function undoubtedly is integrable according to Cauchy's definition, but these conditions are too restrictive for applications of Fourier's series in such branches of analysis as the theory of numbers.

Riemann began by making a change in the Cauchy definition, which appears to be slight but is profoundly significant. As before, he divided the domain of integration $[a,b]$ by a finite number of points,

$$a = x_0 < x_1 < x_2 < \ldots < x_n = b, \quad x_{k+1} - x_k < \varepsilon,$$

but he calculated the sum

$$I_n = \sum_{k=0}^{n-1} (x_{k+1} - x_k) f(s_k),$$

where s_k is *any* point in the interval $[x_k, x_{k+1}]$ and not necessarily the end-point x_k as in Cauchy's definition.

Next Riemann defined (see section 11.2) the 'oscillation' D_k of the function $f(x)$ in the interval $[x_k, x_{k+1}]$ as the difference between the greatest and least values of $f(x)$ in this interval. The sums I_n clearly depend on the choice of the points s_k, but the difference between the greatest and least values of I_n cannot exceed

$$\sum_{k=0}^{n-1} (x_{k+1} - x_k) D_k$$

when the points of subdivision, x_k, are fixed. He then deduced that if J_σ is the total length of the intervals in which $D_k > \sigma$ then J_σ must tend to zero as ε tends to zero, in order that the sums I_n should converge to a definite limit and thus define the integral of $f(x)$ over the interval $[a,b]$. To illustrate this general theory he constructed the function

$$f(x) = \sum_{n=1}^{\infty} \frac{(nx)}{n^2}$$

where (x) denotes the excess of x over the nearest integer, or zero if x is half an odd integer, but this is almost the only function to which Riemann's criterion of integrability is easily applicable.

We note that the limit of D_k as the interval $[x_k, x_{k+1}]$ shrinks to a single point s is called by Baire (1899) the 'oscillation' $w(s)$ of the function $f(x)$ at the point $x = s$.

11.5. The Darboux Integral

The deficiencies of Riemann's definition of an integral needed to be repaired by the discovery of a more convenient test for the integrability of a function and by a deeper analysis of what is involved in the 'convergence' of the Riemann sums I_n. The first desideratum was provided by Jordan, the second by Darboux.

Darboux (1875) modified the Riemann sum by replacing $f(s_k)$ by either U_k, the greatest value of $f(x)$ in the interval $[x_k, x_{k+1}]$ or by L_k, the least value of $f(x)$ in this interval. He thus obtained an upper limit $\bar{I}_n$ and a lower limit $\underline{I}_n$ to the Riemann sum I_n,

where
$$\bar{I}_n = \sum_{k=0}^{n-1} (x_{k+1} - x_k) U_k,$$

and
$$\underline{I}_n = \sum_{k=0}^{n-1} (x_{k+1} - x_k) L_k.$$

Then, if $f(x)$ is bounded, and if we consider all possible modes of subdividing the interval $[a, b]$, the Darboux sums $\bar{I}_n$ have a definite lower bound $\bar{I}$, while the Darboux sums $\underline{I}_n$ have a definite upper bound $\underline{I}$. These numbers $\bar{I}$ and $\underline{I}$ are the upper and lower Darboux integrals of $f(x)$ over the interval $[a, b]$. The necessary and sufficient condition for the existence of a Riemann integral is that $\underline{I} = \bar{I}$.

This provides a convenient and compact definition of the Riemann integral, from which its properties are easily deduced.

11.6. Riemann and 'Integrable Groups of Points'

In order to make the maximum use of the Riemann integral it is desirable to know the necessary and sufficient conditions for a function $f(x)$ to possess a Riemann integral over an interval $[a, b]$. The condition given by Riemann can be expressed in terms of an 'integrable group of points'—a concept due to P. du Bois-Reymond (1882). Such a set of points can be covered by a *finite* number of intervals which can be chosen so that their total length is less then any assigned positive number. Riemann's condition for the integrability of $f(x)$ is that the set of points at which the oscillation of $f(x)$ is greater than an arbitrary positive number ε should form an integrable set.

11.7. Hankel and the 'Content' of a Set of Points

In his researches on functions which were discontinuous or which possessed an infinite number of maxima and minima in any interval,

Hankel (Section 11.3) had employed the same methods to isolate the 'wild' points of a function and had introduced the concept of the 'content' (*Inhalt*) of a linear set of points E as the lower bound of the sum

$$\sum_{k=1}^{n} |I_k|,$$

where $I_1, I_2 \ldots, I_n$ is a *finite* set of intervals which cover the set of points E. The limitations of this idea are obvious, for if E consists of the rational points in the interval [0,1] the Hankel-content of E is clearly unity, since rational points occur in any finite sub-interval.

11.8. The 'Content' of Harnack and Cantor

This limitation was removed by Harnack (1881, 1885) by defining the 'content' of the set E to be the lower bound of the sum

$$\sum_{k=1}^{\infty} |I_k|,$$

where $I_1, I_2, \ldots$ is an *enumerably infinite* set of intervals covering the set E.

Meanwhile Cantor (1884a, b) invented a similar concept in a form which could be applied to a set of points E in n-dimensional Euclidean space. Let $S(\varrho)$ denote the n-dimensional sphere with radius ϱ and centre at a point p of the set E, and let $V(\varrho)$ be the n-dimensional volume of the union of all the spheres corresponding to all the points of E. Then the Cantor-content of the set E is the limit of $V(\varrho)$ as ϱ tends to zero. This definition has the disadvantage that the rational points in [0, 1] would have Cantor-content equal to unity, when we take the 'sphere' $S(\varrho)$ to be an interval of length 2ϱ centred on the point p.

11.9. The 'Outer and Inner' Content of Peano and Jordan

In their definition of the 'content' of a set of points E lying in a large square R in a Euclidean plane, Peano (1887) and Jordan (1892) introduced the new ideas of 'outer' and 'inner' content. The square R is divided into a finite number of small squares of side ϱ. The 'inner' squares S_1 with total area $S_1(\varrho)$ each have all their points interior to E; the 'outer' squares S_2 of total area $S_2(\varrho)$ each have no points interior to E, the remaining squares S_3 of total area $S_3(\varrho)$ each contain some points interior to E and some points not belonging to E, i.e. they contain the 'frontier' of E.

Peano and Jordan define the limit of $S_1(\varrho)$ as $\varrho \to 0$ to be the 'inner measure' of E, and the limit of $S_1(\varrho) + S_3(\varrho)$ as $\varrho \to 0$ to be the 'outer measure' of E.

The set E was said to be 'measurable' if its outer and inner measures were equal. Unfortunately this definition implied that the rational points in R had a measure equal to the area of R, and that the same was true for the irrational points in R!

However, the concepts of inner and outer measure proved to be of permanent value when generalized by Borel and Lebesgue. In Borel's theory the main instrument of measurement is not a *finite* collection of squares but an enumerably *infinite* collection of rectangles (or intervals). In Lebesgue's theory the outer measure $m^*(E)$ is the lower bound of the total area of any disjoint *enumerable* collection of rectangles covering E, but the inner measure is defined as

$$m_*(E) = m^*(R) - m^*(F)$$

where R is any rectangle covering E, and F is the complement of E with respect to R. This shifts the interest from the Jordan inner measure defined as the limit of $S_1(\varrho)$ as $\varrho \to 0$ to the limit of $|R| - S_2(\varrho)$ as $\varrho \to 0$.

11.10. Borel and 'Well-defined' Sets

Hitherto all writers on measure theory had envisaged an abstract set of points E without giving any serious consideration to the precise manner in which such a set was defined. It was left to Borel (1898; 1914, pp. 217–56) to make a theory of measure simple, rigorous and widely applicable by restricting it to sets E, which he described as 'well-defined' and which are 'constructable' according to a definite set of rules. Just as Kronecker had restricted analysis to finite operations on the natural numbers, so Borel restricted measure theory to enumerable operations on intervals.

The sets of points which alone are considered by Borel to be 'well-defined' are the sets which can be obtained from the intervals $[a,b]$ on the real axis by two operations repeated indefinitely:

(1) By forming the union of a finite or enumerably infinite number of disjoint intervals.
(2) By forming the complement of any 'well-defined' set E with respect to any 'well-defined' set F which covers E.

The Borel measure $B(I)$ of an interval $I = [a,b]$ is its length $b - a$, or its volume if the 'interval' is the set of points x

$$a_k < x_k < b_k, \quad k = 1, 2, \ldots, N$$

in N-dimensional Euclidean space. The Borel measure of the union of the enumerable disjoint intervals I_n, $(n = 1, 2, \ldots)$ is the sum of the series

$$\sum_{n=1}^{\infty} B(I_n).$$

The Borel measure of the complement of E with respect to F is

$$B(F) - B(E).$$

These rules provide a definition of the Borel measure $B(E)$ of any 'well-defined' or 'Borel' set E.

These principles are enunciated even more forcefully in the Note VI added to the second edition of Borel's *Leçons sur la Théorie des Fonctions* (1914), but even then Borel gave no formal proof that they were mutually consistent, but contented himself with a reference to the famous covering theorem which bears his name.

11.11. *The Heine–Borel–Lebesgue Covering Theorem*

This theorem can be compactly expressed in modern terminology by defining

(1) a collection of sets $\{A\}$ to be a 'covering' of a set S if each point of S belongs to at least one set of $\{A\}$, and

(2) a subcollection $\{B\}$ of $\{A\}$ to be a 'refinement' of $\{A\}$ if it is still a covering of S.

Borel's covering theorem (Borel 1895) is that any covering of a bounded, closed interval in Euclidean space by enumerable open sets has a refinement consisting of a finite set of open sets.

This theorem first appeared in Heine's proof (Heine 1872) that if $f(x)$ is continuous in a bounded, closed interval $[a,b]$ then it is also 'uniformly' continuous, i.e. to ensure that $|f(x) - f(x)| < \varepsilon$ where x and y are any points of $[a,b]$, it is only necessary to choose x and y so that $|x - y| < \delta$, where δ depends *only* on ε and not on x or y. Its next appearance is in a paper by Cousin (1895).

Subsequently H. Lebesgue (1904; Chap. VII, § II) and Lindelöf (1903) showed that Borel's theorem is still valid if the number of sets in the collection $\{A\}$ is non-enumerably infinite.

11.12. *The Borel Integral*

The restrictions which Borel (1914) insisted were necessary in order that a function should be effectively calculable led him to limit his definition of an integral to functions 'asymptotically equivalent' to polynomials. If $f(x,y,z)$ is such a bounded function defined in a domain D, then to any pair of positive numbers, ε and α (however small), there corresponds a polynomial $P(x,y,z; \varepsilon,\alpha)$ such that

$$|f - P| < \varepsilon,$$

except in a set of points of 'Borel measure' less than α. The Borel integral of f in the domain $a \leqslant x \leqslant A$, $b \leqslant y \leqslant B$, $c \leqslant z \leqslant C$ is then asymptotically equivalent to the polynomials

$$\int_a^x \mathrm{d}\xi \int_b^y \mathrm{d}\eta \int_c^z \mathrm{d}\zeta P(\xi,\eta,\zeta; \varepsilon,\alpha).$$

To define the integral of f over a Borel measurable domain Δ contained in such a rectangular domain D, the polynomials P are chosen to be asymptotically equivalent to f in Δ, and to zero in the complement of Δ with respect to D.

Unbounded functions are integrated by a method of excluding the points at which $f(x)$ becomes infinite by a set of intervals whose total volume is less than any assigned positive number ε.

11.13. Lebesgue's Theory of Measure

Inspired by the work of Borel, H. Lebesgue proceeded to give a theory of measure which should be applicable to any abstract set of points and independent of its 'constructability'—a concept which was repudiated by Borel as devoid of meaning.

In his famous thesis, published in 1902, H. Lebesgue made a most significant advance on the theory of outer and inner measure given by Peano and Jordan by considering the covering of a linear (or plane) set of points by an *enumerably infinite* set of intervals (or triangles!), instead of the finite set employed by the previous authors (H. Lebesgue 1902, 1904, 1928). The Lebesgue outer measure of E was then defined as the lower bound $m^*(E)$ of the sum

$$\sum_{n=1}^{\infty} B(I_n)$$

for all possible enumerable coverings of E.

To obtain the Lebesgue inner measure $m_*(E)$ of a set E lying in a bounded interval I, he considered the complementary set $I - E$ and gave the definition

$$m_*(E) = m^*(I) - m^*(I - E).$$

If the outer and inner measures of E are equal, their common value is the (Lebesgue) measure, $m(E)$ of E. If E is 'well-defined' then its Lebesgue measure $m(E)$ and its Borel measure $B(E)$ are equal.

Despite the greater generality of Lebesgue's definition, he was able to prove that if a set E has a Lebesgue measure $m(E)$, then there exist two sets F and G, each constructable by the methods of Borel, and such that F is covered by E, which is itself covered by G, and such that

$$B(F) = m(F) = m(E) = m(G) = B(G).$$

H. Lebesgue also extended his definition of measure to plane sets of points by covering them with enumerable sets of triangles.

The great advantage of H. Lebesgue's definition of measure is that it is 'completely additive' in the sense that if $\{E_n\}$ is a bounded enumerable collection of disjoint measurable sets, then their union E is also measurable and

$$m(E) = \sum_{n=1}^{\infty} m(E_n).$$

11.14. *De la Vallée-Poussin's Exposition of Lebesgue's Theory*

H. Lebesgue never gave a formal and rigorous account of his theory of measure, but this was provided by de la Vallée-Poussin (1915, 1916), who proved, for example, that the inner measure of a set defined in terms of an interval I covering E, is really independent of I!

The main instruments in the exposition by de la Vallée-Poussin are the systematic use of open and closed sets and the pair of 'union-intersection theorems', viz.

$$m^*(\sigma) + m^*(\tau) \geqslant m^*(\sigma \cup \tau) + m^*(\sigma \cap \tau)$$

and
$$m_*(\sigma) + m_*(\tau) \leqslant m_*(\sigma \cup \tau) + m_*(\sigma \cap \tau).$$

de la Vallée-Poussin gave the definitive form for H. Lebesgue's definitions of outer and inner measure on the real line by writing

$$m^*(E) = \inf\, m(A)$$

$$m_*(E) = \sup\, m(B)$$

for all open sets A covering E and for all closed sets B covered by E.

However, the extension of these definitions to higher dimensional space was not entirely satisfactory, because it is then no longer true that an open set is an enumerable collection of disjoint intervals.

11.15. *The Lebesgue Integral as a Lebesgue Measure*

To obtain the (two-dimensional) measure of a set of points E in a Euclidean plane, H. Lebesgue used the same technique as for a linear set, covering E by a collection of triangles. This led him to give a new definition of the integral of a bounded non-negative function $f(x)$ over an interval $[a,b]$ as the two-dimensional measure of the set of points (x,y) such that

$$a \leqslant x \leqslant b \quad \text{and} \quad 0 \leqslant y \leqslant f(x),$$

(provided this set is indeed measurable!). To integrate a bounded function $f(x)$ which can take both positive and negative values, H. Lebesgue expressed $f(x)$ as the sum of its positive and negative parts,

$$f^+(x) = \tfrac{1}{2}f(x) + \tfrac{1}{2}|f(x)|$$

and
$$f^-(x) = -\tfrac{1}{2}f(x) + \tfrac{1}{2}|f(x)|,$$

and defined the integral as

$$\int_a^b f(x)\, \mathrm{d}x = \int_a^b f^+(x)\, \mathrm{d}x - \int_a^b f^-(x)\, \mathrm{d}x.$$

Unbounded functions and integrals over an infinite interval were dealt with by the familiar limiting processes introduced by Cauchy. However, the splitting of the integrand $f(x)$ into its positive and negative parts implied that $f(x)$ has a Lebesgue integral only if $|f(x)|$ has a Lebesgue integral.

11.16. The Young–Lebesgue Integral

In his famous thesis H. Lebesgue (1902) defined the integral of a non-negative function $f(x)$ over an interval $[a,b]$ as the two-dimensional measure of the set of points (x,y) for which

$$a \leqslant x \leqslant b \quad \text{and} \quad 0 \leqslant y \leqslant f(x),$$

but in the second edition of his *Leçons sur l'Intégration* (Lebesgue 1928) he gave another definition in which the 'Lebesgue' integral was expressed as a Riemann integral. Meanwhile W. H. Young had also grappled with the problem of integration and in a series of papers had independently arrived at the same definition (W. H. Young 1905).

Both definitions apply to a bounded, measurable function $f(x)$, i.e. a function such that the set of points for which $f(x) > t$ has a Lebesgue measure for each value of t. If $A < f(x) \leqslant B$ the interval $[A,B]$ is divided at a finite number of points $t_0, t_1, \ldots, t_n$ such that

$$A = t_0 < t_1 < t_2 < \ldots < t_n = B.$$

Let $m(t)$ be the measure of the set of points x at which $f(x) > t$.

H. Lebesgue considers the two sums

$$\sigma = \sum_{p=0}^{n-1} t_p \{ m(t_p) - m(t_{p+1}) \}$$

and

$$\Sigma = \sum_{p=0}^{n-1} t_{p+1} \{ m(t_p) - m(t_{p+1}) \}$$

and shows that the upper bound of σ and the lower bound of Σ coincide when we consider all possible subdivisions of the interval $[A,B]$. This common bound furnishes the analytical definition of the Lebesgue integral

$$\int_a^b f(x) \, dx.$$

The integral defined by Young is

$$A(b - a) + \int_A^B m(t) \, dt$$

which is exactly equivalent to H. Lebesgue's analytical definition.

It is odd that H. Lebesgue never alludes to the fact that the common limit of his two sums, σ and Σ, is expressible at once as the Riemann integral of the monotone, non-increasing measure function $m(t)$.

H. Lebesgue (1928) also approached the problem of integration from the standpoint of a descriptive definition, in which the integral of a function $f(x)$ is characterized as a linear functional $L(f)$ which is

(1) positive, i.e. such that $f \geqslant 0$ implies that $L(f) \geqslant 0$;
(2) absolute, i.e. if $L(f)$ exists, so also does $L(|f|)$;

(3) monotonely convergent, i.e. such that if $L(f_n)$ exists for a bounded, monotone sequence of functions $\{f_n\}$, so also does $L(f)$, where $f = \lim f_n$, and, moreover

$$L(f) = \lim L(f_n);$$

(4) normalized, i.e. if $f \equiv 1$, its integral over (a,b) is $b - a$.

The proof, sketched by H. Lebesgue, shows that the necessary and sufficient condition for the existence of $L(f)$ is that f should be measurable, and that in this case $L(f)$ is the Lebesgue–Young integral.

The Lebesgue integral of bounded functions is therefore the most general integral which can be constructed, subject to the conditions listed above as 'characteristic' of integration. In extending the definition to unbounded functions f, H. Lebesgue was constrained to restrict himself to functions whose integrals were 'absolutely' integrable, i.e. such that the integrability of $f(x)$ implied the integrability of $|f(x)|$.

It is then sufficient to consider separately the integration of the 'positive' and 'negative' parts of an unbounded function $f(x)$ and to consider the 'truncated' functions

$$f_n^+(x) = \begin{matrix} f_n^+ & \text{if} & f^+(x) \leqslant n, \\ 0 & \text{if} & f^+(x) > n, \end{matrix}$$
$$f_n^-(x) = \begin{matrix} f_n^- & \text{if} & f^-(x) \leqslant n, \\ 0 & \text{if} & f^-(x) > n. \end{matrix}$$

Then the integrals are defined as

$$\int f^+(x) \, dx = \lim_{n \to \infty} \int f_n^+(x) \, dx$$

and

$$\int f^-(x) \, dx = \lim_{n \to \infty} \int f^-(x) \, dx.$$

Both limits necessarily exist, but the definition is significant only if the limits are finite. The function $f(x)$ is than said to be 'summable'.

The great advantages of the Young–Lebesgue integral are that

(1) it can be easily generalized for integration over a measurable set E, by taking the integrand $f(x)$ to be zero on the complement of E;
(2) if $\{f_n(x)\}$ is any sequence of measurable functions defined in a measurable set E, which has 'dominated' convergence to a limit function $f(x)$, i.e. if

$$|f_n(x)| \leqslant \phi(x) \text{ for each } n$$

where $\phi(x)$ is summable in E, and if $f_n(x)$ converges to $f(x)$ as $n \to \infty$ at each point of E, then $f(x)$ is summable in E and

$$\int_E f_n(x) \, dx \to \int_E f(x) \, dx \quad \text{as} \quad n \to \infty;$$

(3) if $f(x)$ is summable over the interval $[a,b]$ then the Lebesgue integral

$$\phi(x) = \int_a^x f(t)\ dt$$

possesses a derivative $\phi'(x)$ equal to $f(x)$ at all points of $[a,b]$ except for those of a certain set of measure zero.

11.17. Dini Derivatives

Shortly after Weierstrass had startled the mathematical world by constructing a non-differentiable continuous function (Weierstrass 1876), Dini (1878, p. 190) showed that any real function $f(x)$ of a real variable x possesses at each point x four generalized derivatives. The incrementary ratio

$$I(x,k) = [f(x+k) - f(x)]/k$$

may not converge to a unique limit as $k \to 0$, but there can be no doubt about the existence of the greatest lower bound and the least upper bound of $I(x,k)$ as k tends to zero through positive values or negative values. Hence we obtain the derivatives, denoted in Dini's lambda-notation or in Scheefer's D-notation (Scheefer 1884), as

$$\lambda'_x = D_-(x) = \varliminf_{k<0} I(x,k),$$

$$\Lambda'_x = D^-(x) = \varlimsup_{k<0} I(x,k),$$

$$\lambda_x = D_+(x) = \varliminf_{k>0} I(x,k),$$

$$\Lambda_x = D^+(x) = \varlimsup_{k>0} I(x,k).$$

11.18. Stieltjes and the Moment Problem

In his researches on continued fractions Stieltjes (1894, 1895) was led to consider two distributions of mass along the real axis which possessed the same moments of order k at the origin. In one distribution masses μ_i were situated at the points $x = \lambda_i$, $(i = 1, 2, \ldots)$ and in the other distribution masses ν_i were situated at the points $x = \theta_i$. Their moments of order k were

$$c_k = \sum_{i=0}^{\infty} \mu_i \lambda_i^k = \sum_{i=0}^{\infty} \nu_i \theta_i^k .$$

The question arises: are these two mass distributions identical, i.e. is a mass distribution uniquely specified by its moment?

This question suggests the study of the moments of any distribution of mass whatever along the real axis. In general Stieltjes noted that such a distribution is 'perfectly determined' by the total mass $\phi(x)$ distributed over a segment Ox, but he did not specify if this segment was to be an open or a closed interval.

To calculate the moments of such a distribution over the interval (a,b) Stieltjes formed the sum

$$\sum_{p=1}^{n} f(\xi_p)[\varphi(x_p) - \varphi(x_{p-1})]$$

where
$$x_{p-1} \leq \xi_p \leq x_p$$
$$a = x_0 < x_1 < x_2 < \ldots < x_n = b,$$

and
$$f(x) = x^k.$$

This expression is obviously a generalization of the sums employed in defining the Riemann integral (Section 11.4, where $\varphi(x) = x$), and by proceeding to the limit in the same manner as Riemann, Stieltjes obtained the value of the kth moment of the mass distribution about the origin, which he expressed in the form

$$c_k = \int_a^b f(u) \, \mathrm{d}\varphi(u).$$

In fact Stieltjes formulated this definition for any bounded monotone non-decreasing function $\varphi(u)$, and any continuous function $f(u)$ and thus greatly extended the theory of Riemann integration.

11.19. *Dieudonné and the Regulated Integral of Functionals*

The fundamental principles and methods of classical analysis discuss the mappings of the space of one real variable into the space of another real variable, but they are capable of immediate extension to the mappings of the space of functions of a real variable into a Banach space, and indeed Dieudonné (1960) has shown that this is the most economical way of expressing the foundations of modern analysis.

11.20. *The Integrals of Denjoy and Perron*

The demonstration of the classical property of the reciprocity of integration and differentiation was achieved by Lebesgue for primitives which possess a finite summable derivative, but the more general theorem for non-summable derivatives remained unsolved until Denjoy (1912) invented the process of 'totalization'. This great achievement gave the most general definition of integration, and moreover a definition which can be called 'constructive', although it involves the use of transfinite induction.

A little later Perron (1914), who later introduced major and minor integrals of ordinary differential equations (Perron 1915), introduced the same concept into the theory of integration.

de la Vallée-Poussin (1916) later defined major and minor functions, φ and ψ, of an integral

$$I = \int_a^x f(t) \, dt \qquad (a \leqslant x \leqslant b)$$

by the conditions[1] $\bar{D}\psi \leqslant f \leqslant \underline{D}\varphi$ and established the existence of such functions for any absolutely continuous function f and any 'tolerance' ε, with the conditions that

$$\varphi - I < \varepsilon, \; I - \psi < \varepsilon.$$

Perron proposed to describe the integral of f by considering the totality of all major and minor functions φ and ψ, and by defining f to be integrable if the infimum of $\varphi(b) - \varphi(a)$ is equal to the supremum of $\psi(b) - \psi(a)$ (for all such functions). It appears that Perron did not propose to generalize Lebesgue's integral, and was happy to propose a definition of integration, which led to a very simple theory.

However, Bauer (1915) extended Perron's definition to multiple integrals and proved that Perron integration is equivalent to Lebesgue integration. Also Alexandroff (1924) and Looman (1925) proved the astonishing result that Perron integration is equivalent to Denjoy totalization.

This result was not welcomed by Denjoy, and his criticism of the Perron integral is a vigorous piece of denunciation (Denjoy 1941). Perron's integral by no means supercedes the process of totalization, which is essential to establish that Perron integration is valid for a wide class of non-summable derivatives.

11.21. The Radon Integral

The Stieltjes integral assumed a much greater significance when Riesz (1909b) discovered that any continuous linear functional on the space of continuous functions could be expressed in the form

$$F(x) = \int f(x) \, d\varphi(x).$$

This led H. Lebesgue (1910) to show that the Stieltjes integral of a continuous function f, i.e.

$$\int_a^b f(x) \, d\varphi(x)$$

could be expressed as a Lebesgue integral of the form,

$$\int_\alpha^\beta f\,[x(t)] \, \frac{d\varphi(t)}{dt} \, dt$$

where $t(x)$ is the total variation of the function φ on $[a,x]$ and $x(t)$ is the inverse function of $t(x)$.

---

[1] $\bar{D}$ is the maximum of D^+ and D^-, $\underline{D}$ is the minimum of D_+ and D_- (Section 11.17).

A much more direct approach was opened up by the introduction of 'set functions', introduced by H. Lebesgue in the special case of integrals in n-dimensional Euclidean space.

Radon (1913) considered a 'σ-ring' T of sets of points in n-dimensional Euclidean space, i.e. a class of subsets which is closed under the operations of countable union and countable intersection, and a non-negative additive set function which takes the place of the Lebesgue measure.

The Lebesgue measurable functions are replaced by functions f which are Φ-measurable, i.e. the sets $E\{x: f(x) > c\}$ are Φ-measurable, and the Radon or Lebesgue–Stieltjes integral $\int f(x)\, d\Phi(x)$ is the common limit of the upper and lower sums

$$\Sigma = \sum y_k \Phi\{(y_{k+1} \leqslant f(x) < y_k)\},$$

$$\sigma = \sum y_{k-1} \Phi\{(y_{k-1} \leqslant f(x) < y_k)\}.$$

A further step in the process of generalization was carried out by Fréchet (1915) by replacing the n-dimensional space of Radon by an abstract topological space; and a final step was taken by Nikodym (1930), who replaced the σ-ring of sets by 'un corps d'ensembles', H, such that if $E_1, E_2, \ldots$ is any enumerable sequence of elements of H, then the union of $E_1, E_2, \ldots$ also belongs to H, and if E belongs to H, so also does its complement.

12. Distributions

12.1. The Problem of Vibrating Strings

The theory of 'distributions' (i.e. generalized functions) arose from the desire of mathematical physicists to employ the derivatives of non-differentiable functions.

The fundamental equations of mathematical physics, whether in continuum mechanics or in electromagnetic theory, were originally expressed as partial differential equations, and they were therefore applicable only to phenomena in which the physical variables (such as elastic displacement or electric intensity) were not only continuous functions of space and time, but also possessed sufficient partial derivatives to make the differential equations intelligible. From the very beginning it gradually appeared that there was a marked difference between static phenomena, usually described by elliptic equations, and dynamic phenomena, usually described by hyperbolic equations. The physical variables in static problems were not only continuous and differentiable, but often infinitely differentiable. The physical variables in dynamic problems could exhibit discontinuities which might be an essential feature.

Thus potential functions are infinitely differentiable (in empty space), but wave functions can exhibit a discontinuity at a front advancing into an undisturbed region. For example, a violin string plucked at its mid-point vibrates initially according to a law of the form

$$g(x,t) = a - \tfrac{1}{2}(a/l)[\,|x - ct| + |x + ct|\,] \qquad (0 \leqslant ct \leqslant l),$$

and the derivative of the displacement $y(x,t)$ is discontinuous at $x = \pm ct$. The problem is therefore insoluble if we adhere rigidly (as did D'Alembert) to differentiable functions, and can only be solved if we are prepared to waive this restriction (as apparently was Euler (1748, 1749).

The methods which have been devised to solve this problem may be broadly characterized as follows:

(1) The fundamental differential equations can be replaced by integral equations in which the physical variables are subject to less onerous restrictions than was the case for differential equations. This is the method extensively developed.

(2) The concepts of a function and of differentiation can be generalized so as to allow for generalized differentiation even of discontinuous generalized functions. This is the method implicitly employed by Kirchoff and Heaviside, and in the axiomatic exposition of Sebastião e Silva.

(3) These two methods can be combined, as in the work of Dirac and Sobolev, and in the definitive form expounded by Laurent Schwartz.

12.2. *Kirchhoff and the Delta Function*

The first appearance of a generalized function which I have been able to trace is in the researches of Kirchhoff (1882, 1883) on the equation of wave propagation

$$\frac{\partial^2 F}{\partial x^2} + \frac{\partial^2 F}{\partial y^2} + \frac{\partial^2 F}{\partial z^2} = \frac{\partial^2 F}{\partial t^2}.$$

In static potential problems, where F is independent of the time t, the fundamental solution of this equation is

$$F = (4\pi R)^{-1} \qquad (R > 0),$$

where
$$R = [(x - a)^2 + (y - b)^2 + (z - c)^2]^{\frac{1}{2}}.$$

In the dynamic problem, where F depends on t, Liouville (1856) had employed wave-functions of the form

$$F = f(R + t)/4\pi R,$$

but it was left to Kirchoff to take the bold step of introducing the fundamental solution

$$F = \delta(R + t)/4\pi R,$$

where $\delta(x)$ is a function which vanishes if x is positive or negative, but behaves at the origin $x = 0$, in such a way that

$$\int_a^b \delta(x)\, dx = 1 \quad \text{for} \quad a < 0 < b.$$

In order to persuade his readers that such a function as $\delta(x)$ really existed Kirchoff referred to the function

$$\delta_n(x) = \pi^{-\frac{1}{2}} n \exp(-n^2 x^2),$$

which he described as 'vanishingly small for large n and for any positive or negative value of x' while

$$\int_{-\infty}^{\infty} \delta_n(x)\, dx = 1,$$

Thus, in some sense, as yet undefined, $\delta(x)$ was regarded as a generalized limit of the sequence $\{\delta_n(x)\}$, $n = 1, 2, \ldots$.

There are many other sequences with similar properties to Kirchoff's sequence. Thus Jeans (1925) uses the sequence

$$\delta_n(x) = (n/\pi)(1 + n^2 x^2)^{-1},$$

and L. Schwartz (1950) the sequence

$$\delta_n(x) = \alpha x^{\alpha-1} \qquad (x \geqslant 0)$$
$$\text{or} \quad 0 \qquad (x < 0),$$

where $\alpha = 1/n$.

12.3. Heaviside and Generalized Differentiation

In studying the response of a mechanical system to 'a continued constant force of unit strength, commencing when $t = 0$', Heaviside (1892, 1893a) noted that if this force is represented as a Fourier integral

$$f(t) = \frac{1}{2} + \frac{1}{\pi} \int_0^\infty \frac{\sin nt}{n} \, dn,$$

then the problem is reduced to the response of the system to a sinusoidal force. He continues 'we see that a unit impulse is represented by

$$\frac{1}{\pi} \int_0^\infty \cos nt \, dn,$$

acting at the moment $t = 0$,' this result being obtained by the differentiation of $f(t)$.

This suggests that what Heaviside had in mind was that the sequence

$$\delta_n(t) = \frac{1}{\pi} \int_0^n \cos st \, ds = \frac{\sin nt}{\pi t} \qquad (n = 1, 2, \ldots)$$

converged in some generalized way to Kirchhoff's delta function $\delta(x)$. Heaviside also gave another generalized limit for the delta function, which may be written as

$$\delta_n(x) = \lim_{n \to -1} \frac{x^n}{\Gamma(n+1)},$$

together with a rich variety of conjectural results for divergent, alternating series and for generalized differentiation of fractional order, the latter being developed in ignorance of the work of Riemann (1847).

But the outstanding feature of Heaviside's creative imagination was the calm assurance with which he differentiated the function

$$f(t) = \begin{cases} 1 & \text{if } t \geqslant 0, \\ 0 & \text{if } t < 0. \end{cases}$$

to give the impulse function $\delta(t)$ which is zero if $t > 0$ or $t < 0$ and infinite if $t = 0$.

In ignorance of the work of Liouville (1832) and Riemann (1847), Heaviside also invented a naïve theory of fractional integration, which was later given a definition by M. Riesz (1949).

The earlier writers had noted that the integral

$$I^\alpha f(x) = (1/\Gamma(\alpha)) \int_a^x f(t)(x - t)^{\alpha-1} \, dt$$

represented the integral of $f(x)$,

$$\int_a^x f(t)\, \mathrm{d}t$$

when $\alpha = 1$, and that for all positive values of α and β

$$I^\alpha I^\beta f(x) = I^\beta I^\alpha f(x),$$

$$\frac{\mathrm{d}}{\mathrm{d}x}\{I^{\alpha+1}f(x)\} = I^\alpha f(x).$$

Hence $I^\alpha f(x)$ may be regarded as a fractional integral of $f(x)$ of order α, and the definition is valid for all positive values of α. (Heaviside was particularly interested in the case when $\alpha = \frac{1}{2}$.)

The great advance made by M. Riesz was to take the parameter α to be a complex number with its real part positive. The function $I^\alpha f(x)$ then becomes a holomorphic function of α. Moreover, if $f(x)$ possesses derivatives of all orders $k \leqslant p$ then the function $I^\alpha f(x)$, defined for complex values of α with real part positive, possesses an analytic continuation for all complex values of α with real part greater than $-p$.

M. Riesz also generalized this integral for applications to potential theory, wave theory and electron theory and thus obtained existence theorems for a wide variety of linear partial differential equations with constant or variable coefficients.

12.4. Weak Convergence

The direct approach to a theory of generalized differentiation initiated by Heaviside was not developed until the researches of Sobolev (1936), but meanwhile the ground was prepared by researches into the convergence of Fourier series and into generalized solutions of partial differential equations.

In view of the difficulty of a direct discussion of the convergence of the Fourier series of a function $f(x)$, summable in the interval $[0, 2\pi]$, the following theorem due to Fatou (1906) (especially p. 356) marks a notable advance in the theory:

If $f_n(x)$ is the sum of the first n terms of the Fourier series of $f(x)$, then

$$\int_0^x f_n(t)\, \mathrm{d}t \to \int_0^x f(t)\, \mathrm{d}t \quad \text{as} \quad n \to \infty,$$

i.e. whether or not the Fourier series of $f(x)$ converges, it can always be integrated term by term to yield the indefinite integral of $f(x)$.

This is a special example of the 'weak convergence' of a sequence of functions $\{f_n(x)\}$—a concept introduced in its full generality by Fischer (1907) and F. Riesz (1910), and defined by the conditions that all the functions $f_n^2(x)$ are integrable over an interval $[a,b]$, all the integrals

$$\int_a^b f_n^2(x)\, \mathrm{d}x$$

are less than the same finite number and the sequence of integrals

$$\int_0^x f_n(t)\, dt$$

converges to a limit function of the form

$$\int_0^x f(t)\, dt$$

for each x in $[a,b]$.

There is a similar theory of the weak convergence of Fourier integral transforms (see Section 12.10).

However, in these early approaches the theory was subject to the limitation that the sequence $\{f_n(x)\}$, whose weak convergence was studied, did actually possess a 'weak' limiting function, $f(x)$, such that the integrated sequence converged pointwise to the integral of $f(x)$, i.e.

$$\int_0^x f_n(t)\, dt \to \int_0^x f(t)\, dt \quad \text{as} \quad n \to \infty.$$

12.5. Weak Solutions of Differential Equations

In the theory of linear partial differential equations, as exemplified by the potential equation, the heat equation or the wave equation, there is an identity, first employed by Green (1828) in electrostatic theory in the familiar form

$$v \Delta u - u \Delta v = \operatorname{div}(v \operatorname{grad} u - u \operatorname{grad} v),$$

where Δ is the Laplacian operator.

In general if $L(u)$ is any linear function of u and its first and second derivatives, with coefficients which are functions of the independent variables x, y, ... then there exists an 'adjoint' linear function $M(u)$ of the same type, such that for any functions u and v with continuous second derivatives

$$u M(v) - v L(u) = \frac{\partial R}{\partial x} + \frac{\partial S}{\partial y} + \ldots$$

where $R, S, \ldots$ are bilinear functions of $u, v, \ldots$ and their first derivatives.

In the classical theory v is taken to be the 'fundamental' solution of the adjoint equation $M(v) = 0$, i.e. what we now call the generalized solution of $M(v) = \delta(P)$, the right-hand side being the Dirac delta function with the point P as its support. However, Wiener (1926) suggested taking v to be an infinitely differentiable function which vanishes on and outside the boundary of the region D in which a solution is required for the equation $L(u) = 0$.

On integrating Green's identity over all space it is then found that the integral of the right-hand side vanishes and that

$$\int_D vL(u) \;=\; \int_D uM(v).$$

Wiener suggested that the function u should be regarded as a 'weak' solution of $L(u) = 0$, if the relation

$$\int_D uM(v) = 0$$

was true for all functions v with the properties listed above.

The simplest example of such a 'test' function v is that given by Bochner and Martin (1948) in the form

$$v = \begin{array}{ll} \exp\,(r^2 - 1)^{-1} & \text{if}\quad 0 \leqslant r < 1, \\ 0, & \text{if} \qquad r \geqslant 1 \end{array}$$

for the domain D, $0 \leqslant r \leqslant 1$, where r is the distance from the origin. However, Bochner (1952) has pointed out that Riemann used a simple form of test function in his researches on Fourier series, viz. the functions which he denoted by $\lambda(x)$ and $\varrho(x)$ (Riemann 1854; Section XII, Subsection VIII and IX).

In Wiener's definition the support of the test function v was the domain D of the problem under consideration, and the explicit construction of such functions would in general be as difficult as the original problem. Sobolev (1936) greatly simplified the definition by allowing the test functions to be any functions with continuous derivatives of order less than some integer s and with compact support, i.e. such that they vanished outside *some* bounded domain.

In this and in a subsequent paper the generalized derivative, of a function $f(x_1, x_2, \ldots, x_n)$, corresponding to the classical derivative

$$\mathscr{D}f = \partial^{a_1 + \cdots + a_n} f(x_1, x_2, \ldots, x_n)/\partial x_1^{a_1}\,\partial x_2^{a_2}\ldots\partial x_n^{a_n},$$

was defined to be a function $g(x_1, x_2, \ldots, x_n)$ such that

$$\int f\mathscr{D}\varphi + (-1)^{a+1}\varphi g = 0$$

for all test functions φ, where $a = a_1 + \ldots + a_n$.

A systematic theory of weak solutions of differential equations was given by Friedrichs (1939, 1944).

12.6. *Hadamard's Finite Part of an Improper Integral*

In his researches on linear hyperbolic differential equations in general (Hadamard 1932, 1952) and on the wave-equation (Hadamard 1903b) in particular, Hadamard was led to the invention of the 'finite part' of certain divergent integrals. Thus the integral

$$\int_\varepsilon^1 \frac{A(x)\,dx}{x^{p+\frac{1}{2}}}$$

where p is a positive integer, is separated into two parts

$$F(\varepsilon) = \int_{\varepsilon}^{1} \frac{A(x) - B(x)}{x^{p+\frac{1}{2}}}\, dx \quad \text{and} \quad I(\varepsilon) = \int_{\varepsilon}^{1} \frac{B(x)\, dx}{x^{p+\frac{1}{2}}}$$

where $\quad B(x) = A(0) + xA'(0) + \ldots + x^{p-1}A^{(p-1)}(0)/(p-1)!$.

The 'infinite part' $I(\varepsilon)$ is the sum of a number of terms in $\varepsilon^{p-\frac{1}{2}}$, $\varepsilon^{p-\frac{3}{2}}$, ..., $\varepsilon^{-\frac{1}{2}}$ while the finite part $F(\varepsilon)$ converges to a finite limit as $\varepsilon \to 0$. The 'finite part' of the integral

$$\int_{0}^{1} \frac{A(x)\, dx}{x^{p+\frac{1}{2}}},$$

now often written as

$$*\int_{0}^{1} \frac{A(x)\, dx}{x^{p+\frac{1}{2}}},$$

is defined to be the limit of $F(\varepsilon)$ as $\varepsilon \to 0$. It remains invariant under any (differentiable) change of the variable of integration x.

In the case of the wave-equation in two spatial dimensions,

$$\frac{\partial^2 u}{\partial x^2} + \frac{\partial^2 u}{\partial y^2} = \frac{\partial^2 u}{\partial t^2},$$

the elementary solution is $u = \Gamma^{-\frac{1}{2}}$,

where $\qquad \Gamma = (t - \tau)^2 - (x - \xi)^2 - (y - \eta)^2$,

and the general solution involves integrals over the region $\Gamma \geqslant 0$, and over surfaces which intersect the cone $\Gamma = 0$, while the integrands contain terms in $\Gamma^{-\frac{1}{2}}$, $\Gamma^{-\frac{3}{2}}$, ... To isolate the finite part of these integrands the domains of integration are restricted to their intersections with $\Gamma > \varepsilon$ and the integrands are expressed in the form

$$A_0\varepsilon^{-\frac{1}{2}} + A_1\varepsilon^{-\frac{3}{2}} + F(\varepsilon),$$

where $F(\varepsilon)$ converges to a finite limit as $\varepsilon \to 0$.

Hadamard's general theory is applied to linear hyperbolic equations with variable coefficients. The theory was adapted to problems of supersonic aerodynamics by A. Robinson (1948) and A. Robinson and Laurmann (1956) [see also the monograph by Ward (1955) and the reports by Heaslet, Lomax and Jones (1947), Heaslet and Lomax (1948) and Lomax and Sluder (1949)].

12.7. Bochner's Generalized Trigonometric Integrals

Riemann (1854; pp. 227–65) and Fatou (especially p. 386) had shown that even if the Fourier series of an integrable function $f(x)$ failed to converge, nevertheless the series integrated, once or twice, term by term, would converge to

$$g(x) = \int_{0}^{x} f(t)\, dt \quad \text{or to} \quad h(x) = \int_{0}^{x} g(t)\, dt.$$

In Chapter VI of his famous lectures on Fourier integrals Bochner (1932) developed a much more general theory for the Fourier transforms of functions $f(x)$ which behave at infinity as polynomials in x of degree k.

For such functions the integral

$$\int_{-\infty}^{\infty} \frac{|f(x)|\ \mathrm{d}x}{1+|x|^k}$$

is finite but the Fourier transform

$$E(\alpha,0) = \frac{1}{2\pi} \int_{-\infty}^{\infty} f(x) \exp(-i\alpha x)\ \mathrm{d}x$$

does not converge. Formal integration of $E(\alpha,0)$ repeated $(k-1)$ times, yields the well-defined function

$$E(\alpha,k) = \frac{1}{2\pi} \int_{-\infty}^{\infty} f(x) \frac{\exp(-i\alpha x) - L_k}{(-ix)^k}\,\mathrm{d}x,$$

where

$$L_k = \sum_{n=1}^{k-1} \frac{(-i\alpha x)^n}{n!} \quad \text{for} \quad |x| \leq 1,$$

$$\text{or} \qquad\qquad 0 \quad \text{for} \quad |x| > 1,$$

together with a polynomial of degree not greater than $k-1$. Thus, for $f(x) = 1$, $E(\alpha,2) = \frac{1}{2}|\alpha| + cx$.

It only needs a definition of generalized differentiation to define the Fourier transform of $f(x)$ as the kth generalized derivative of $E(\alpha,k)$, but this further step had to await the researches of Sobolev (1936).

12.8. The Dirac Delta Function

The theory of generalized functions received a great impetus when Dirac (1930) introduced his famous 'delta function', $\delta(x)$, as a 'convenient notation' in the mathematical formulation of quantum theory.

In the (classical) form of quantum theory due to Dirac the state of an atomic system is represented by a (complex) function ψ of the spatial coordinates of its constituent proton and electrons, and the probabilities that the system may be in the eigenstate corresponding to an energy-level E are derived from an expansion

$$\psi = \Sigma a_n \psi_n + \int a_p \psi_p\ \mathrm{d}p,$$

where the symbols ψ_n refer to energy levels E_n in the discrete spectrum and the symbols ψ_p refer to energies in the continuous spectrum.

The 'eigenvectors' ψ_n of the discrete spectrum may be regarded as vectors in a Hilbert space, forming an orthonormal system such that

$$(\psi_m, \psi_n) = \delta_{mn} = \begin{array}{l} 1 \quad \text{if} \quad m = n, \\ 0 \quad \text{if} \quad m \neq n, \end{array}$$

the parentheses indicating the scalar product in the Hilbert space. But the eigenvectors ψ_p of the continuous spectrum have to satisfy an analogous relation which Dirac writes in the form

$$(\psi_p, \psi_q) = \delta(p - q),$$

where the generalized function $\delta(s)$ is constrained to satisfy the conditions that

$$\delta(s) = 0 \quad \text{if} \quad s \neq 0,$$

while

$$\int_a^b \delta(s) \, ds = 1 \quad \text{if} \quad a < 0 < b.$$

The principal properties assigned to the delta function were

$$\int_{-\infty}^{\infty} f(x) \, \delta(x - a) \, dx = f(a),$$

for any continuous function $f(x)$,

$$\int_{-\infty}^{\infty} \delta(a - x) \, \delta(x - b) \, dx = \delta(a - b)$$

and

$$\int_{-\infty}^{\infty} f(x) \, \delta'(x - a) \, dx = -f'(a),$$

where $f(x)$ is any differentiable function and $\delta'(s)$ represents a generalized derivative of $\delta(s)$ such that

$$\delta'(-x) = -\delta'(x)$$

and

$$x\delta'(x) = -\delta(x).$$

In fact Dirac had rediscovered Kirchoff's delta function and Heaviside's 'unit impulse'.

12.9. *The Functional Analysis of Sobolev*

The concepts which were implicit in Dirac's bold, heuristic theory received explicit expression in the work of Sobolev (1936), which seems to have been developed independently of Dirac's investigations, and which is primarily concerned with the existence of solutions of linear hyperbolic partial differential equations. We have already noted (Section 12.4) that, even in the simplest problems of wave propagation, a physically significant solution may be represented by a function which is not everywhere differentiable. The general theory must certainly take cognizance of such a generalized solution, and the necessary technique was provided by Sobolev.

In Sobolev's work the concept of 'test functions' is given almost in its present classical form, viz. a test function $\varphi(x_1, x_2, \ldots, x_n)$ of n independent variables possesses everywhere partial derivatives of all orders, not exceeding some finite integer s, but nevertheless vanishes outside its 'support', viz. some bounded region. A sequence of test functions $\{\varphi_n\}$ is said to converge to a limit test function φ if the support of each function φ_n is contained in the same fixed finite domain, and if $D^p\varphi_n$ converges uniformly to $D^p\varphi$ everywhere, for each partial derivative D^p of order not less than s viz.

$$D^p\varphi = \frac{\partial^{p_1+p_2+\cdots+p_n}\varphi}{\partial x_1^{p_1}\partial x_2^{p_2}\cdots\partial x_n^{p_n}}, \quad p_1+p_2+\ldots+p_n \leq s.$$

A functional in the space Φ of test functions is said to be linear and 'continuous of class s', and is written as (ϱ,φ) if it is linear in the test functions φ and if $(\varrho,\varphi_n) \to (\varrho,\varphi)$ for any sequence of test functions $\{\varphi_n\}$ which converge to φ. A sequence of functionals (ϱ_n,φ) is said to converge to a functional (ϱ,φ) if $(\varrho_n,\varphi) \to (\varrho,\varphi)$ for each test function φ in Φ. The notation is suggested by the simplest example of a functional in which (ϱ,φ) is the scalar product,

$$(\varrho,\varphi) = \iint \ldots \int \varrho(x_1, x_2, \ldots, x_n)\, \varphi(x_1, x_2, \ldots, x_n)\, dx_1\, dx_2 \ldots dx_n.$$

For functionals of this special type

$$(D^p\varrho,\varphi) = (-1)^p(\varrho,D^p\varphi), \quad (p \leq s),$$

and $$(\omega\varrho,\varphi) = (\varrho,\omega\varphi)$$

where ω is any continuous function, and the function ϱ possesses continuous derivatives of order s. Hence if L is a linear hyperbolic differential operator of the form

$$L = \sum_{i,j=1}^{2k+1} A_{ij}\frac{\partial^2}{\partial x_i\partial x_j} + \sum_{i=1}^{2k+1} B_i\frac{\partial}{\partial x_i} + C - \frac{\partial^2}{\partial t^2}$$

and u has continuous derivatives of order $s = 2k + 1$,

$$(Lu,\varphi) = (u,L^*\varphi),$$

where L^* is the adjoint operator of L, and the scalar product (u,v) is an integral over the space of $x_1, x_2, \ldots, x_{2k+1}, t$.

The classical 'Cauchy problem' is to determine a function u which satisfies the equation,

$$F = Lu$$

where F is a prescribed function, together with initial conditions

$$u = f, \quad \text{and} \quad \partial u/\partial t = g \quad \text{for} \quad t = 0.$$

To formulate the generalized Cauchy problem, Sobolev introduced the dicontinuous function v such that

$$v = u \quad \text{if} \quad t > 0 \quad \text{and} \quad v = 0 \quad \text{if} \quad t \leq 0.$$

The functional (Lv,φ) is then *defined* to be $(v,L^*\varphi)$, although v may not be differentiable and this second functional is expressed in terms of the integral of $F\varphi$ over the space $t > 0$, and the integral of $\{f\partial\varphi/\partial t - g\varphi\}$ over the hyperplane $t = 0$ as a functional (ϱ,φ).

The generalized Cauchy problem can then be expressed in the form $Lv = \varrho$, and $v = 0 = \varrho$ for $t \leqslant 0$, where ϱ and v are 'generalized functions' and L a 'generalized operator'.

In Sobolov's work it will be noted that the definitions of these concepts are given only in an implicit form. The final and explicit definitions were given in two papers by L. Schwartz 1945, 1948) and later in his book, *Théorie des Distributions* (L. Schwartz 1950).

12.10. *The Distributions of Laurent Schwartz*

In 1945 Laurent Schwartz began to publish his researches on generalized functions, which were subsequently fully developed in his magisterial *Théorie des Distributions* (Schwartz 1950), in which all that was implicit, hypothetical and often confused in the earlier researches described above received an explicit, constructive and perfectly clear exposition.

The theory of Schwartz appears to have originated in the attempt to give mathematical precision to the well-known physical concepts of a doublet, and of surface layers of sources and doublets, by generalizing the idea of electric density so that it should be applicable to these well-known distributions of electricity. Hence the name of 'distribution' which Schwartz now gave to the new mathematical entities which he defined with logical precision and topological generality.

The first point to be emphasized is that in the work of Schwartz there is a great variety of different species of distributions or generalized functions, which are defined as functionals in relation to different spaces of test functions on which they operate.

Thus in problems which relate to bounded regions in Euclidean space R^n the test functions φ are *infinitely* differentiable everywhere and have a compact support, outside of which they vanish identically, just as in the work of Bochner (1946). In problems which relate to the whole of R^n the test functions are infinitely differentiable everywhere and decay rapidly towards infinity, i.e. if $D^p\varphi$ is any partial derivative of φ of order p and r the distance from a finite point, then

$$r^k D^p \varphi \to 0 \quad \text{as} \quad r \to \infty$$

for all $k \geqslant 0$. In the theory of Fourier series of functions $f(x_1, x_2, \ldots, x_n)$ defined on the torus

$$0 \leqslant x_k \leqslant 1, \; k = 1, 2, \ldots, n,$$

the test functions can be taken to be linear combinations of the exponentials, $\exp (2\pi i)(l_1 x_1 + l_2 x_2 + \ldots + l_n x_n)$ where $l_1, l_2, \ldots, l_n$ are integers.

A distribution T is defined to be a linear functional operating on one of these spaces $\mathscr{D}$ of test functions, and possessing the strong type of continuity expressed by the relations

$$T(\varphi_n) \to 0 \quad \text{as} \quad n \to \infty$$

if $D^p\varphi_n$ converges uniformly to zero everywhere (and for problems in bounded regions, if $\varphi_n = 0$ outside some bounded region independent of n).

The connection between the functionals of Schwartz's theory and the numerical functions of classical analysis is furnished by two theorems due to Hadamard (1903a) and F. Riesz (1909b).

Hadamard proved that any continuous linear functional must have the form

$$L(\varphi) = \lim_{n \to \infty} \int f_n(x)\varphi(x)\, dx.$$

F. Riesz obtained a more compact representation of L in the form of a Stieltjes integral

$$L(\varphi) = \int \varphi(x)\, df(x),$$

where f is a function of bounded variation. Hadamard's representation leads naturally to the theory of generalized functions developed by Mikusiński (1948) and many others. The representation due to Riesz leads to Schwartz's theory of distributions.

Thus to any continuous function $f(x)$ there corresponds a distribution,

$$T(\varphi) = \int f(x)\varphi(x)\, dx$$

and moreover it is seen at once that in Schwartz's theory the distribution

$$T(\varphi) = \int \varphi(x)\, d(\tfrac{1}{2}\operatorname{sgn} x) = \varphi(0)$$

will take the place of Dirac's delta function $\delta(x)$ and will enable us to dispense with the notorious formula

$$\int \delta(x)\varphi(x)\, dx = \varphi(0).$$

Schwartz takes the derivative of a distribution $T(\varphi)$ to be the distribution $T(-d\varphi/dx)$, x being one of the variables in φ and hence is able to conclude at once that any distribution is infinitely differentiable.

By considering the distributions corresponding to a 'semi-function' $f(x)$, such that $f(x) = 0$ if $x \leq 0$ while $f(x)$ is infinite at the origin but summable in $[0, a]$, Schwartz provided a simple theory of the 'finite part' of an integral due to Hadamard. Thus, if

$$T(\varphi) = \lim_{\varepsilon \to 0} \int_{\varepsilon}^{\infty} \varphi(x)\, x^{-\frac{1}{2}}\, dx,$$

then
$$\partial T(\varphi)/\partial x = \lim_{\varepsilon \to 0}\left[\varphi(0)\varepsilon^{-\frac{1}{2}} - \int_{\varepsilon}^{\infty} \varphi(x)\, \tfrac{1}{2}x^{-\frac{3}{2}}\, dx\right].$$

Here $T(\varphi)$ is the distribution corresponding to the function $H(x)x^{-\frac{1}{2}}$ where $H(x) = 1$ if $x > 0$, or 0 if $x \leq 0$ (Heaviside's unit function), and $\partial T(\varphi)/\partial x$ corresponds to the function

$$x^{-\frac{1}{2}}\delta(x) - \tfrac{1}{2}H(x)\, x^{-\frac{3}{2}},$$

which is the generalized derivative of $H(x)x^{-\frac{1}{2}}$.

Among the numerous important and fertile results obtained by

Schwartz we shall mention the theory of the convolution of two distributions and the theory of Fourier transforms.

With appropriate restrictions on the supports of two distributions S and T we can calculate $T[\varphi(x + s)]$ as a test function $\psi(s)$ and then define the convolution of S and T as

$$U = S \times T \quad \text{where} \quad U(\varphi) = S(\psi).$$

The theory of Fourier transforms applies primarily to 'tempered' distributions, which operate in the space $(\mathcal{S})$ of test functions which 'decay rapidly' at infinity. Such distributions correspond to the generalized derivatives of functions of 'slow growth' at infinity, i.e. continuous functions $f(x)$ such that

$$f(x)(1 + r^2)^{-k}$$

is bounded for all positive k. If $u(x)$ is a test function in the space $(\mathcal{S})$, so also is its classical Fourier transform,

$$v(x) = \int \exp{(-2\pi i x s)} \, u(s) \, ds,$$

and if U is a tempered distribution, its Fourier transform $V = \mathcal{F}U$ is uniquely defined by the relation

$$U[u(-x)] = V[v(x)].$$

Thus
$$\mathcal{F}S = 1, \quad \mathcal{F}1 = \delta.$$

Any linear partial differential equation (or difference equation) can be expressed in the form $A \times T = B$, where T is the unknown distribution, and A, B are given distributions. An 'elementary solution' of this equation is a distribution E which satisfies the equation $A \times E = \delta$, and provides a solution of the form $T = E \times B$.

Perhaps the most original and striking aspect of Laurent Schwartz's theory is the study of the topological space of distributions, resulting in a topological definition of differentiation and in a proof that any distribution in Euclidean space corresponds to the generalized derivative of a continuous function in any prescribed bounded domain.

12.11. *Mikusiński and Weak Convergence*

The publication of the second of Schwartz's papers (L. Schwartz 1948) stimulated Mikusiński to show that the methods of distribution theory could be considerably generalized (Mikusiński 1948).

In its most abstract form the theory of Mikusiński considers three abstract sets F, Φ and C and a 'composition' which maps the product $F \times \Phi$ on to C in such a way that if $f_n \in F$, $(n = 1, 2, \ldots)$ and $\varphi \in \Phi$ then we can define the limit c of the sequence $\{c_n\} = \{f_n\varphi\}$ even if there is no element f of F such that $c = f\varphi$. In such a case Mikusiński says that the sequence $\{f_n\}$ converges 'weakly' to the limit f, if the sequence $\{f_n\varphi\}$ converges in C for all φ in Φ. The sequences $\{f_n\}$ are taken to define the weak limits f, which form the closure of F with respect to the operation of weak convergence.

Clearly the weak limit of a sequence $\{f_n\varphi\}$ is a linear functional in the space Φ which forms a representation of the functional $\{f\varphi\}$ in the manner envisaged by Hadamard (Section 12.10).

12.12. Van der Corput and the Neutrix Calculus

In its simplest form Hadamard's technique for obtaining the finite part of a divergent integral, such as

$$\int_0^a t^{-p-\frac{1}{2}}\varphi(t)\,dt,$$

where p is a positive integer and $\varphi(t)$ is of class C^p in $[0,a]$, is to express the truncated integral

$$J(x) = \int_x^a t^{-p-\frac{1}{2}}\varphi(t)\,dt \qquad (x>0)$$

in the form $I(x) + F(x)$, where $I(x)$ the 'infinite part', is a polynomial in $x^{-\frac{1}{2}}$ and where $F(x)$ tends to a finite limit as $x \to 0$. Then the finite part of the integral is

$$*\!\int_0^a t^{-p-\frac{1}{2}}\varphi(t)\,dt = \lim_{x\to 0} F(x).$$

A slightly different way of describing this technique is to say that in calculating the integral $J(x)$ we neglect any linear combination of the functions $x^{-\frac{1}{2}}$, $x^{-\frac{3}{2}}$, ..., $x^{-p+\frac{1}{2}}$. Indeed, as Hadamard noted, there is no need to restrict p to be integral, and similar techniques can be devised for integrals of the form

$$\int_x^a t^{-p-\mu}\,(\log t)^q \varphi(t)\,dt \qquad (x>0,\ 0<\mu<1),$$

the 'negligible' functions being of the form $x^a(\log x)^b$.

Van der Corput (1952) had noticed that a similar situation arose in the evaluation of an integral which was required for an asymptotic expansion, where in fact some 40 of the 50 'leading' terms dropped out of the final expression, and he systematically developed a technique by which these negligible functions could be recognized as such from the start, so that they could be completely ignored. A necessary condition for the consistency of such a technique is that in the calculation of a given integral or series the set of negligible functions should form a commutative, additive group in which the only constant functions are identically equal to zero. Such a set of functions is called a neutrix.

Thus, for example, the generalized Fourier transform of $\log |x|$, i.e.

$$*\!\int_{-\infty}^\infty e^{-2\pi i x y}\,\log\,|x|\,dx,$$

is obtained as $-\frac{1}{2}|y|^{-1}$ by integration by parts and neglecting the neutrix formed by the set of functions

$$f_y(x) = e^{-2\pi i x y}\,\log|x| \quad \text{for} \quad -\infty<y<\infty.$$

We have thus obtained the same generalized Fourier transform as in Schwartz's theory of distributions by a very simple process.

It appears that many of the results given by the theory of distributions can be obtained by the use of the neutrices consisting of functions $v_n(x)$ [like $f_y(x)$] such that

$$\lim_{n \to \infty} \int_{-\infty}^{\infty} v_n(x)\varphi(x) \; \mathrm{d}x = 0$$

for any test function $\varphi(x)$, and the voluminous papers of Van der Corput contain an inexhaustible store of neutrices suitable for almost every conceivable situation. A completely formalized theory of the vast generalizations given by Van der Corput (1959a,b) is much to be desired.

12.13. *Tempered Distributions*

Schwartz's theory of distributions is admirably simple when the corresponding test functions φ are of compact support, but there are complications when the support is unbounded. Two classes of functions, defined over the whole of n-dimensional Euclidean space $R^{(n)}$, are then of special importance:

(1) 'Slowly increasing' functions f such that each derivative $D^k f$ is bounded by some positive power of r (the distance from the origin to the point which is the argument of f).

(2) 'Rapidly decreasing' functions f such that for each value of k and m,

$$r^m D^k f \to 0 \quad \text{as} \quad r \to \infty.$$

'Tempered distributions' are defined by Schwartz as continuous linear functionals acting on the space of rapidly decreasing test functions. For such tempered distributions Schwartz has given a theory of Fourier transforms. Thus in one dimension, the generalized Fourier transform of x^k is

$$\int x^k e^{i\alpha x} \; \mathrm{d}x = \left(\frac{\mathrm{d}}{\mathrm{i}\mathrm{d}\alpha}\right)^k \int e^{i\alpha x} \; \mathrm{d}x = (-\mathrm{i})^k D^k \delta(x).$$

Tempered distributions can be expanded into series of Hermite polynomials $k_1, k_2, \ldots$ and if

$$f = \Sigma a_n k_n$$

then the Fourier transform of f is $\Sigma i^n a_n k_n$. Thus a sequence of tempered distributions $\{f_\mu\}$ is adequately represented by the matrix with coefficients $a_{\mu n}$ where

$$f_\mu = \Sigma a_{\mu n} k_n.$$

Köthe (1966) has developed a topological theory of spaces of sequences represented by matrices $\|a_{\mu n}\|$ and a simplified theory has been given by Antosik, Mikusiński and Sikorski (1973). Another approach to the theory has been given by H. König (1953).

12.14. Hyperfunctions

In Schwartz's theory of distributions the product of two distributions could be defined only when one factor was a generalized derivative of a continuous function of order p while the other was a function which possessed classical derivatives of order $p-1$. The outstanding problem was to generalize the definition of a distribution so as to ensure the existence of a product in all those problems of quantum theory in which it occurred.

The problem was solved by Sato (1959) by the introduction of 'hyperfunctions', which may be roughly described as the 'boundary values' taken by holomorphic functions $f(z)$ as z approaches the real axis; and, more accurately, in terms of cohomology of sheafs.

A simplified account of the theory was given by Bremermann and Durand (1961) who in effect take a hyperfunction to be a linear functional of the form

$$T(\varphi) = \lim_{\varepsilon \to 0} \int_{-\infty}^{\infty} \{f_1(x + i\varepsilon) + f_2(x - i\varepsilon)\}\varphi(x)\,dx,$$

acting on a test function $\varphi(x)$. Thus if

$$f_1(x + i\varepsilon) = \frac{-i/\pi}{x + i\varepsilon}$$

and $$f_2(x - i\varepsilon) = 0,$$

then $T(\varphi) = f(0)$ so that I corresponds to the Dirac function $\delta(x)$.

The 'nucleus' of the hyperfunction, viz.

$$f_1(x + iy) + f_2(x - iy) \qquad (y > 0),$$

is holomorphic away from the real axis $y = 0$ and hence these nuclei can be freely added and multiplied in an unambiguous manner.

In general, if T is a Schwartzian distribution with compact support the nucleus of the associated hyperfunction is

$$T^0(z) = (1/2\pi i)\, T(x - z)^{-1},$$

and $$T(\varphi) = \lim_{\varepsilon \to 0} \int_{-\infty}^{\infty} \{T^0(x + i\varepsilon) - T^0(x - i\varepsilon)\}\varphi(x)\,dx.$$

Thiess (1968) has shown how such hyperfunctions can be employed in quantum field theory to represent commutator and projector distributions and thus to provide a rational process for the regularization of ultraviolet divergences.

The topological theory of hyperfunctions has been further explored by Martineau (1961) and by Schapiro (1969).

12.15. Sebastião e Silva and Formal Derivatives

In the theory of Laurent Schwartz the possibility of expressing a distribution as (locally) a generalized derivative of a continuous function appears as a theorem. It was an obvious suggestion to use this property as a definition, but Schwartz decided against this temptation because of the indetermination of the order of differentiation of the continuous function in question when several variables are involved. The possibility of overcoming this difficulty was indicated by H. König (1953), the impact of whose researches can be found in the more profound and extensive investigations of Sebastião e Silva (1945, 1960).

If f is a continuous function of k independent variables, x_1, x_2, ..., x_n, the derivative indicated by

$$D^p f \equiv D_{x_1}^{p_1} D_{x_2}^{p_2} \dots D_{x_k}^{p_k} f$$

may not exist ($D_{x_k}^{p_k}$ is the operator $\partial^{p_k}/\partial x_k^{p_k}$, and p represents the ordered set of non-negative integers p_1, p_2, ..., p_k) but we can certainly consider the entity represented by the symbol $[f,p]$. To remove the 'indetermination' in f, remarked by Laurent Schwartz, Sebastião e Silva considers a class of equivalent couples $[f,p]$ such that

$$[f,p] \sim [g,q]$$

if and only if

$$I^q f - I^p g$$

is a function of the class N_{p+q}, the 'nucleus' of the differential operators. Explicitly θ belongs to N_k, if

$$\theta(x) = \sum_{i=1}^{k} \{\gamma_{i,j}(x) x_i^{j-1}\}$$

and $\gamma_{i,j}(x)$ is independent of x_i, and I^p is an inverse of D^p.

In this theory a distribution is simply an equivalence class of couples such as $[f,p]$, with certain restrictions which are imposed by the necessity of defining the restriction of a distribution to one of the set of neighbourhoods which cover a given open set in the space of the variables (x_1, x_2, ..., x_n).

The definition of the sum of two distributions then follows as in the theory of numerical fractions, and the infinite differentiability of distributions becomes immediately evident.

In the case of functions of one variable, the definition in terms of the couples (f,m) had been given independently by Sikorski (1954) and also by Korevaar (1955).

13. Ordinary Differential Equations

13.1. Introduction

Many of the problems of mathematical physics give rise to differential equations, i.e. relations between one or more functions of a variable t and their derivatives of the first and second orders, and the problem is to determine these functions explicitly in terms of t and of such auxiliary conditions as may be prescribed.

In the earlier researches on ordinary differential equations, mathematicians were content to assume, or hope, on the basis of physical intuition, that the problems presented by physics were necessarily soluble and, moreover, soluble by quadratures or by the functions of elementary analysis.

Existence Theorems from Cauchy to Banach

13.2. Cauchy and Dominant Functions

The great advance made by Cauchy (Moigno 1861) was to consider the very general system of differential equations of the form

$$du_a/dt = f_a(u_1, u_2, \ldots, u_n; t), \quad a = 1, 2 \ldots, n$$

where the variables u_a and t are complex and the functions f_a are all analytic in some common domain D. Moreover, he devised a process which he called the 'calculus of limits' which definitely establishes the existence of analytic functions, which satisfy the given system of equations and assume given arbitrary values at a given point in the domain D.

This process depends upon the construction of a 'dominant function'

$$F = M[1 - (t - c)/\varrho]^{-1} \prod_k \left(1 - \frac{v_k - a_k}{r}\right)^{-1}$$

where $|t - c| < \varrho$, $|v_k - a_k| < r$ is a subset of the domain D.

The given system of differential equations determines the initial values of the derivatives $d^n u_k/dt^n$ at the point $t = c$ and hence gives a formal power series for the functions u_a.

Cauchy then considers the dominant system of equations

$$dv_a/dt = F(v_1, v_2, \ldots, v_n, t)$$

which possess a solution of the form

$$[1 - (v_n - a_n)/r]^{n+1} = A + B(n + 1)(\varrho/r) \, M \log \, [1 - (t - c)/\varrho]$$

where M is the maximum of $|f_a|$ in the domain D.

174

The functions v_k thus constructed are analytic if

$$|t| < \varrho\{1 - \exp[-a(n+1)^{-1}M^{-1}\varrho^{-1}]\} = T.$$

Finally Cauchy proves that each coefficient in the formal power series for u_k is less, in absolute terms, than the corresponding coefficient in the power series for v_k. Hence the formal power series actually converge and provide a solution of the original problem.

The limitations of Cauchy's method of dominant functions are two-fold. In the first place, the power series for the solution u_k may have a much larger radius of convergence than the radius T determined by the dominant function F. Secondly, although the method establishes the existence of unique analytic solutions, it cannot disclose the existence of irregular solutions.

Cauchy's results have been completed by Poincaré (1882b) who has determined the conditions under which the solution of a system of differential equations can be expressed, not only as a convergent power series in the independent variable, but as a power series in the initial values of the dependent variables and also as a series in a parameter occurring in the equations.

Moreover, Poincaré (1882b) has shown how to obtain power series solutions of the equations

$$dx_k/dt = X_k, \qquad k = 1, 2, \ldots n,$$

valid for *all* values of t, when the functions X_k are polynomials in x_1, x_2, ..., x_n. He introduces a new variable s, defined by the equations $ds/dt = X_1^2 + X_2^2 + \ldots + X_n^2$, and obtains power series solutions in the variable

$$(e^{\alpha s} - 1)/(e^{\alpha s} + 1)$$

where α is an appropriate real number.

13.3. *Briot and Bouquet and Irregular Equations*

The success of Cauchy's method of dominant functions in establishing the existence of analytic solutions naturally leads to the examination of 'irregular' differential equations whose solutions possess a singularity.

Well-known examples have been provided by Legendre's equation and Bessel's equation. A systematic examination of equations of the form

$$x \, dy/dx = qx + py + \varphi(x,y)$$

where f is analytic near $x = 0$, $y = 0$ was undertaken by Briot and Bouquet (1856), and a painstaking account is given by Forsyth (1890–1906) (Volume I). One of the most general equations considered is of the form

$$\frac{dy}{dx} = x^s y^r \frac{q(x,y)}{r(x,y)} \exp G(x,y)$$

where r and s are integers, and the initial conditions are $y = 0$, $x = 0$, the functions p,q and G being analytic near the origin.

Briot and Bouquet's researches provide abundant examples, but their investigations are essentially *local* in character, i.e. they elucidate the behaviour of the integrals near a prescribed point.

13.4. *The Equations of Mathieu and Hill*

Mathieu (1868) investigated the problem of vibration of an elliptic membrane which is governed by the two-dimensional wave-equation

$$\partial^2 u/\partial x^2 + \partial^2 u/\partial y^2 = c^{-2}\partial^2 u/\partial t^2.$$

The appropriate coordinates ξ,η are defined by the equation

$$x + iy = h \cos h(\xi + i\eta),$$

and the classical method of the 'separation of the variables' provides solutions of the form

$$u = F(\xi)G(\eta) \cos pt,$$

where $\qquad$ $d^2F/d\xi^2 + (k^2 \cosh^2\xi - A)F = 0,$

$$d^2G/d\eta^2 + (-k^2 \cos^2\eta - A)G = 0.$$

Both equations can easily be reduced to the form

$$d^2u/dz^2 + (a + 16q \cos 2z)u = 0.$$

The function $G(\eta)$ must be periodic with period 2π and this condition is sufficient to determine a set of values of a in terms of q, q itself is determined by the fact that $F(\xi) = 0$ on the boundary, $\xi = $ constant, of the membrane.

In 1877 Hill constructed an equation of the form

$$d^2u/dz^2 + p(z)u = 0$$

where p is a periodic function of the form

$$p(z) = \sum_{n} c_n \cos 2nz,$$

in connection with his researches on the determination of the motion of the Lunar Perigee (Hill 1886).

In astronomical problems the relevant solutions of this equation are not periodic, but nevertheless Floquet (1883) showed that the general solution of Hill's equation has the form

$$u = ae^{\mu z}\varphi(z) + be^{-\mu z}\varphi(-z),$$

where a,b are constants and $\varphi(z)$ is periodic with period 2π.

In the paper published in 1886 Hill showed that the value of the parameter μ is given by an equation of the form

$$\sin^2 (\tfrac{1}{2}\pi i\mu) = \Delta(0) \sin^2 (\tfrac{1}{2}\pi \sqrt{c_0})$$

where $\Delta(0)$ is a determinant whose elements depend on the coefficients c_n.

The researches of Mathieu and Hill are noteworthy as a splendid example of the classical analytic approach to the theory of differential equations. Their work has been of lasting value as providing 'exact' solutions of certain vibration problems in which the 'restoring force' is periodic in time.

Mathieu functions have attracted numerous investigators, partly because of the challenging analytical problems to which they give rise, partly because of their applications to physical problems, and partly because of the need for numerical tables. A very full account is given by McLachlan (1947).

The main contributors to the analytic theory were Sieger, E. T. Whittaker and his pupil, Ince. Sieger (1908) gave numerous properties of Mathieu functions, including their orthogonality and the integral equations which the functions satisfy. These results were obtained independently by Whittaker, whose address to the International Congress of Mathematicians at Cambridge in 1912 (E. T. Whittaker 1914) gave a fresh impetus to the subject.

During the years 1915 to 1939 Ince published some eighteen papers on Mathieu functions (which are listed in McLachlan's book cited above). His outstanding achievement was the calculation of the characteristic numbers, the constants in the Fourier series for the functions, together with the zeros and turning points of the functions.

Hill's analysis was applied by Rayleigh (1887) to the classical experiment of Melde, in which one end of a horizontal thread is fixed and the other attached to a prong of a tuning fork, and subharmonic transverse oscillations are produced in the thread.

Perhaps the most important and influential studies were made by Jeffreys (1924) in a series of four papers, which are remarkable not only for their contributions to the theory of the Mathieu equation, but also for their revival of what has unfortunately become known as the 'W-K-B' method for obtaining approximate solutions of Schrödinger's wave-equation.

13.5. The 'W-K-B' Method

Although the so-called 'W-K-B' method really belongs to the theory of hyperbolic partial differential equations, it had such an influence on the solution of ordinary differential equations in one-dimensional problems of quantum wave-mechanics that it demands a summary account in this chapter.

This method has its roots in the relation between wave-theory and geometrical optics. The wave-equation for light of frequency c/λ (c being the speed of propagation) is

$$\nabla^2 \psi + \psi/\lambda^2 = 0$$

with the special solution for a parallel beam

$$\psi = A \sin (ct - x)/\lambda, \quad A = \text{const.}$$

The form of this solution suggests that the general solution should have the form

$$\psi = A \sin (ct - S)/\lambda$$

where the 'eikonal' function S satisfies the equations

$$\lambda^2 \nabla^2 A - A(\operatorname{grad} S)^2 + A = 0,$$

$$2\lambda^2 \operatorname{grad} A \operatorname{grad} S + A\nabla^2 S = 0.$$

For waves of length small compared with the linear dimensions of the optical apparatus under consideration, it appears to be reasonable to search for series solutions in ascending power of λ, the leading terms being given by

$$(\operatorname{grad} S_0)^2 = 1, \qquad \nabla^2 S_0 = 0.$$

Unfortunately the power series is only asymptotically convergent, but the first few terms often yield valuable approximations to the full solution.

Green (1838) (of potential theory fame) appears to have been the first to use this method to obtain useful approximations, but the analytical theory was not developed until the researches of Horn (1899) and Schlesinger (1906, 1907). The hydrodynamical applications seem to have remained in abeyance until the researches of Jeffreys (1915) and Fowler and coworkers (1920).

The method was formalized in the papers by Jeffreys (1924), which consider an equation of the form

$$y'' = (h^2\chi_0 + h\chi_1 + \chi_2)y,$$

with a large parameter h. The approximate solution is of the form

$$y = \varphi \exp h\omega(1 + f_1/h + f_2/h^2 + \ldots)$$

where $\quad \omega'^2 = \chi_0, \quad \varphi'/\varphi = (\chi_1 - \omega'')/2\omega', \quad 2\omega_1'f_1 = \chi_2 - \varphi''/\varphi$

and $\quad\quad\quad\quad \varphi = (\omega')^{-\frac{1}{2}} \exp \frac{1}{2}\int \chi_1/\omega' \, dx.$

Jeffrey's method was developed independently by Wentzel (1926), Kramers (1926) and Brillouin (1926) in a famous series of papers which elucidate the relation between the early quantum theories of Bohr and de Broglie and the wave-mechanical theory of Schrödinger.

In the one-dimensional theory the wave-equation for a single particle of total energy E moving in a field of potential V is

$$d^2\psi/dx^2 + k(E - V)\psi = 0$$

where $k = 8\pi^2\mu/h^2$, μ being the mass of the particle and h being Planck's constant. The 'W-K-B' method is to search for solutions of the form

$$\psi = A \exp \{(2\pi i/h)\int y(x,E) \, dx\}$$

where y must satisfy the Ricatti equation

$$(h/2\pi i) y' + y^2 = 2\mu(E - V).$$

Now h is a small parameter, and if y is expanded in a power series in h,

$$y = y_0 + hy_1 + \ldots$$

the leading terms are

$$y_0 = \pm [2\mu(E - V)]^{\frac{1}{2}},$$

and

$$y_1 = \tfrac{1}{4}V'/(E - V).$$

Clearly the solution takes periodic or exponential forms accordingly as $E - V$ is positive or negative, and the main problem is to connect these different forms at the points where $E = V$.

There is an immense literature on this problem which has important relations with the general theory of asymptotic expansions and an admirable introductory account is given by Kemble (1937).

13.6. *The Lipschitz Condition*

The theory of differential equations in the complex domain is considerably simplified by the fundamental principle that an analytic function of a complex variable necessarily possesses derivatives of all orders within the domain of analyticity. In the corresponding theory in the real domain we know (Section 11.3) that even a continuous function does not necessarily possess a first derivative. Some restriction on the continuity of $f(x,y)$ must be imposed in order to ensure the existence of a unique solution of even the simplest equation

$$dy/dx = f(x,y).$$

The necessary and sufficient condition was discovered by Lipschitz (1876) and forms the basis of the method of solution by successive approximation due to Picard. Curiously enough, Lipschitz had earlier introduced a condition of the form

$$|f(\beta + \delta) - f(\beta)| < B\delta^\alpha, \quad B, \alpha > 0$$

in a paper on Fourier series (Lipschitz 1864). This condition is usually attributed to Hölder.

The well-known 'Lipschitz condition' was first introduced by Lipschitz in a paper (Lipschitz 1876) which obtained the solution of a system of simultaneous ordinary differential equations

$$dy_\alpha/dx = f_\alpha(x, y_1, y_2, \ldots, y_n), \quad \alpha = 1, 2, \ldots, n,$$

by replacing them by a system of simultaneous algebraic equations,

$$\Delta y_\alpha = f_\alpha \Delta x.$$

Lipschitz appears to be unaware of the much earlier work of Cauchy and Coriolis and regards it as a new and challenging problem to obtain a solution of the differential equation when $x, y_1, y_2, \ldots, y_n$ are restricted to be *real*. He was aware that Cauchy had solved the problem by his method of dominant functions when the variables are allowed to be complex and the functions are analytic, but he needed some other property of these functions when they are merely continuous. The necessary and sufficient condition he expressed in the form

$$[f_\alpha(k, k_1, k_2, \ldots, k_n) - f_\alpha(l, l_1, l_2, \ldots, l_n)] < \sum_\beta c_{\alpha,\beta} [k_\beta - l_\beta]$$

where the quantities $c_{\alpha,\beta}$ are positive and the symbol $[w]$ represents the absolute value of w. This is the famous Lipschitz condition, which is more restrictive than mere continuity.

13.7. Picard and the Method of Successive Approximations

We owe to Liouville (1836–38) the most important and fertile concept that a differential equation together with boundary conditions can be transformed into an integral equation. Liouville's researches were limited to linear equations of the second order, but the extensions to linear equations of the order n are obvious generalizations. The theory was perfected by Picard (1893) and is adequately exemplified by the simplest problem, viz.

$$dy/dx = f(x,y)$$

and $y = b$ when $x = a$.

The essential features of Picard's method are (1) the restriction to real values of x and y and (2) the conditions whose importance was discovered by Lipschitz (Section 13.6) that in some domain D containing the initial point (a,b),

$$|f(x,y)| < M$$

and
$$|f(x,y_2) - f(x,y_1)| < k|y_2 - y_1|,$$

when the points (x_1,y_1), (x_2,y_2) belong to D. This condition takes the place of the condition of analycity imposed in Cauchy's theory.

The differential equation, as given above, is equivalent to the integral equation

$$y(x) = b + \int_a^x f[t,y(t)]\, dt,$$

which leads to the sequence of successive approximations

$$y_{n+1}(x) = b + \int_a^x f[t,y_n(t)]\, dt, \quad n = 1, 2, \ldots.$$

This sequence converges to the solution of the differential equation.

A notorious paper by Peano (1890a) shows that the Lipschitz condition is essential for the unicity of the solutions. This paper, written in the mathematical symbolism invented by Peano, which mercilessly expounds every detail of the argument, is almost unreadable, but, happily it has been summaried in good German by Mie (1893).

Thus Clairaut's equation

$$y = px + p^2 \quad \text{or} \quad (p + \tfrac{1}{2}x)^2 = y + \tfrac{1}{4}x^2$$

where $$p = dy/dx$$

possess the regular solutions, $y = cx + c^2$ and the so-called 'singular' solution $y = -\tfrac{1}{4}x^2$.

It is easily verified that the equation

$$p + \tfrac{1}{2}x = (y + \tfrac{1}{4}x^2)^{\frac{1}{2}}$$

does not satisfy the Lipschitz condition at points on the parabola

$$x = c, \quad y = -\tfrac{1}{4}c^2.$$

Unfortunately the significance of the Lipschitz condition for 'singular' solutions has not yet percolated into elementary textbooks.

13.8. Banach and Contraction Operators

The integral equation employed by Picard has been the genesis of a new chapter in functional analysis, which may be introduced by considering the relation

$$\eta(x) = U[y(x)] = \int_a^x f[t,y(t)] \, dt$$

which maps the region D onto itself. If the differential equation $dy(x)/dx = f[x,y(x)]$ possesses a solution $y(x)$, then such a solution is a 'fixed' point of the mapping, i.e.

$$y(x) = U[y(x)].$$

Now the Lipschitz condition implies that there is a positive constant α less than unity such that if

$$y(x) = U[y(x)]$$

then $$\left|\eta(x_2) - \eta(x_1)\right| < \alpha \left|y(x_2) - y(x_1)\right|,$$

i.e. that U is a 'contraction' operator, which systematically reduces the 'distances' $\left|y_n(x_2) - y_n(x_1)\right|$ in the function space of $y(x)$ as n steadily increases, the successive functions $y_n(x)$ being given by

$$y_{n+1} = Uy_n.$$

We owe to Banach (1922) the theorem that the mapping given by any such contraction operator U has a unique fixed point $y(x)$ and that $y_n \to y$. This theorem has initiated a new era in functional analysis.

The transformation of a differential equation, or of a system of differential equations, into the search for fixed points of a mapping in function space is an important technique which has been extensively developed by Cronin (1964).

Poincaré's Researches

13.9. The Originality of Poincaré

The historian is often at a loss to discover the true sources of the inspiration which appears in the researches of the great mathematicians. This difficulty does not arise in the case of Poincaré. It is scarcely an exaggeration to say that this great innovator was not indebted to any of his colleagues or predecessors for any of the great ideas which flourish in his works.

At the celebrations in 1954 which marked the centenary of Poincaré's birth, Freudenthal (1955) seized the opportunity to reveal Poincaré's complete ignorance of such classical subjects as Dirichlet's principle in potential theory, the work of Jacobi and Riemann on the problem of the inversion of hyperelliptic integrals, and the theory of Riemann surfaces. This astonishing limitation in his mathematical education was perhaps due in part to his unfamiliarity with languages other than French (and perhaps Danish?) and to the relative immobility of young scientists in the days before international conferences and symposia. But perhaps it was also due to the modest and retiring character of a mathematician occupied with his own thoughts, which were only invaded at intervals by the outside world.

13.10. Local and Global Solutions

The earliest investigations of differential equations were restricted to the examination of a few special equations for which the solution was expressible in finite terms, by means of elementary and well-known functions such as x^n, $\sin x$, $\cos x$ and e^x, and by definite integrals involving these functions. But the number of equations which are integrable by these methods is extremely restricted and further progress along these lines required the introduction of new 'transcendentals', such as the elliptic functions of Jacobi.

In general we do not know in advance what class of functions will be required to represent the solution of a differential equation, nor do we know which equations possess an integral. It was Cauchy who first penetrated this *terra incognita* of differential equations. He invented a powerful and ingenious method ('le calcul des limites') by which he could establish the existence of solutions, in the form of convergent power series in x, for differential equations of the form

$$dy/dx = f(x,y)$$

when the function f itself is holomorphic in the complex variables x and y.

These investigations, which were essential for a rational theory of differential equations, were, however, limited in scope, for they provided the solution of a differential equation

$$y' = f(x,y)$$

only in a region in which the function $f(x,y)$ was holomorphic, and they were thus strictly *local* in character. To complete these investigations it would be necessary to 'match' the solutions which are valid in different regions and thus extend the solution to the whole of the complex x,y plane. Thus, for example, the power series expressions for the Legendre functions, valid near the origin, have to be matched with the solutions valid at infinity. We should thus obtain a number of power series, each valid in some finite region, and representing in their totality a solution of a differential equation.

But, although such an analytic solution is admirably adapted to the calculation of the numerical value of the solution at any given point, considerable research may still be required to determine certain important general characters of the solution. Thus, for example, it by no means leaps to the eye that the series

$$y = \sum_n (ix)^n/n! \quad \text{for} \quad \exp ix$$

which satisfies the equation $y'' = -y$ is a periodic function of x.

It was Poincaré who was the first to recognize clearly the need for a 'global' theory of differential equations, which should exhibit the qualitative features of all the solutions of a given equation throughout the whole domain of the dependent variables. It seems that he was led to investigate this question from the consideration of certain problems of celestial mechanics, especially the problem of three bodies (treated as point-masses) moving under Newtonian attraction. A number of problems of considerable importance in astronomy, such as the existence of periodic orbits, the capture of comets and the possible escape of planets, can only be examined with great difficulty, if at all, by means of solutions expressed as convergent series. The analytic solutions are too detailed and microscopic in character to furnish the synoptic view demanded by these global problems. The basic techniques for the solution of these problems were invented by Poincaré (1881).

13.11. The Problems of Celestial Mechanics

In 1892, 1893 and 1899 Poincaré published his three volumes on *Les Méthodes Nouvelles de la Mécanique Céleste*—a work which marks an epoch in this subject, because the author not only gives a masterly summary of preceding investigations, but also describes his own important contributions. The historian can scarcely do better than paraphrase the text of the introduction.

According to Poincaré, the final objective of celestial mechanics is to solve the 'three-body problem' and to determine whether Newton's law of attraction is sufficient to account for all astronomical phenomena. The 'two-body problem' of the motion of a single planet or comet around a sun is exactly soluble in terms of the well-known elliptic, parabolic or hyperbolic orbits and has long been reduced to an exercise in particle dynamics, as the attracting bodies can be treated as point masses. But the three-body problem, say of the Sun, Earth and Moon, has proved most intractable, even when one of the bodies has an 'infinitesimal' mass compared with the other two.

Vibration Theory

13.12. *The Phase Plane*

The simplest problems in the theory of vibrations relate to the movement
of a particle of mass m under the action of a restoring force, depending on
its displacement x; the equation of motion is therefore of the form

$$\ddot{x} = f(x, \dot{x}),$$

since, by Newton's laws of motion, these forces are balanced by what the
engineer calls the 'reversed effective force'.

At some time long before the present history begins some unknown
genius tore this second-order equation into two first-order equations of
the form

$$\dot{x} = u$$

$$\dot{u} = f(x, u).$$

The state of the vibrating system at any time t is completely specified
by the displacement $x(t)$ and the velocity $u(t)$, and the vibrations can be
represented geometrically by plotting the curve in the x,u-plane with the
parametric equations

$$x = x(t), \qquad u = u(t).$$

The x,u-plane is called the 'phase' plane, a term which Gibbs borrowed
from the astronomers, who had long referred to the varying appearances
of the moon as its different 'phases'. In his book on the *Elementary
principles in statistical mechanics* (Gibbs 1902) he had to consider a gas
consisting of a 'great number of independent systems, identical in nature,
but differing in phase, that is, in their condition with respect to
configuration and velocity'. The use of the word 'phase' is, of course, not
found in Poincaré's paper, but the concept is fundamental in his
researches.

The simplest example is clearly the 'simplest harmonic motion' described by the equations

$$\ddot{x} = -\omega^2 x \quad \text{or} \quad \dot{x} = u, \ \dot{u} = -\omega^2 x,$$

and represented in the phase plane by the ellipses

$$\omega^2 x^2 + u^2 = \text{constant}.$$

The concept of the phase plane had also been of great service in the
discussion of first-order equations of the form

$$dy/dx = f(x, y),$$

where the function f, although not holomorphic for all finite values of x
and y, could nevertheless be expressed in the form

$$f(x, y) = Q(x, y)/P(x, y),$$

where P and Q have this desirable character. The equations

$$\dot{y} = Q, \qquad \dot{x} = P$$

have solutions which are valid even at the poles of f (i.e. the zeros of P), and thus reveal global properties of the solutions of the original equation $dy/dx = f$.

13.13. Perturbation Theory

The technique of the phase plane has also been applied in perturbation theory. In the simplest case we have a differential equation

$$dy/dx = f(x,y,\varepsilon)$$

which involves a parameter ε, which 'perturbs' the original equation

$$dy/dx = f(x,y,0).$$

If the function f is holomorphic in ε near $\varepsilon = 0$ the perturbation is 'regular' and there is little difficulty in obtaining a solution $y = y(x,\varepsilon)$ expressed as a convergent power series in ε. But if f regarded as a function of ε, has a singularity at $\varepsilon = 0$ the perturbation is 'irregular' and it may be possible to write f in the form $f = Q(x,y,\varepsilon)/P(x,y,\varepsilon)$ when P and Q are holomorphic in ε at the origin. The equations $\dot{x} = P$, $\dot{y} = Q$ can then be solved in powers of ε.

The study of irregular perturbations received a new impetus from the work of Lighthill (1949) who made numerous applications to physical problems, especially in aerodynamic theory. Lighthill's method, which was brilliantly empirical in character, was subsequently rationalized by Wasow (1955). An alternative approach can be illustrated by the equation

$$(x + \varepsilon y)dy/dx + (2 + x)y = 0$$

with the condition $y = \varepsilon^{-1}$ at $x = 1$.

All solutions of the unperturbed equation (with $\varepsilon = 0$) are of the form $y = Cx^{-2}e^{-x}$, where C is a constant. However the equations

$$dy/dt = -(2 + x)y, \quad dx/dt = x + \varepsilon y$$

possess solutions which can be expressed as power series in ε in the form

$$x = z + \varepsilon\varphi(z) + \varepsilon^2 + \dots \quad (t = \log z)$$

$$y = z^{-2}e^{-z} - \varepsilon z^{-2}e^{-z}\int_1^z \varphi(t) \, dt,$$

where
$$\varphi(z) = \int_1^z s^{-4}e^{-s} \, ds.$$

Hence, near $z = 0$, $\qquad x = z - \tfrac{1}{3}\varepsilon z^{-2} + 0(\varepsilon^2/z^4)$

and $\qquad\qquad\qquad y = z^{-2} - \tfrac{1}{6}\varepsilon z^{-4} + 0(\varepsilon^2/z^6),$

and, at $z = 0$ $\qquad\qquad y = (3/\varepsilon)^{\frac{2}{3}} + 0(\varepsilon^{-\frac{1}{3}}).$

13.14. Poincaré and the Phase Portrait

In his famous 'Mémoire sur les courbes définis par une équation différentielle' Poincaré (1881) concentrated his attention on the simplest type of non-linear differential equation. As a result of his studies in celestial mechanics his studies were directed to the pair of dynamical equations

$$\dot{x} = P(x,y), \quad \dot{y} = Q(x,y)$$

rather than to the single equation $dy/dx = Q/P$. The pair of equations are 'autonomous', inasmuch as the functions P and Q are independent of the independent variable t, and depend only on two real variables x and y, and are restricted to be polynomials.

The totality of the curves defined by the equation

$$dy/dx = Q/P$$

constitutes the 'phase-portrait' of the solutions, and Poincaré's method enables us to obtain the general features of all these curves (the 'allure' of the whole family of paths). In fact he showed that there are only two species of solutions:

(1) the closed paths of 'limit cycles', and
(2) the paths which tend as $t \to \pm\infty$, either to a limit cycle, or to a critical point at which $P = 0$ and $Q = 0$.

Lefschetz (1967) noted that Poincaré repeatedly refers to 'general' situations and that he implicitly anticipates the definition of 'generic points'. In fact for Poincaré a point M is general if the functions $P(x,y)$ and $Q(x,y)$ have the same relevant properties in some neighbourhood of M as they have at the point M itself.

Poincaré, who seems to be unaware of the earlier work of Kronecker (1869) or of Picard's account (Picard 1891a, b, c) developed a theory of the index of a vector field in two dimensions. If (P,Q) are the Cartesian components of the vector field at a point M, Poincaré considers the variations in the expression Q/P as the point M describes a closed curve C in the positive direction, i.e. so that the interior of C is always on the left-hand side of the direction of motion of M. If h is the number of times the expression Q/P suddenly changes from $-\infty$ to $+\infty$, and if k is the number of times in which it changes from $+\infty$ to ∞, then the integer $i = \frac{1}{2}(h - k)$ is the Poincaré index of the curve C for the given vector field.

The critical points of the vector field are those at which both P and Q vanish. If a critical point A is isolated then all the simple closed curves which enclose no other critical point than A have the same Poincaré index, which may therefore be called the index of A. If $\varphi(x,y) = $ constant is the equation of the trajectories of the differential equation

$$dx/P = dy/Q$$

then the phase portrait may be considered to provide the contour lines, $\varphi(x,y) = $ constant, of the surface $z = \varphi(x,y)$ and the simplest classification of the critical points, at which $\partial\varphi/\partial x = 0$, $\partial\varphi/\partial y = 0$ are (1) the maxima, (2) the minima, (3) the cols or saddle points of the surface.

Poincare's classification is much deeper and depends on the character of the matrix

$$M = \begin{pmatrix} a & b \\ c & d \end{pmatrix}$$

formed by the leading terms of the power series for P and Q, viz.

$$P = ax + by + \ldots,$$
$$Q = cx + dy + \ldots.$$

(the critical point being taken at the origin). The simplest types of critical points (for which the matrix M is not singular), are exemplified by the following five cases:

(1) a 'centre', with $\dot{x} = -y$, $\dot{y} = x$, $x^2 + y^2 = $ const. (index $= +1$),
(2) a 'focus', with $\dot{x} = -x-y$, $\dot{y} = x - y$, with $r = (x^2 + y^2)^{\frac{1}{2}} = Ae^{-t}$, (index $= +1$),
(3) a 'node', with $\dot{x} = x$, $\dot{y} = y$, $y/x = $ const. (index $= +1$),
(4) an 'improper node', with $\dot{x} = x$, $\dot{y} = \lambda y (\lambda \neq 1)$, $y = Cx^\lambda$ (index $= +1$),
(5) a 'col' or 'saddle point', with $\dot{x} = x$, $\dot{y} = -y$, $xy = $ const. (index $= -1$).

In the delineation of the phase portrait of a differential system it is necessary first of all to sketch the 'outlines', i.e. the trajectories which terminate at critical points (which must be saddle points). These curves are now known as 'separatrices'.

But it is only in the simplest problems that the 'outlines' are sufficient to determine even the main features of the phase portrait, and Poincaré therefore had recourse to an auxiliary 'topographical system' consisting of a family of closed curves, one of which passes through each point of the x,y plane. The points in which the paths of the differential equation are tangential to the curves of the topographical system can be determined without solving the equation. These points of contact lie on the 'contact curve', which is a valuable accessory in sketching the phase portrait.

13.15. Bendixson and Higher Critical Points

Fifteen years after Poincaré's pioneer investigations, Bendixson (1901) completed the analysis of critical points by making a more detailed classification.

The neighbourhood of a critical point is divided into sectors by the paths (separatrices) which actually attain the critical point. In each sector the paths may exhibit a hyperbolic character, with the separatrices as asymptotes, or an elliptic character with each path emerging from and returning to the critical point. The Bendixson index is

$$i = 1 + \tfrac{1}{2}(e - h)$$

where e and h are the numbers of elliptic and hyperbolic sectors at the critical points. When the functions P and Q are polynomials, the only three possible types are nodes with $i = 1$, saddle-points with $i = -1$, half-node/half-saddle with $i = 0$.

13.16. *Trajectories on Arbitrary Surfaces*

Poincaré also studied differential manifolds and established the fine theorem that the sum of the indices of the critical points (when finite in number) is equal to the Poincaré–Kronecker 'characteristic' $2 - 2p$ of the surface (p being its 'genus'). The generalization of the result to n-dimensional manifolds is due to Lefschetz (1963).

13.17. *Liapunov and Stability*

Earlier studies on the stability of points of equilibrium of a conservative dynamical system had been confined to the examination of the leading linear terms in the Lagrangian equations,

$$\ddot{x}_k = - \sum_j V_{kj} x_j \quad k = 1, 2, \ldots n.$$

The nature of the general solution

$$x_k = \sum_j A_j e^{\lambda_j t}$$

depends upon the characteristic roots, $\lambda_1, \lambda_2, \ldots, \lambda_n$, of the matrix associated with the equation

$$\sum_j V_{kj} A_j = \lambda^2 A_n,$$

and for complete stability the real parts of the characteristic roots must be negative or zero.

This investigation is clearly insufficient, and the only alternative seemed to be the construction of explicit solutions of the full and exact equations of the system.

The great achievement of Liapunov (1892) was a complete theory of stability, based on the full equations of movements and a technique which did not require the explicit knowledge of the exact solutions.

To discuss the system described by the vector equation $\dot{x}_k = f_k(x,t)$. Liapunov's technique consists of the construction (by imaginative intuition) of a function $V(x,t)$ which is positive definite, such that $V(0,t) = 0$ for $t > 0$, and whose Eulerian derivative

$$DV = \sum_k f_k \partial V/\partial x_k + \partial V/\partial t$$

is negative definite. The existence of such a function is a necessary and sufficient condition for the uniform asymptotic stability of the point $x_k = 0$.

13.18. *Non-Linear Vibrations—the Experimental Foundation*

Linear vibrations are exemplified by the small oscillations of a simple pendulum, where the angular displacement θ satisfied the equations

$$\ddot{\theta} = n^2 \theta, \qquad \theta = A \cos n(t - t_0),$$

where $n^2 = g/l$, g being the acceleration due to gravity and l the length of the pendulum. Such linear vibrations have two remarkable characteristics, viz. the frequency, $n/2\pi$, is independent of the initial conditions, but the amplitude A is entirely dependent on the initial conditions,

$$\theta = A \cos nt_0 \quad \text{and} \quad \dot\theta = -An \sin nt_0, \text{ at } t = 0.$$

These properties sharply differentiate vibrations controlled by a linear restoring force (here proportional to θ) and those controlled by a non-linear restoring force. The latter case is realized in the exact equations of the simple pendulum

$$\ddot\theta = -n^2 \sin\theta, \quad \sin\tfrac{1}{2}\theta = \sin\tfrac{1}{2}A \operatorname{sn} n(t - t_0)$$

where sn is Jacobi's elliptic function, with modulus $k = \sin\tfrac{1}{2}A$ and frequency

$$\tfrac{1}{2}\pi n / K \quad \left[K = \int_0^{\frac{1}{2}\pi} (1 - k^2 \sin^2 \varphi)^{-\frac{1}{2}} \, d\varphi \right].$$

Thus the frequency varies with the amplitude, and the amplitude is determined by the initial conditions.

The theory of the small motion of the pendulum was studied by Galileo in 1632 and the theory of finite motion was studied by Euler in 1736. The vibrations of the 'sleeping' top were also studied by Euler in 1758, and the general motion of a rigid body under no external forces was studied by Jacobi in 1849. But these problems mainly afford exercises in the use of elliptic functions, and the general motion of a rigid body under gravity, with one arbitrary point fixed, requires the theory of Fuchsian functions.

While pure mathematicians were studying vibration problems from the standpoint of elliptic and hyperelliptic functions, astronomers had developed a general theory of the non-linear oscillations which are produced by the perturbing influence of, say, the sun on the orbit of the moon around the earth, or of one planet on the orbit of another planet. The remarkable success of these investigations is due in part to the fact that the mathematical expressions for these perturbations are controlled by small parameters, and that adequate solutions can be found by methods of successive approximations. Poincaré's treatise, *Les Méthodes Nouvelles de la Mécanique Céleste* (Poincaré 1892a), provides a masterly account of these investigations.

Another powerful impetus to the study of non-linear oscillations was provided by the experimental work of Lord Rayleigh on vibrating sources of sound and by the observations of engineers on the vibrations of heavy machinery (Rayleigh 1877). The engineering approach to the subject is exemplified by Duffing (1918).

A new phase of research was inaugurated by van der Pol's investigations of the oscillations in any electric system containing a 'vacuum tube' or triode valve (see Section 13.29).

13.19. The Method of the 'Variation of Constants'

In view of the immense volume of literature on celestial mechanics it is hazardous to attempt a summary of its influence on 'pure' mathematics, but two methods for the investigation of differential equations do seem to merit especial attention, viz. the method of the 'variation of constants' and the related method of parametric representation. Both of these methods were devised for application to the problem of perturbation. The simplest example is the non-linear oscillations controlled by an equation of the form

$$\ddot{x} + n^2 x = f(x,t).$$

The vibrations, unperturbed by the non-linear term $f(x,t)$ are given by the relation

$$x = A \cos nt + B \sin nt,$$

where A and B are constants. A systematic method of allowing for the influence of the non-linear term is to search for a solution of the form

$$x = A(t) \cos nt + B(t) \sin nt.$$

This leads to differential equations for the two functions $A(t)$ and $B(t)$, and these equations determine these functions definitely only when some other condition is imposed. It is this arbitrariness in the method which gives it such flexibility.

The most convenient auxiliary condition is

$$\dot{A} \cos nt = \dot{B} \sin nt$$

which leads to the equation

$$f(x,t) = -n\dot{A} \sin nt + n\dot{B} \cos nt$$

whence
$$x = n^{-1} \int^t f(x,s) \sin n(t-s) \, ds.$$

13.20. *The Method of 'Parametric Expansion'*

In calculating the perturbations of the small planets there arises a difficulty that a straightforward if naïve process of successive approximation gives rise to 'secular' terms in the expressions for the coordinates of the planet. To avoid this difficulty Linstedt (1883) transformed the equations of motion by introducing a new variable in the place of time. Thus the coordinates of the planet are no longer determined as functions of the time t, but these coordinates, and the time t itself, are determined as functions of a new parameter τ. Poincaré (1892b) gave a critical account of Linstedt's method.

The difficulty and complex problems of celestial mechanics are essentially the problems of the existence and stability of periodic solutions of differential equations, and such questions lead Poincaré to examine the simpler problems of non-linear vibrations of one-dimensional systems (Section 13.14).

The astronomical examples of Linstedt's method are rather too complex and of technical interest to be cited here, but the method is adequately described by the application made by Rayleigh to the non-linear vibrations of a tuning fork (Rayleigh 1894).

The vibrations of the tuning fork are governed by a non-linear equation of the form

$$\ddot{u} + n^2 u + \beta u^3 = 0$$

where β is small, and the first approximation is given by

$$\ddot{u}_1 + n^2 u_1 = 0, \qquad u_1 = A_1 \cos nt.$$

It is essential to the Linstedt–Rayleigh method to note that the effect of the non-linear term βu^3 is not only to introduce harmonics in addition to a fundamental $C \cos vt$, but also to change the frequency of the fundamental from $2\pi/n$ to a value $2\pi/v$, which is to be determined later on.

Neglect of this effect would lead us to determine a second approximation from the equation

$$\ddot{u}_2 + n^2 u_2 = -\beta u_1^3$$
$$= -\tfrac{1}{4}\beta A_1^3 \cos 3nt - \tfrac{3}{4}\beta A_1^3 \cos nt$$

whence u_2 would contain a 'secular' term of the form $t \sin t$.

The correct procedure is to express u_2, in the term βu_2^3 in the form $A \cos vt$ with $v = n + \beta p + 0(\beta^2)$ and solve the equation

$$\ddot{u}_2 + n^2 u_2 = -\tfrac{1}{4}\beta A^3 \cos 3vt - \tfrac{3}{4}\beta A^3$$

The method of 'parametric expansion' requires for its formal justification a rigorous existence theorem and this was provided by Poincaré (1892a) for equations of the form

$$\ddot{u} + u = \mu f(u, \dot{u}).$$

Poincaré's proof depends upon the transformation suggested by the work of Linstedt (1883), which introduces the variable $\psi = \omega t$ instead of the time, ω being the *unknown* frequency of the periodic solution. Poincaré then shows that, under appropriate restrictions on f, there exists a periodic solution of the form

$$u = u_0(\psi) + \Sigma \mu_n u_n(\psi)$$

where u_n has the period 2π in ψ and $u_0 = A \cos \psi$.

Many problems of non-linear oscillations require the solution of an equation of the form

$$\ddot{x} + ax = \varepsilon f(x, \dot{x}, \omega t)$$

where the function f is periodic in t with period $2\pi/\omega$ and where ε is a small parameter. The fundamental existence theorem is due to Friedrichs and coworkers (1942–3) and Stoker (1950) who developed a method adumbrated by Poincaré (1892a). If f is an analytic function of the complex variables $x, \dot{x}$ then there exists a periodic solution $x(t, \varepsilon)$ with period $2\pi/\omega$, which is an analytic function of t and ε, and such that $x(t, 0) = A_0 \cos \omega t$.

This periodic solution must have the form

$$x(t, \varepsilon) = \sum_{0}^{\infty} A_n \cos n\omega t + \sum_{1}^{\infty} B_n \sin n\omega t$$

where
$$A_0 = C_0 + \varepsilon C_1(\varepsilon),$$
$$A_1 = C + \varepsilon C_1(\varepsilon),$$
$$A_n = \varepsilon C_n(\varepsilon) \quad (n > 1),$$
and $\quad B_n = \varepsilon K_n(\varepsilon) \quad (n \geqslant 1).$

Influence of Planetary Theory

13.21. E. T. Whittaker's Criterion for Periodic Orbits

One of the earliest criteria for the existence of periodic orbits was given by E. T. Whittaker (1902, 1937). In the simplest case of a particle of unit mass and Cartesian coordinates x,y moving in a plane under the action of forces with a potential $V(x,y)$ the equation of energy is

$$\tfrac{1}{2}(\dot{x}^2 + \dot{y}^2) + V(x,y) = h.$$

The 'action' A around any simple closed curve C for particles of total energy h is given by the integral

$$A(C) = \oint_C \{h - V(x,y)\}^{\frac{1}{2}} \{(dx)^2 + (dy)^2\}^{\frac{1}{2}}.$$

If $A + \delta A$ is the action around any simple closed curve C' enclosing C, then

$$\delta A = \oint_C \{h - V(x,y)\}^{-\frac{1}{2}} K(C) \{(dx)^2 + (dy)^2\}^{\frac{1}{2}},$$

where $K(C) = \varrho^{-1}\{h - V(x,y)\} - \tfrac{1}{2}\partial V/\partial n$, $\partial V/\partial n$ being the derivative of V along the outward normal to C and ϱ is the radius of curvature of C at (x,y).

Whittaker considers the aggregate of simple closed curves situated in the ring-shaped space bounded by two curves C_1 and C_2, such that $K(C_1)$ is negative and $K(C_2)$ is positive. Then the values of the action for such curves are bounded and must possess a minimum. In view of the values of $K(C_1)$ and $K(C_2)$ this minimum is not attained on either of the boundary curves C_1 or C_2. Whittaker, following the precedent of Riemann in the potential problem, assumes that the minimum is actually attained on some curve C in the annulus bounded by C_1 and C_2. Then it follows from the properties of the action function that the curve C is a possible orbit, and it is manifestly periodic.

13.22. Problems of Convergence

The method of successive approximations is rarely pursued beyond the second or third term and it is evident that the value of these results is entirely dependent on an investigation of the convergence of the whole infinite series of approximations.

Thus the exact solution of the equation

$$\ddot{x} + x = (2\varepsilon - \varepsilon^2)x, \quad x(0) = A, \quad \dot{x}(0) = 0,$$

is
$$x = A \cos(1 - \varepsilon)t$$
$$= A(1 - \tfrac{1}{2}\varepsilon^2 t^2 + \ldots) \cos t + A(\varepsilon t - \tfrac{1}{6}\varepsilon^3 t^3 + \ldots) \sin t.$$

The first approximation is $x_1 = A \cos t$ and the second, derived from the equation

$$\ddot{x}_2 + x_2 = 2\varepsilon x_1$$

is
$$x_2 = A \cos t + A\varepsilon t \sin t.$$

The presence of the secular term, $A\varepsilon t \sin t$, is disturbing, but it gives correctly the *initial* trend of the exact solution, although it is useless for large values of t. Linstedt's technique introduces the new variable $\tau = \tau(t)$ with $\tau = 1 - \varepsilon$ and thus obtains a solution $A \cos \tau$ valid for large τ (and in fact for all t).

Poincaré (1886) noted that many of the solutions in series obtained by astronomers were in fact divergent, but that they possessed the valuable property of *asymptotic* convergence, i.e. for such a series

$$\sum_{k=1}^{\infty} u_k(t)$$

the partial sums

$$s_n(t) = \sum_{k=1}^{n} u_k(t)$$

approach asymptotically a sum function $f(t)$ i.e. the ratio

$$\{f(t) - s_n(t)\}/t^n \to 0 \quad \text{as} \quad n \to \infty.$$

He proved that such asymptotic series can be added, multiplied and integrated term by term to yield new asymptotic series, and, what is far more important, he identified the characteristic property of the asymptotic series most commonly obtained in analysis, viz. that the successive terms begin by decreasing with great rapidity and then increase beyond all limit. The most familiar example is Stirling's series for $\log x$!

13.23. *The Problem of 'Small Divisors'*

The mathematical problems of celestial mechanics required the solution of the equations of motion of N mass points attracting one another according to Newton's law of gravitation. The differential equations of motion have the form (in an obvious notation)

$$m_k \ddot{x}_k = \partial U / \partial x_k, \qquad k = 1, 2, \ldots, N,$$

where
$$U = \sum m_k m_l |x_k - x_l|^{-1}$$

and each x_k is a three-dimensional Cartesian vector. The main problem is to discuss the stability of solutions, i.e. are there solutions which do not experience collisions and do not escape? For such solutions

$$\max_{k<l} (\tau_{kl}, \tau_{kl}^{-1})$$

is bounded for all t, where

$$\tau_{kl} = |x_k - x_l|.$$

Even in the case in which the mass ratios, m_k/m_N, of $N-1$ planets compared with a sun, are small, the problem is of great difficulty and has only recently received a definitive solution. In general the solution (if stable) will not be simply periodic but quasi-periodic, in the sense that each vector function $x_k - x_N$ can be represented by series of the form

$$\sum_j c_j \exp i(j_1\omega_1 + j_2\omega_2 + \ldots + j_s\omega_s)$$

where the periods $2\pi/\omega_k$ are all real, the j_k's are integers and $\sum_j |c_j|$ is convergent, and $j = (j_1, \ldots, j_n)$.

Unfortunately the usual analytic process for the determination of the coefficients c_j leads to fractions in which the denominators are of the form $\sum_k j_k\omega_k$. Even if the ω_k are rationally independent, these expressions can become arbitrarily close to zero, and thus become 'small divisors', which make any discussion of convergence extremely delicate and difficult.

A magisterial account of the whole subject has been given by Moser (1968) to whom we are indebted for the following extracts.

Laplace had attempted to construct a proof of convergence and Weierstrass claimed that he was able to construct solutions but could not establish their convergence. Poincaré attempted unsuccessfully to solve this problem, but was thereby stimulated to initiate a number of entirely new lines of research, such as the qualitative or geometric theory of differential equations and a number of 'fixed-point' theorems for the existence of periodic solutions.

The successful solution of the problem of small divisors is due to Siegel, Kolmogorov, Arnold and Moser. One of the most interesting techniques which can be used in the proof is the famous 'twist theorem', stated by Poincaré, later proved by G. D. Birkhoff (1913) and further developed and applied by Moser (Section 13.24).

This theorem concerns the one–one mapping of the annulus $a < r < b$ into itself. We use polar coordinates (r,θ) and write the mapping in the form

$$r \to r_1 = f(r,\theta), \quad \theta \to \theta_1 = \theta + g(r,\theta).$$

The functions f and g are continuous and of period 2π in θ. Moreover the mapping preserves the area of corresponding regions. But the crucial property of the mapping is that the boundaries of the annulus, $r = a, r = b$, are mapped into themselves in opposite directions, i.e. $g(a,\theta) g(b,\theta) < 0$. Then, under these conditions the mapping possesses at least one fixed point in the interior of the annulus.

13.24. Mappings, Flows and Poincaré's 'Consequent Points'

The dynamical manifold or phase space has as coordinates the generalized coordinates $q_1, q_2, \ldots, q_n$ and momenta $p_1, p_2, \ldots, p_n$ of a conservative dynamical system with n degrees of freedom. The motion of an individual system is represented by a curve or orbit or trajectory in the phase space, and, since the time of Liouville, the totality of all these trajectories has been regarded as the set of paths of particles of a continuous incompressible fluid.

The close connection between such flows and certain area-preserving mappings is easily envisaged for systems with two degrees of freedom. All systems with the same total energy E move on the same three-dimensional region $E(q_1, q_2, p_1, p_2) = $ constant. Let this region be intersected by some two-dimensional manifold σ. Let the trajectory which interests σ in a point P_0 next intersect it in the consequent point P_1. Then the flow determines a mapping T of σ into itself, such that $P_1 = TP_0$. Clearly periodic motions correspond to fixed points of the mapping T.

Since the flow in the phase space or on the energy manifold $E = $ constant is incompressible, 'tubes of flow' will always have the same cross-section and the mapping T will preserve areas on the surface σ. Thus, as Poincaré was the first to show, area-preserving mappings provide a powerful technique for the determination of periodic motion.

13.25. Poincaré's Geometric Theorem

Poincaré's famous geometric theorem (Poincaré 1912b) has a special interest as one of his last researches on the 'three-body problem' of mathematical astronomy, as an incomplete investigation which remained as a challenge to mathematicians until it was taken up by G. D. Birkhoff (1913), and as an important contribution to the theory of geodesics.

The theorem relates to 'twist' transformations T of an annulus. Let T be a one–one differentiable transformation of the ring,

$$0 < a \leqslant x \leqslant b, \qquad 0 \leqslant y \leqslant 2\pi$$

into itself, where x, y are the usual polar coordinates r, θ i.e. $T: x \to X$, $y \to Y$. Then the boundary $x = a$ must transform into $X = a$ and $x = b$ into $X = b$ (otherwise T would be biunifom). Suppose, moreover, that under the transformation T the points of $x = a$ advance in the clockwise sense so that $Y < y$ and that the points of $x = b$ retreat in the counter-clockwise sense so that $Y > y$. Then the annulus will suffer a species of twist or rotatory shear. Finally, suppose that the transformation T preserves areas so that, if any region σ is transformed into a region $\Sigma = T\sigma$ then the area of Σ is equal to the area of σ.

Then Poincaré asserted that the transformation T always possesses two fixed points in the interior of the annulus $a \leqslant x \leqslant b$.

There may be more than two fixed points. In fact the transformation

$$X = x, \quad Y = y + \lambda(2x - a - b), \quad \lambda > 0$$

carries each point of the curve $x = \frac{1}{2}(a - b)$ into itself, and it is clear that, if $x = a$, $Y = y - \lambda(b - a)$, and if $x = b$, $Y = y + \lambda(b - a)$. Moreover

$\partial(X,Y)/\partial(x,y) = 1$ so that the transformation preserves the magnitude of areas.

Poincaré never succeeded in proving this theorem, but he verified it in a number of special cases, he was well aware of its importance for the proof of the existence of periodic solations of the 'three-body' problem, and he gave cogent reasons for his conviction that the theorem is true in general.

The first proof of Poincaré's geometric theorem was given by G. D. Birkhoff (1913, 1926), and followed by generalizations and related theorems due to Moser (1962).

Poincaré (and George D. Birkhoff) applied the geometric theorem to establish the existence of closed geodesics on a closed convex surface, but the simplest application is to Birkhoff's example of the motion of a billiard ball on a perfectly smooth convex billiard table with a perfectly elastic boundary. The billiard ball projected from a point P_0 on the boundary strikes the boundary again at a point $P_1 = TP_0$ and a simple modification of the geometric theorem shows that the transformations T^2, T^4, T^6 ... possess two invariant 'points' corresponding to two closed polygonal paths. Even in this case the analysis, although straightforward, is somewhat complex.

13.26 *George D. Birkhoff and Dynamical Systems*

The *Colloquium Lectures* which George D. Birkhoff delivered in 1927 provide a masterly survey of the advances made since the time of Poincaré and a modest summary of the contributions made by the author himself (G. D. Birkhoff 1927).

The earlier chapters introduce the generalization of Hamiltonian systems which arise from the more general Pfaffian variational principle

$$\delta \int_{t_0}^{t_1} \left\{ \sum_j P_j p'_j + Q \right\} dt = 0 \quad (p'_j = dp_j/dt)$$

where P_j, Q are functions of $p_1, \ldots, p_n$, and n is even. These systems have the advantage that they retain their form under any general point transformation of the system.

The properties of these systems, in the neighbourhood of a point of equilibrium or of a periodic solution, which are invariant under the group of analytic point transformations, are investigated in a purely formal manner, all questions of convergence being systematically ignored.

The final chapters attack the general problem of the 'qualitative determination of all possible types of motions' and their interrelations. The theory (which in essence is due to Poisson and was greatly extended by Poincaré) considers the motion of the representative points in the dynamical manifold as the motion of an incompressible fluid and concentrates attention on the small 'molecule' of fluid initially surrounding a point P_0. It is easy to prove that (if the dynamical manifold is bounded and closed) at some subsequent time this molecule must occupy a position overlapping its initial position. Thus any trajectory eventually returns to points which are arbitrarily near its original position.

Birkhoff greatly extended this theory to arbitrary dynamical systems and showed that they always possess a closed set of 'central motions' which possess this property of regional recurrence, towards which all other motions of the system in general tend asymptotically.

The later chapters, which contain the main achievements of the author, deal with the various general methods by which the existence of periodic motions can be established. These are the 'method of the minimum', the 'minimax method', and the fixed-point method.

The first method is a generalization of the well-known principle that a stretched elastic loop slipped over a smooth surface such as a dumb bell will take up a position of minimum length and form a geodesic corresponding to a periodic motion. In the second method Birkhoff considers the same problem for a 'torus-shaped surface' with any number of constrictions. The positions of equilibrium correspond to maxima, minima or minimaxes of the length of the elastic loop. This problem of global analysis has been the subject of extensive and detailed treatment by Morse (1934, 1969, 1973).

The third method, due to Poincaré, considers the intersections of the trajectory of a representative point with a suitable 'surface of section'. These successive intersections determine a transformation of the surface of section into itself.

13.27. Moser's Twist Theorem

The twist theorems of Poincaré and George D. Birkhoff established the existence of two fixed points, but for stability studies it is more important to discover invariant curves which separate the boundaries of the annulus. Moser (1973) noted that up to that date the existence of such curves had been established only for small perturbations of the elementary twist mappings

$$r_1 = r, \quad \theta_1 = \theta + \alpha(r).$$

These perturbations are of the form

$$r_1 = r + \varepsilon^\sigma f(r,\theta,\varepsilon)$$

$$\theta_1 = \theta + \alpha + \varepsilon^\varrho \gamma(r) + \varepsilon^\sigma g(r,\theta,\varepsilon)$$

where $0 \leqslant \varrho < \sigma$. A trivial example is the 'rotary shear'

$$r_1 = r, \quad \theta_1 = \theta + \lambda(2r - a - b)$$

with the invariant curve $r = \frac{1}{2}(a + b)$.

Non-linear Springs and Damping

13.28. Helmoltz and Combination Tones

Helmholtz observed in his studies of physiological acoustics that the ear often 'hears' sounds of frequencies which are not present in the incident sounds and he showed that this effect is due to the non-linearity of the forces which control the vibrations of the tympanic membrane (Helmholtz 1862).

Helmholtz gives the example of the system governed by an equation of the form

$$\ddot{x} + a^2 x + bx^2 + \varepsilon f \sin pt + \varepsilon g \sin (qt + c) = 0$$

and constructs the first few terms of a solution in powers of ε, viz.

$$x = \varepsilon x_1 + \varepsilon^2 x_2 + \ldots.$$

The first term contains only terms in

$$\sin at, \ \cos at, \ \sin pt \quad \text{and} \quad \sin (qt + \varepsilon)$$

but the second term contains terms in

$$\cos 2pt, \ \cos 2(qt + c), \ \cos [(p - q)t + c], \ \cos [(p + q)t + c]$$

and thus exhibits combination tones of frequencies $(p \pm q)/2\pi$.

13.29. *Duffing's Equation*

Duffing had the happy idea of constructing an exact analytic solution of the equation

$$\ddot{y} + Ay + By^3 = 0$$

by introducing elliptic functions Duffing (1918). Unfortunately he chose to work with Weierstrassian functions which are notoriously unwieldly in application. Duffing's result can be exhibited more concisely with the Jacobian functions *snt*, *cnt* and *dnt* which satisfy the equations

$$\begin{array}{ll} d\textit{snt}/dt = \textit{cnt dnt}, & \textit{sn}^2 t + \textit{cn}^2 t = 1, \\ d\textit{cnt}/dt = -\textit{snt dnt}, & h^2 \textit{sn}^2 t + \textit{dn}^2 t = 1, \\ d\textit{dnt}/dt = -k^2 \textit{snt cnt}, & \textit{sn}(0) = 0. \end{array}$$

From these equations it easily follows that the function $x = \textit{snt}$ satisfies a Duffing equation

$$\ddot{x} + (1 + k^2)x - 2k^2 x^3 = 0$$

with a period equal to $4K$ where K is the complete elliptic integral with modulus k.

The possibility of subharmonic resonance is illustrated by the rather artifical equation

$$\ddot{x} + 4\pi^2 x + 2\lambda x^2 = \lambda(1 + \cos 4\pi t)$$

in which the exerting force $\lambda(1 + \cos 4\pi t)$ with frequency 2 can produce a resonant vibration with frequency 1. Such subharmonic resonance has been observed in electric circuits with triode valves, but is extremely difficult to obtain—a fact not unrelated to the very special character of the example given above.

13.30. *Van der Pol and Relaxation Oscillations*

The experimental researches of Appleton and Greaves (1923), Appleton and van der Pol (1922) and Greaves (1923a, b) on the oscillations of an electric circuit containing a triode valve focused attention on the non-linear equation

$$\ddot{v} - \varepsilon(1 - v^2)\dot{v} + v = f(t)$$

where $f(t)$ is zero for free oscillations and of the form $A \sin \omega t$ for forced oscillations. This equation exhibits the effect of non-linear damping (proportional to $\dot{v}$) on the well-known isochronous system $\ddot{v} + v = 0$. The damping is negative for small values of v and positive for large values so that we may expect stable oscillations of some definite finite amptitude.

The methods of solution invented by van der Pol (1926, 1927) (first published in Dutch in 1920) are remarkable for their simplicity, their reliance on inspired guesswork, and their extraordinary agreement with experimental observations and subsequent more rigorous theory.

To discuss free oscillations, van der Pol assumed a solution of the form

$$v = a \sin t$$

where a is a 'slowly varying' amplitude and neglected terms in $\dot{a}$ or $\ddot{a}$ in comparison with a, and ignored terms in the higher harmonic, $\cos 3t$, to obtain the simple approximate equation

$$2a\dot{a} - \varepsilon(a^2 - \tfrac{1}{4}a^4) = 0$$

whence $\quad a^2 = 4\{1 + \exp(-\varepsilon t)\}.$

This represents the gradual development of a stable oscillation,

$$v = 2 \sin t$$

with an amplitude independent of the initial conditions.

To provide further evidence for this conclusion van der Pol devised an 'averaging' method, in which the original equation for v is multiplied by v or by $u = \int v\, dt$ and integrated over the unknown period T of the oscillations $v = a \sin (t/T)$. This yields the exact equations

$$\overline{\dot{v}^2} = \overline{v^2}$$

and
$$\overline{v^2} = \tfrac{1}{3}\overline{v^4}$$

whence $T = 1$ and $a^2 = 4.$

Finally the argument is clinched by actually plotting the isoclines of the exact equation

$$dz/dv - \varepsilon(1 - \dot{v}^2) + v/z = 0 \qquad (z = \dot{v})$$

for various values of ε and thereby constructing graphical solutions which confirm the bold approximations of the previous arguments.

13.31. Van der Pol and Forced Oscillations

To discuss the forced oscillations in an electric circuit containing a 'vacuum tube', which are governed by an equation of the form

$$\ddot{v} - \alpha\dot{v} + 3\gamma v^2\dot{v} + \omega_0^2 v = B\omega_0^2 \sin \omega_1 t,$$

van der Pol adopted the same bold step of assuming a solution of the form $v = b_1(t) \sin \omega_1 t + b_2(t) \cos \omega_1 t$ with 'slowly varying' functions $b_1(t), b_2(t)$, so that $\dot{b}_i$ and $\ddot{b}_i$ are small of the first and second order respectively in α, γ and B. If the 'detuning'

$$\Delta = 2(\omega_0 - \omega_1)$$

is also of the first order, we obtain the approximate equations

$$2\dot{b}_1 + b_2\Delta - \alpha b_1(1 - b^2/a_0^2) = 0,$$

$$2\dot{b}_2 - b_1\Delta - \alpha b_2(1 - b^2/a_0^2) = B\omega_0,$$

where $\frac{3}{4}\gamma a_0^2 = \alpha$ and $b^2 = b_1^2 + b_2^2$.

Van der Pol proceeded to analyse the stability of the periodic oscillations corresponding to $\dot{b}_1 = 0$, $\dot{b}_2 = 0$ by replacing b_1, b_2, by $b_1 + \delta b_1$, $b_2 + \delta b_2$, but a more satisfactory method was introduced by Andronov and Wit (1930) who studied van der Pol's equations for b_1 and b_2 by the methods given by Poincaré. A rigorous proof that the van der Pol equation does actually possess a stable, periodic solution has been given by Levinson and Smith (1942).

13.32. *Liénard's Equation*

In order to discuss the equation

$$\ddot{x} + f(x)\dot{x} + x = 0$$

Liénard (1928) introduced the 'kinetic' energy, $\frac{1}{2}y^2$, where

$$y = \dot{x} + F(x), \quad F(x) = \int_0^x f(s) \, ds,$$

and the potential energy $G(x) = \frac{1}{2}x^2$. Liénard imposed on the function $f(x)$ a number of restrictions, which are usually satisfied in practice, the most important of which is that $f(x)$ is a continuous function. It follows that if $x(t)$, $y(t)$ is a solution of the equations $\dot{x} = y - F(x)$, $\dot{y} = -x$, then so is $-x(t)$, $-y(t)$. By tracing the solution qualitatively from a point A, $x = 0$, $y = a$ to a point B where $\dot{x} = 0$ and on to C where $y = 0$. Liénard proves that there is a unique point $A(0,a)$ such that $\frac{1}{2}y^2 + \frac{1}{2}x^2$ has the same value at A and C whence there exists a periodic solution.

13.33. *Minorsky and the 'Stroboscopic Method'*

The rigorous and systematic examination of periodic solutions of non-linear differential equations requires complex and tedious calculations. It is therefore worthwhile to record the 'stroboscopic' method invented by Minorsky (1962), which appears to merit a rigorous theory.

If a vibrating system, say

$$\ddot{x} + x = \mu f(x, \dot{x}, t)$$

governed by a small parameter μ, has a period T which differs little from 2π, then the successive values of the displacement x at times $t = 2\pi n$,

$n = 0, 1, 2, \ldots$, will exhibit only small differences. These are the values which would be observed if the system were illuminated 'stroboscopically' by flashes of light at these instants of time.

Thus, let the general solution of the equation

$$\dot{x} = y, \quad \dot{y} = -x + \mu f(x,y,t)$$

with the initial conditions $x = \xi$, $y = \eta$ be

$$x(t) = x(\xi,\eta,\mu,t), \quad y(t) = y(\xi,\eta,\mu,t).$$

If the parameter μ is small the initial point ξ,η will be near a point ξ^*,η^* on a periodic orbit. Let

$$u = \xi - \xi^*, \quad v = \eta - \eta^*.$$

After the lapse of a time 2π, u and v become $u + \Delta u$ and $v + \Delta v$. These small increments Δu and Δv will vanish if $\mu = 0$, $u = 0$ or $v = 0$. Hence

$$\Delta u = \mu(bu + b'v) + 0(\mu^2)$$

$$\Delta v = \mu(cu + c'v) + 0(\mu^2)$$

neglecting higher terms in u and v.

Minorsky took the bold step of replacing these difference equations by the differential equations

$$du/d\tau = bu + b'v,$$

$$dv/d\tau = cu + c'v,$$

regarding the parameter μ as an increment in a quasi-temporal variable τ. The stability of the periodic orbit can then be discussed in terms of the eigenvalues of the matrix

$$\begin{pmatrix} b & b' \\ c & c' \end{pmatrix}.$$

13.34. Fixed-point Theorems

The theory of differential equations has been transformed by the study of 'fixed-point theorems', i.e. by the study of the mappings M of a space S into itself and of the existence of fixed points x in S such that $M(x) = x$.

The remote origins of the theory can be found in the 'method of successive approximation' of Cauchy (Moigno 1861) and Liouville (1836–8), but the importance of the method was only recognized after the elegant exposition of Picard (Section 13.7).

It was also Picard who recognized the importance of the Kronecker index which gives the number of oriented intersections of a number of curves

$$f(x,y) = 0, \quad g(x,y) = 0.$$

Poincaré was confronted with the same problem in his studies (Poincaré 1881) of the ordinary differential equations

$$\dot{x} = P(x,y), \quad \dot{y} = Q(x,y)$$

and, in ignorance of the work of Kronecker, introduced the concept of the 'rotation of the vector field' on a simple closed curve.

The decisive step of incorporating these fragments into a complete topological theory was taken by Brouwer (1910, 1912a, b) who defined the 'degree' of a mapping of closed orientable manifolds onto orientable manifolds, and proved the crucial theorem that if an n-cell C is mapped onto itself then there exists at least one fixed point.

Apparently in ignorance of the earlier work, J. W. Alexander (1922) gave Kronecker's index integral and hence obtained Brouwer's index for a continuous transformation of an n-sphere into itself.

Extensions of Brouwer's theorem by G. D. Birkhoff and Kellogg (1922), Schauder (1930) and Tychonoff (1935) are all included in Lefschetz's fixed point theorem (Section 9.22) or in its extensions.

These theorems are *global* in character, i.e. they give the algebraic number of fixed points of a continuous mapping f of a space X into itself, but, for applications to partial differential equations, what is needed is the number of fixed points on open subsets of X. A second new era in the theory was opened by Leray (1945) and Leray and Schauder (1934) who provided the necessary *local* theorems for continuous mappings of certain 'convexoid' spaces. The essential feature of this theory was the definition of the *local* topological degree of a mapping, relative to some open set G of X.

The work of Leray and Schauder is based on the use of simplicial approximation to the given mapping f. A much simpler approach has been given by Nagumo (Section 13.36), and Browder (Section 9.22) has extended the Leray-Schauder theorem to more general spaces.

The bare fact of the existence of periodic solutions of a certain large class of differential equations has been established by Lefschetz (1943) by a simple application of Brouwer's fixed-point theorem. These equations have the form

$$\ddot{x} + g'(x)\dot{x} + f(x) = e(t)$$

where $e(t)$ has the period T, and the differentiable functions, $e(t)$, $f(x)$ and $g(x)$ satisfy the conditions that

$$x^{-1}f(x) \quad \text{and} \quad x^{-1}g(x) \to \infty \quad \text{with} \quad |x|,$$

while for some constants b and B,

$$|x^{-1}g(x) - bx^{-1}f(x)| \leq B.$$

Lefschetz shows that, if

$$2u = ax^2 - 2 + by^2, \qquad (ab > 1, a > 0)$$

then $\dot{u}$ is negative outside a circle $x^2 + y^2 = n^2/u^2$ whence there is an elliptic region, $u = $ constant, in the phase plane which is mapped into itself by the transformation

$$x(0) \to x(\tau) \qquad \dot{x}(0) \to \dot{x}(\tau),$$

when $x(\tau)$ is a solution of the original differential equation.

13.35. Cartwright and Littlewood

To the delight of the historian, it is possible to give an exact date for the reflorescence of the theory of non-linear oscillations. In January 1938 the Radio Research Board of the Department of Scientific and Industrial Research of the United Kingdom issued an appeal to pure mathematicians for help in obtaining solutions of non-linear equations of the form

$$\varphi(D)v + \psi(D)i = 0, \quad D = d/dt$$

$$i = i(v) = -\alpha v + \beta v^2 + \gamma v^3 + \ldots, \quad (\alpha, \gamma > 0)$$

when φ and ψ are polynomials in D. The problems arose from the use of thermionic valves or vacuum tubes and the important questions relate to the existence, frequency, amplitude and stability of their oscillations, especially as they are affected by the physical parameters occurring in the questions. The study of these problems has had a great influence on the mathematical theory. The memorandum from the DSIR was sent to the London Mathematical Society and came into the hands of Dame Mary Cartwright and aroused her interest in the theory. The renaissance of the non-linear theory is undoubtedly due in a large measure to her researches in this field.

The existence results obtained by Cartwright and Littlewood (1945) on forced oscillations in non-linear systems are mainly concerned with van der Pol's equation.

$$\ddot{x} - k(1 - x^2)\dot{x} - x = bk \cos(\lambda t + \alpha),$$

and with the solution $f(t)$ for which $\dot{x} = b > 0$, $x = 0$, when $t = 0$. Their researches operate in the x,t-plane rather than the phase plane of x and $\dot{x}$, and depend on the fact that the curve $x = f(t)$ crosses the t-axis, $x = 0$, again, with $\dot{x} = b' > 0$ at time $t = 2\pi + k\tau$. Hence there is another solution of van der Pol's equation with α replaced by

$$\alpha' = \alpha + 2\pi(\lambda - 1) + k\lambda\tau,$$

and there is a (1–1) transformation from the point (b, α) to the point (b', α'). By a study of this transformation of the difference equations connecting b and b', it is found that solutions of period $2\pi/\lambda$ are given approximately by the intersection of the curves

$$b/p = \tfrac{1}{2}k \sin \alpha/(\lambda - 1),$$

$$b(\tfrac{1}{4}b^2 - 1) = p \cos \alpha.$$

13.36. Cronin and Local Topological Degree

The fixed-point theorem, which states that a continuous mapping of a closed solid 'sphere' in n dimensions into itself necessarily has at least one invariant or fixed point, was undoubtedly of great value in the early history of differential equations, and it was in fact the basis of the existence theorems due to Liouville and Picard. But the fixed-point theorem suffers from the disadvantage that it gives no information about the exact number of fixed points.

This disadvantage is removed or mitigated by the theory of the local topological degree of a mapping. Broadly speaking, the problem is to estimate the multiplicity of a mapping $f:S \rightarrow T$ from a space S to a space T, i.e. to estimate the number of points of the domain S which are mapped into a single point p of the codomain T. The reason for emphasizing that the theory provides only an 'estimate' is that the topological methods employed necessarily yield results which are invariant under small 'continuous' changes of the point p in T or in the mapping f.

The simplest example is provided by the classical problem of the zeros of a real polynomial

$$f(x) = a_0 + a_1 x + \ldots + a_n x^n$$

which has a long history which includes the work of Descartes and Sturm, well known in Victorian textbooks of algebra. In a sense the problem is considerably simplified by the consideration of the equation $f(x) = k$ where k is a real number. The total number of solutions $n(k)$ of this equation may vary with the parameter k, but the *parity* of $n(k)$ remains constant. This result depends on the fact that, in general, the sign of $f'(x)$ at a zero of $f(x) - k$ remains unaltered for small changes in k.

There is an analogous theorem in the theory of integral equations as developed by Arzela and Schmidt (1907a, b, 1908).

Leroy (1933) had the happy idea of generalizing this result to certain non-integral functional equations and of defining the 'total index' which characterizes the (algebraic) number of the solutions of an equation of the type

$$F(x) = x$$

where the functional $F(x)$ is completely continuous and x is a point of a certain Banach space.

The Leray–Schauder theory can be applied at once to a non-linear equation, say $G(x) = x$, provided that we can introduce a parameter k, such that the functional $F(x,k)$ varies continuously with k, $G(x) = F(x,1)$ and the 'total index' N of $F(x,0)$ can be determined and is not zero. Then the total index N of $F(x,k) = x$ is independent of k, and the total index of $G(x)$ must also be N. It follows that the non-linear equation $G(x) = x$ must have at least one solution.

This total index is of course simply the 'topological degree' of the mapping $y = x - G(x)$ at the origin $y = 0$, i.e. the algebraic number of points x which are mapped into $y = 0$. In the case when x and y are points in an n-dimensional space Brouwer (1912a) had already succeeded in defining the topological degree by approximating to the mapping by 'simplicial' transformations $y = \phi(x,\varepsilon)$ where the transformation ϕ operates in n-dimensional space, and Leray and Schauder (1934) extended this definition to mappings in functional space.

The Leray–Schauder theory is justly famous for the number of existence theorems which it has established, especially in the domain of non-linear partial differential equations, but it is only just to draw attention to a number of difficulties which have to be surmounted.

The formal definition of the topological degree of a transformation was first given by Kronecker in terms of an integral (Kronecker 1869), but this definition is restricted to differentiable mappings, and is of very little use in actually computing the numerical value of the degree. Nevertheless in two dimensions it does serve excellently to provide a clear geometrical definition of the Poincaré index of a vector field.

Another simpler definition has been given by Nagumo (1951a). This avoids all topological theory and is based entirely on classical calculus, but, once again, it is said to be unsatisfactory for actual computation.

A thoroughgoing topological definition in terms of cohomology has given by Rado and Reichelderfer (1955). The definition adopted by Cronin (1964) is that given by Alexandroff and Hopf (1935) and is carefully constructed on geometric lines from the elementary concept of the order of a point relative to a cycle or to a continuous image of a cycle.

After the formal definition of topological degree the next problem in the theory of differential equations is to reformulate the equation as a fixed-point problem for a mapping.

In order to establish the existence of solutions with given initial conditions it is often sufficient to replace the differential equation by an integral equation as in the classical theory of Picard (Section 13.7) and to apply a fixed-point theorem. But to establish the existence and stability of periodic solutions is a much more difficult problem to which Cronin has successfully applied the theory of the topological degree.

The calculations are lengthy and tedious, but the general idea can be illustrated by the two-dimensional problem of the form

$$\dot{x} = y + \mu f(x,y,t)$$
$$\dot{y} = -x + \mu g(x,y,t)$$

where f and g have a period $2\pi/v$ and $\mu > 1$. The solution $x = \xi(t)$, $y = \eta(t)$ such that

$$x = \alpha, y = \beta \quad \text{at} \quad t = 0$$

becomes $\quad x = \alpha^1 = \xi(\tau), \; y = \beta^1 = \eta(\tau) \quad \text{at} \quad t = \tau = 2\pi/v,$

and thus determines a mapping $\alpha \to \alpha^1$, $\beta \to \beta^1$ as in Poincaré's theory of 'consequent' points. If the topological degree of this mapping is not zero then there exists at least one periodic solution. Moreover, Cronin has shown that the *sign* of the local degree can provide information about the stability of the system, and has thus extended and made rigorous the theory of Andronov and Witt (1930).

14. Calculus of Variations

14.1. Introduction

In the preface to his *Calculus of Variations* Forsyth (1927) rightly remarks: '[This subject] has attracted a rather fickle attention at more or less isolated intervals in its growth. Its progress has been neither steady nor consecutive . . . [and] it has not secured an abiding interest.'

Although problems of the calculus of variations, such as Newton's problem of the missile of least resistance, Bernoulli's problem of the brachistochrone and Bernoulli's isoperimetric problem, had been considered at the end of the seventeenth century, and, although Euler and Lagrange had discovered the 'method of variations', to be further developed by Jacobi and Clebsch, it was Weierstrass who first gave a satisfactory foundation to the theory in his lectures at Berlin between 1865 and 1890, and thus inaugurated the Victorian period of the calculus of variations.

This period ended gloriously when the calculus of variations burst into full flower with the advent of functional analysis, a 'second spring' whose herald was Hadamard and whose manifesto was his *Leçons sur le Calcul des Variations* (Hadamard 1910).

During the earlier period the problems of the calculus of variations were reduced to questions of the existence of solutions of differential equations (ordinary or partial) until Hilbert devised a new method in which the existence of a minimizing function was established directly as the limit of a sequence of approximations (Hilbert 1900).

By employing the resources of functional analysis and the theory of Hilbert space, Weyl showed how Hilbert's methods could be simplified and strengthened so as to establish the solution of Dinichlet's problem in a general differentiable manifold (Weyl 1940).

Finally Morse succeeded in freeing the calculus of variations from the limitations imposed by the restriction to 'small, or local, variations', and gave a general global theory of 'variations in the large' (Morse 1934).

14.2. The Eulerian Equations

In the simplest case the problem is to determine which functions $y(x)$ minimize (or maximize) an integral of the form

$$F(y) = \int_a^b f(x,y,y')\,\mathrm{d}x$$

where $y' = dy/dx$ and the function $y(x)$ takes prescribed values, say α and β at the extremities of the interval of integration. Here we have to consider curves, $y = y(x)$, in the two-dimensional space of the variables x and y, and the problem is exemplified by the motion of a particle of mass m moving under gravity in a vertical plane. The actual motion of the particle is determined by the stationary values of the 'action' which is the integral of twice the kinetic energy

$$A(C) = \int_C 2T \, dt$$

taken along paths C from (a,α) to (b,β) subject to the condition that

$$T = \tfrac{1}{2}m(\dot{x}^2 + \dot{y}^2) = E - V, \quad V = mgy$$

$\dot{x}$ and $\dot{y}$ being the components of the velocity and T,V the kinetic and potential energies. The action integral is easily transformed into the standard type by writing

$$A(C) = \int \sqrt{(2T)} \, \sqrt{2(E - V)} \, dt = \int_a^b f(x,y,y') \, dx$$

where $f = \sqrt{m}[2(E - mgy)]^{\frac{1}{2}}(1 + y'^2)^{\frac{1}{2}}$.

The general theory seeks to determine which curves in $(n + 1)$-dimensional Euclidean space, of coordinates, x, y_1, y_2, ..., y_n, give minimum values to an integral of the form

$$J = \int_a^b f(x, y_1, y_2, \ldots, y_n, y_1', y_2', \ldots, y_n') \, dx.$$

If there is such a curve C, say $y_k = \overline{y}_k(x)$, we must compare the integral J taken along C with the value of J taken along neighbouring curves, and we shall certainly obtain *necessary* conditions for J to be a minimum along C, if we limit ourselves to neighbouring curves $C(\varepsilon)$ of the form

$$y_k = \overline{y}_k(x) + \varepsilon \eta_k(x), \quad \eta_k(a) = 0 = \eta_k(b), \quad k = 1, 2, \ldots, n,$$

where ε is a small parameter, and the functions $\eta_k(x)$ have as many derivatives as we shall find convenient. The value of J along $C(\varepsilon)$ is easily obtained in the form

$$J(\varepsilon) = J(0) + \varepsilon \sum_k \int_a^b \{\eta_k' \partial f/\partial y_k' + \eta_k \partial f/\partial y_k\} \, dx + 0(\varepsilon^2),$$

the derivatives being evaluated for $\varepsilon = 0$, i.e. along the curve C. Integration by parts expresses the integral in the form

$$\varepsilon \sum_k \int \eta_k \{\partial f/\partial y_k - d(\partial f/\partial y_k')/dx\} \, dx.$$

Since the small parameter ε can be positive or negative, if the value of J is to be a minimum (or maximum) for the curve C, it is essential that the terms of the first order in ε in $J(\varepsilon) - J(0)$ should vanish. And this will undoubtedly be the case if f satisfies the Eulerian equations

$$\partial f/\partial y_k = \mathrm{d}(\partial f/\partial y'_k)/\mathrm{d}x, \quad k = 1, 2, \ldots, n.$$

These conditions are therefore *sufficient* to ensure that a certain *necessary* condition, viz. $J(\varepsilon) = J(0) + 0(\varepsilon^2)$ is satisfied. But there remain a whole host of questions to be answered.

In the first place we need to know if the Eulerian equations are necessary to ensure that J is stationary for the curve C.

Then we must consider more general types of curves which are neighbours to the curve C. We have considered small variations in the functions y_k and by implication small variations in their derivatives y'_k i.e. we have considered neighbouring 'smooth' curves. What is the generalization of the Eulerian equations for curves with sharp corners? Again, what is the generalization if the comparison curves have infinite derivatives? And when we have settled the condition for the curve C to make the integral J stationary, we still have to consider the really interesting question—what are the conditions that C should make J not merely stationary but a genuine maximum or minimum? We therefore proceed to summarize the history of these problems. Although the earlier history lies outside our period, a rapid summary of the 'prehistory' is necessary to make intelligible the great advances due to Weierstrass.

14.3. The Fundamental Lemma

The 'fundamental lemma', on which all the earlier researches on the calculus of variations depend, relates to the conditions under which the vanishing of a continuous function $\varphi(x)$ in an interval $[a,b]$ can be inferred from the vanishing of the integrals

$$I = \int_a^b \varphi(x)\eta(x) \, \mathrm{d}x$$

for all functions $\eta(x)$ of a certain class F, which vanish at $x = a$ and $x = b$.

The earlier writers thought that it was evident that it was sufficient if the functions $\eta(x)$ were 'arbitrary', and it was a great advance when it was realized that the class F should be carefully defined and a formal proof provided.

The question of priority is of antiquarian interest, but it appears that the first serious investigation is due to Stegmann (1854). All the proofs depend upon the construction of a class of functions F, which vanish at $x = a$ and $x = b$, which are arbitrarily small except in the neighbourhood of an arbitrary point $x = c$ ($a \leqslant c \leqslant b$), and which make the integral

$$\int_a^b \eta(x) \, \mathrm{d}x$$

arbitrarily large.

The real interest of these proofs is that they unconsciously anticipate the idea of 'test functions' and of the notorious Dirac function $\delta(x)$

(Section 12.2) which stimulated Laurent Schwartz to invent the theory of distributions. In fact it is necessary and sufficient for the validity of the 'fundamental lemma' that the class F should contain a sequence of functions $\{\varphi_n(x)\}$ such that

$$\int_a^b \varphi_n(x)\ dx = 1 \quad \text{and} \quad \varphi_n(x) \to 0, \text{ except at a single point } c,\ a < c < b.$$

Thus Stegmann takes F to include the functions

$$C(x-a)^{\mu}(b-x)^{\nu}$$

and Zermelo (1894) following the indications given by Weierstrass in his Berlin lectures, takes F to include the functions

$$C(x-a)^m(b-x)^n \exp\left[-\varrho^2(x-c)^2\right]$$

which were also used by Kirchoff in his researches on Huyghens' principle (Section 12.2).

A closely related result is due to du Bois-Reymond (1879) who showed that, if the integral

$$\int_a^b \varphi(x)\eta'(x)\ dx$$

is zero for a continuous function $\varphi(x)$ and all functions $\eta(x)$ which vanish at $x = a$ and $x = b$ and which possesses continuous derivatives in the interval $[a,b]$, then $\varphi(x)$ is constant in this interval. The use of this lemma gives a greater generality to the Eulerian equations.

14.4. Legendre's Condition for a Maximum

The earlier work can be adequately illustrated by the simplest problem of minimizing the integral

$$J(y) = \int_a^b f(x,y,y')\ dx,$$

where $y' = dy/dx$. To compare this integral along a curve $y = \bar{y}(x)$ with its value for a neighbouring curve $y = \bar{y}(x) + \varepsilon\eta(x)$, with $\eta(a) = 0 = \eta(b)$, we expand $J(y)$ in powers of ε and thus obtain

$$J(y) = J_0(y) + \varepsilon J_1(y) + \tfrac{1}{2}\varepsilon^2 J_2(y) + 0(\varepsilon^3),$$

where
$$J_0(y) = \int_a^b f(x,\bar{y},\bar{y}')\ dx,$$

$$J_1(y) = \int_a^b \left\{\eta'\frac{\partial f}{\partial y'} + \eta\frac{\partial f}{\partial y}\right\} dx,$$

$$J_2(y) = \int_a^b \left\{P\eta^2 + 2Q\eta\eta' + R\eta'^2\right\} dx,$$

and
$$P = \partial^2 f/\partial y^2, \quad Q = \partial^2 f/\partial y\partial y', \quad R = \partial^2 f/\partial y'^2,$$

the integrals being evaluated on the curve $y = \bar{y}(x)$.

The vanishing of the first variation $J_1(y)$ is necessary if the curve $y = \bar{y}(x)$ minimizes (or maximizes) the integral. A sufficient condition that this curve should minimize the integral is that the second variation $J_2(y)$ should be positive.

By the device of adding the expression

$$2\eta\eta'w(x) + \eta^2 w'(x) = \mathrm{d}[\eta^2 w(x)]$$

to the argument of the integral $J_2(y)$, and by choosing the function $w(x)$ to make the new argument a perfect square, Legendre (1786) reduced $J_2(y)$ to the form

$$J(y) = \int_a^b \{\eta' + (Q + w)\eta/R\}^2 R \ \mathrm{d}x.$$

He then concluded that to make $J(y)$ a minimum it is necessary that

$$R = \partial^2 f/\partial y'^2$$

should be positive in the range $a \leqslant x \leqslant b$.

This ingenious process depends upon choosing the auxiliary function w to satisfy the equation

$$R(P + w') = (Q + w)^2.$$

14.5. The Jacobi Equation

Jacobi's contributions to the Calculus of Variations fell into two parts— the analytical part contained in a few paragraphs of his letter to Encke dated 29 November 1836 (Jacobi 1837), and the geometrical part briefly noticed in his lectures on dynamics (Jacobi 1866).

The analytical theory consists in the reduction of Legendre's equation

$$(Q + \alpha)^2 = R(P + \alpha')$$

by the Riccati transformation

$$Q + \alpha = -Ru'/u$$

to the linear form

$$(Ru')' + (Q' - P)u = v,$$

and the general solution of this equation in terms of a complete integral of the Euler equations

$$\frac{\partial F}{\partial y} = \frac{\mathrm{d}}{\mathrm{d}x} \frac{\partial F}{\partial y'}.$$

This complete integral has the form $y = f(x, a, b)$ and involves two independent arbitrary constants, a and b. The general integral of the equation for u is given by Jacobi as

$$u = A\partial f/\partial\alpha + B\partial f/\partial\beta,$$

where A and B are constants of integration. Jacobi gave no indication how he obtained this extraordinarily simple and useful result.

The effective use of Jacobi's integral depends crucially upon the choice of A and B so as to prevent u from vanishing in the range of integration $a \leqslant x \leqslant b$ of the original integral of $F(x,y,z')$. If such an integral exists then Legendre's criterion for a minimum, viz. $R = \partial^2 F/\partial y'^2 > 0$ is *sufficient*.

Jacobi also gave, without proof, a generalization of these methods to integrals of the form $\int F(x,y,y', \ldots, y^{(n)}) \, dx$ involving derivatives of $y(x)$ of orders 1, 2, ..., n.

14.6. Jacobi and Conjugate Points

If an integral u of Jacobi's equation vanishes at $x = x_1$ and $x = x_2 > x_1$ and does not vanish at any point x between x_1 and x_2 then x_2 is called the 'conjugate point' of x_1. Jacobi stated, without formal proof, that the necessary and sufficient condition for the function u not to vanish in the range $a \leqslant x \leqslant b$ is that a', the conjugate point of a, must be beyond b, i.e. $a' > b$.

Jacobi (1866) stated without proof that the conjugate point of a is the point a' at which an extremal through a touches the envelope of all the extremals through a, and he gave the simple but exceptional example of geodesics on a sphere.

The classical example of the trajectory of a particle moving under gravity is summarized in Section 14.8.

14.7. The Commentators on Jacobi

From Jacobi to Weierstrass the history of the calculus of variations consists of commentaries and elucidations of the terse and pregnant notes left by Jacobi. The earliest commentators on his work were V.-A. Lebesgue (1841), Delauncy (1841) and Minding (1858), but the most complete investigation is due to Hesse (1857), who expressed Jacobi's equation

$$(P - Q')u - (Ru')' = 0$$

in the form

$$\psi(\eta) = \frac{\partial \Omega}{\partial \eta} - \frac{d}{dx} \frac{\partial \Omega}{\partial \eta'} = 0$$

where

$$2\Omega = P\eta^2 + 2Q\varphi\eta\eta' + R\eta'^2.$$

The general theory was expounded by Mayer (1868).

It must be confessed that some readers find these investigations complicated and wearisome, even when they are digested in standard expositions. Indeed, Hedrick (1902) has gone so far as to protest, with special reference to 'the only modern textbook on the theory' (viz. Kneser's *Variationsrechnung*), against 'the extreme complication recently introduced in some quarters into the essentially simple subject of the calculus of variations'.

Rather than risk such a criticism I shall follow the example of Darboux (1889) in his elegant introduction to this subject by a reference to the well-known problem of the motion of a projectile under terrestrial gravity.

14.8. Darboux and Projectile Theory

The x and y coordinates being respectively horizontal and vertically upwards, the potential energy of a particle of mass m is $V = mgy$ and its kinetic energy is $T = \frac{1}{2}m(\dot{x}^2 + \dot{y}^2)$, g being the gravitational acceleration. For a particle moving with total energy $E = T + V$, the 'action' A for any path C,

$$x = x(t), \qquad y = y(t)$$

(not necessarily the actual trajectory according to Newton's equations,

$$m\ddot{x} = -\partial V/\partial x, \qquad m\ddot{y} = -\partial V/\partial y,)$$

is defined to be

$$A(C) = \int_C 2T \mathrm{d}t = \int_C F(x,y,y')\,\mathrm{d}x,$$

where

$$F(x,y,y') = (2T)^{\frac{1}{2}}[2(E - V)]^{\frac{1}{2}}(1 + y'^2), \quad y' = \mathrm{d}y/\mathrm{d}x.$$

The first necessary condition that the path C should make the action a minimum is given by Euler's equation

$$\frac{\mathrm{d}}{\mathrm{d}x}\frac{\partial F}{\partial y'} = \frac{\partial F}{\partial y}$$

which has the complete integral

$$y = x \tan \alpha - \tfrac{1}{4}(x^2/c)\sec^2 \alpha, \quad c = E/mg.$$

This equation represents the familiar parabolic orbits springing from the origin and touching the envelope

$$y = c - \tfrac{1}{4}(x^2/c)$$

at the point P, $x = 2c \cot \alpha$, $y = c - c \cot^2 \alpha$.

This point is the Jacobi 'conjugate' of the origin. There are two parabolic paths from the origin to a prescribed target point (ξ, η) with parameters α given by $\xi \tan \alpha = 2c \pm (4c^2 - 4c\eta - \xi^2)^{\frac{1}{2}}$. One parabola (with the greater elevation α), touches the envelope before hitting the target point P. The other parabola touches the envelope after hitting the target point. According to Jacobi's criterion, it is the latter trajectory which minimizes the action, along the curve from the origin to P.

We note that the Weierstrass Excess Function (Section 14.9) is

$$E(x,y,p,q) = m\sqrt{2g}\,(c - y)M(1 + p^2)^{\frac{1}{2}}$$

where $M = (1 + q^2)^{\frac{1}{2}}(1 + p^2)^{\frac{1}{2}} - (1 + pq)$.

Since $(1 + q^2)(1 + p^2) - (1 + pq)^2 = (p - q)^2$ the Excess Function is positive if $y < c$.

14.9. Weierstrass and the Excess Function

The second great period in the history of the calculus of variations opens with the work of Weierstrass, whose main contributions were the introduction of 'strong variations' and of the 'excess function'. Weierstrass never published these outstanding discoveries but announced them in his lectures. We are indebted to those who had the good fortune to hear his lectures for the accounts of his theory (Kneser 1900; Bolza 1904, 1909), although some of the lectures were published in his collected works.

The conditions for a minimum of the integral

$$J = \int_a^b F(x,y,y') \, dx$$

given by Legendre and Jacobi were obtained by considering only weak variations of the form

$$y = f(x) + \varepsilon \omega(x),$$

where ε was a small parameter, $\omega(a) = 0 = \omega(b)$, and $\omega(x)$ was a 'smooth' function.

More straight conditions were obtained by Weierstrass by considering variations of the form

$$y = \frac{(x-a)}{(\alpha-a)} [\gamma(\alpha) - f(\alpha)] + f(x),$$

from $[a,f(a)]$ to $[\alpha,\gamma(\alpha)]$ and $y = \gamma(x)$ from $[\alpha,\gamma(\alpha)]$ to $[b,f(b)]$ the function $\gamma(x)$ being differentiable and such that $\gamma(b) = f(b)$.

If $J(\alpha)$ denotes the integral J taken along this curve (which may present a very sharp corner at $[\alpha,\gamma(\alpha)]$) and $J(b)$ is the integral along an extremal obtained by making α equal to b, then the condition $J(\alpha) \geqslant J(b)$ is necessary for $J(b)$ to be a minimum. Since $\alpha \leqslant b$, this implies that $J'(b) \leqslant 0$. Also since $[b,f(b)]$ is any point on the extremal, we replace $x, f(b)$ and $f_x(b)$ by x, y, b to obtain Weierstrass' condition in the standard form for the 'excess' function

$$E(x,y,p,q) \equiv F(x,y,q) - F(x,y,p) - (q-p)\partial F(x,y,p)/\partial p \geqslant 0.$$

14.10. The 'Corner Conditions'

In deriving the Eulerian equations, which express the vanishing of the first variation of the integral

$$\int_a^b f(x,y,y') \, dx$$

it is usually assumed that the function $y = y(x)$ possesses a second derivative $y''(x)$ at all points in the interval $[a,b]$.

By making use of the lemma due to du Bois-Reymond it is, however, possible to relax this condition, and to allow the function $y(x)$ to possess a second derivative only at the interior points of a finite set of intervals $[a,x_1], [x_1,x_2], \ldots [x_n b]$, it being understood that $y(x)$ is everywhere continuous, but $y'(x)$ may be discontinuous at $x = x_1, x_2, \ldots, x_n$.

The continuous curve $y = y(x)$ may then present sharp corners at these points, and the conditions for the vanishing of the first variation now consist of the Eulerian equations in each open interval

$$(a, x_1), \ (x_1, x_2), \ \ldots, \ (x_n, b),$$

and the 'corner' conditions, that $\partial f/\partial y'$ must be continuous at each such corner. These corner conditions were first given explicitly by Weierstrass in 1877 and Erdmann (1877) [see also Whittemore (1901)], but certain special problems were solved by Todhunter (1871). Without these conditions we cannot compare the length of the straight line joining two points with the length of a curvilinear polygon joinging the same points.

Such solutions of the variational problem in which $y'(x)$ is only piece-wise continuous are commonly known (by abuse of language) as 'discontinuous' solutions.

14.11. *The Parametric Equations of Weierstrass*

In locating and computing the minima of a function $f(x)$ which is continuous in a closed interval $[a, b]$ it is only necessary to compare the value of $f(x)$ at an arbitrary point $x = c$ with its value at points in *any small* interval

$$a \leqslant c - \eta \leqslant x \leqslant c + \varepsilon \leqslant b.$$

In the corresponding problem for an integral

$$J = \int_a^b F(x, y, y') \ \mathrm{d}x$$

we compare the value of J for an arbitrary smooth curve joining the points $[a, y(a)]$, $[b, y, (b)]$ with its value for any 'neighbouring curve', joining the same points.

What may be called the 'strength' of the minimum thus obtained will depend entirely upon the class of neighbouring curves which are considered. The earlier writers—Euler, Lagrange, Legendre, Jacobi and Clebsch—had limited themselves to extremals given by equations of the form $y = f(x)$ and to neighbouring curves of the form $y = f(x) + \varepsilon\omega(x)$ where f and ω possessed second derivatives and ε was a small numerical parameter.

In his lectures given in 1872 (or perhaps earlier) Weierstrass greatly extended the class of extremals and neighbouring curves by allowing a parametric representation, in which extremals are given in the form

$$x = x(t), \quad y = y(t)$$

and neighbouring curves in the form

$$x = x(t) + \varepsilon\xi(t), \quad y = y(t) + \varepsilon\eta(t).$$

The integral J now takes the form

$$J = \int F(x, y, x', y') \ \mathrm{d}t$$

where $x' = dx/dt$ and $y' = dy/dt$, and the vanishing of the first variation of the integral J is now given by a pair of Eulerian equations

$$\frac{\partial F}{\partial x} = \frac{d}{dt}\frac{\partial F}{\partial x'}, \quad \frac{\partial F}{\partial y} = \frac{d}{dt}\frac{\partial F}{\partial y'}.$$

If the value of the integral J is independent of the particular parametric representation employed (e.g. $x = t$, $y = t^2$, or $x = u^3$, $y = u^6$), then F is positively homogeneous in x' and y', i.e.

$$F(x,y,kx',ky') = kF(x,y,x',y')$$

if $k > 0$ and the two Eulerian equations are no longer independent.

The great value of the parametric representation is that it allows us to consider neighbouring curves for which $y' = dy/dt \div dx/dt$ may attain infinite values. A valuable comparison of the parametric method with the older method [which considers only neighbouring curves of the form $y = f(x)$] is given by Bolza (1903).

14.12. *Variational Problems with Variable End-points*

From the very beginning of its history, the calculus of variations had been applied by the Bernoulli's, Euler and Lagrange to problems in which the upper and lower extremeties, x_2 and x_1, of the integral

$$J = \int_{x_1}^{x_2} F(x,y,y') \, dx$$

were not fixed but allowed to move along certain fixed curves $y = \varphi_1(x)$, $y = \varphi_2(x)$. The first variation of the integral J is then obtained by replacing y by $y + \delta y$ and by replacing the initial point x_1, y_1 by $x_1 + \delta x_1$, $y_1 + \delta y_1 = y_1 + \varphi_1'(x) \, \delta x_1$, and the final point x_2, y_2 by $x_2 + \delta x_2, y_2 + \delta y_2 = y_2 + \varphi_2'(x_2) \, \delta x$. The first variation therefore has the form

$$\int_{x_1}^{x_2} A\delta y \, dx + B\delta x_1 + C\delta x_2,$$

where A is the usual Eulerian expression

$$\frac{\partial F}{\partial y} - \frac{d}{dx}\frac{\partial F}{\partial y'}.$$

The other terms, first enunciated explicitly by Lagrange (1760), are

$$B\delta x_1 = \{F(x_1,y_1,y_1') - y_1' \partial F(x_1,y_1,y_1')/\partial y_1'\} \, \partial x_1 + \partial F(x_1,y_1,y_1')/\partial y_1' \, \delta y_1$$

and

$$C\delta x_2 = \{F(x_2,y_2,y_2') - y_2' \partial F(x_2,y_2,y_2')/\partial y_2'\} \, \partial x_2 + \partial F(x_2,y_2,y_2')/\partial y_2' \, \delta y_2.$$

In order that the first variation shall vanish it is clear that not only must the curve $y = f(x)$ be an extremal (so that $A = 0$), but also that we must have

$$F(x_1,y_1,y_1') + \{\varphi_1'(x_1) - y_1'\} \partial F(x_1,y_1,y_1')/\partial y_1' = 0$$

and

$$F(x_2,y_2,y_2') + \{\varphi_2'(x_2) - y_2'\}\partial F(x_2,y_2,y_2')/\partial y_2' = 0.$$

The Eulerian equation, $A = 0$, determines a two-parameter system of extremals, and, in general, the other two equations just given determine one or more pairs of points on the curves $y = \varphi_1(x)$ and $y = \varphi_2(x)$.

The discrimination of maxima and minima naturally requires the construction of the second variation, explicit expressions for which have been given by Erdmann (1878).

14.13. *Isoperimetric Problems*

The analytical problem of determining the *relative* maxima or minima of a function $f(x,y,z, \ldots)$ of many variables, $x,y,z \ldots$, subject to conditions of the form

$$g(x,y,z, \ldots) = 0, \quad h(x,y,z, \ldots) = 0, \ldots,$$

was examined by Lagrange by the method of 'undetermined multipliers', $\lambda, \mu, \ldots$. The necessary conditions for a relative extreme value of f are given by the equations

$$\partial F/\partial x = 0, \quad \partial F/\partial y = 0, \quad \partial F/\partial z = 0, \ldots$$

where $$F = f + \lambda g + \mu h + \ldots$$

and where the multipliers $\lambda, \mu, \ldots$ are determined from the equations $g = 0, h = 0, \ldots$.

Lagrange also considered the similar problem in the calculus of variations, where it is a question of finding the extreme value of an integral

$$F = \int f(x,y,y') \, dx,$$

subject to 'isoperimetric conditions'

$$\int g(x,y,y') \, dx = G, \quad \int h(x,y,y') \, dx = H, \ldots$$

where $G, H, \ldots$ are given constants. The necessary condition for a minimum, given by Euler and by Lagrange, is that the integral

$$\int (f + \lambda g + \mu h + \ldots) \, dx$$

should have an extreme value, the undetermined multiples $\lambda, \mu, \ldots$ being found from the isoperimetric condition.

The most famous of such problems is to find the form $y = f(x)$ of a heavy uniform chain, hung in the vertical (x,y) plane, from two fixed points $y = 0$, $x = \pm a$. The problem is to minimize the height of the centre of gravity,

$$c = \frac{1}{l} \int y(dx^2 + dy^2)^{\frac{1}{2}}$$

subject to the condition that the chain has a length

$$l = \int (dx^2 + dy^2)^{\frac{1}{2}}.$$

Here $F = \int y(1 + y'^2)^{\frac{1}{2}} \, dx$ and $G = \int (1 + y'^2)^{\frac{1}{2}} \, dx$.

Since $l \geqslant 2c$, the form of the chain must be the catenary

$$ky = \cosh kx - \cosh kc,$$

where

$$kl = 2 \sinh kc.$$

14.14. Kneser and Fields of Extremals

The third period in the development of the calculus of variations was marked by its emancipation from the restrictions of 'weak variations'—almost completely effected by Kneser (1900) who developed the idea of a 'field' of extremals, due originally to Weierstrass, and generalized the idea of 'transversals', first conceived by Darboux.

A field of extremals is a one-parameter system of extremal curves, say $y = \varphi(x,\alpha)$, such as, for example, all the extremals which issue from the same point in different directions. The field is also to be restricted to a region R in which no two different extremals intersect.

Darboux made extensive use of the fields of extremals in his lectures at the Collège de France in 1866/7 and published his results in Volume 2 of his *Théorie des Surfaces* (Darboux 1889). Darboux was concerned with the problem of determining geodesics, first on a surface in Euclidean space, and secondly on any Riemannian manifold, and with the closely related problem of Least Action and the Hamilton–Jacobi theory of generalized dynamics.

On a Euclidean surface Darboux constructed a system of geodesics ($q = $ const.) and their orthogonal trajectories ($p = $ const.). If the coordinate p is chosen to be the length of a geodesic arc drawn from a fixed trajectory, then the element of length, ds, on the surface is given by an expression of the form

$$ds^2 = dp^2 + m^2 dq^2,$$

which exhibits at once the minimum character of the geodesics.

The coordinates p and q are the 'geodesic' coordinates on a surface, introduced by Gauss and generalized by Riemann to any Riemannian manifold. The great contribution of Kneser was to discover the corresponding intrinsic coordinates for problems in the calculus of variations.

Weierstrass had restricted himself to the one-parameter system of extremals issuing from the same fixed point, but Kneser considered any one-parameter system of extremals, say $y = f(x,\alpha)$. In order to generalize the concept of the orthogonal trajectories ($p = $ const.) of a system of geodesics ($q = $ const.), Kneser considered the terms in the first variation of

$$J = \int_{x_1}^{x_2} F(x,y,y') \, dx$$

which arise when the initial point x_1, y_1 is allowed to move along a fixed curve $y = \varphi(x)$.

In a one-parameter family of extremals, $y = f(x, \alpha)$, the derivative $f'(x, \alpha)$ of the unique extremal which passes through a point (x, y) is a single-valued function of x and y, say $u(x, y)$. The equation

$$F(x, y, u) + \{\varphi'(x) - u\} \, \partial F(x, y, u,)/\partial u = 0$$

then defines a system of curves, $y = \psi(x)$, which are the 'transversals' of the extremals $y = f(x, \alpha)$.

Kneser showed the extremals, $y = f(x, \alpha)$, and their associated transversals can be used to define a system of curvilinear coordinates, intrinsically related to the function $F(x, y, y')$. It is sufficient to take as coordinates the parameter α of the extremal passing through the point (x, y) and the value $w(x, y)$ of the integral

$$J = \int F(x, y, y') \, dx$$

taken along the extremal from some fixed transversal T to the field point $(x.y)$.

This system of coordinates is effective only in the region R in which no two extremals of the family intersect. The region R is bounded by the envelope of these extremals and, if the extremal from a point P on the basic transversal T touches the envelope at Q then Q is called by Kneser the focal point of P [for the family $y = f(x, \alpha)$]. If T generates into a point circle the focal point of P becomes the conjugate point of P as defined by Jacobi.

14.15. Hilbert and the Auxiliary Integral

At the conclusion of his famous lecture on 'Mathematical problems' (Hilbert 1900), Hilbert gave a new method for the discussion of the minimum value of the integral

$$J = \int_0^1 F(x, y, y') \, dx, \quad y' = dy/dx,$$

by introducing the auxiliary integral

$$J^* = \int_0^1 \{F + (y' - p) \, F_p\} \, dx$$

where $F = F(x, y, p)$ is now a function of three *independent* variables x, y, p.

This integral J^* will have a value independent of the path of integration given by $y = y(x)$ from $x = 0$ to $x = 1$ if

$$\frac{\partial F_p}{\partial x} + \frac{\partial (pF_p - F)}{\partial y} = 0.$$

By taking the path to be an extremal of J, Hilbert obtained another derivation of Weierstrass' function and a new approach to Jacobi's problem of determining necessary and sufficient conditions for the existence of a minimum of J—and all this without the introduction of the second variation of J!

In fact if J_G is the integral J taken along an extremal G joining the points $(0,y(0))$, $(1, y(1))$ and if J_Γ is the integral J taken along any smooth curve joining the same points, then

$$J_\Gamma - J_G = \int_\Gamma E \, dx$$

where $E = F(x,y,y') - F(x,y,u) - (y' - u) \, F_u(x,y,u)$ and $y' = u$ on the extremal G.

14.16. Beltrami and the 'Unabhängigkeitssatz'

Hilbert's method rests upon the fact that the integral $\int \{f(x,y,Y) + (y' - Y) f_Y(x,y,Y)\} \, dx$ is independent of the form of the curve $y = y(x)$ joining two fixed end-points, provided that the function Y is suitably chosen. In fact this theorem was discovered by Beltrami (1868b) but had been forgotten until Hilbert drew attention to its importance.

Beltrami was primarily concerned with geodesics, but he noticed that the Eulerian equations for the integral

$$\int f(x,y,y') \, dx$$

can be put into a 'Hilbertian' form, once we know a first integral of the Eulerian equations in the form

$$y' = \varphi(x,y).$$

(In Kneser's theory such an integral refers to a 'field' of extremals, $y = \varphi(x,\alpha)$, and gives the value of y' at a point (x,y) for this field.)

Beltrami explicitly wrote the Eulerian equation in the form

$$\frac{\partial}{\partial y} \left\{ f - y' \frac{\partial f}{\partial y'} \right\} = \frac{\partial}{\partial x} \left\{ \frac{\partial f}{\partial y'} \right\}$$

where y' is to be replaced by $\varphi(x,y)$, and noted the existence of a function $F(x,y)$ such that

$$\frac{\partial F}{\partial x} = f - y' \frac{\partial f}{\partial y'}, \quad \frac{\partial F}{\partial y} = \frac{\partial f}{\partial y'}$$

and such that the elimination of y' leads to a partial differential equation of the first order.

Thus for the geodesics in a plane

$$f = (1 + y'^2)^{\frac{1}{2}}$$

and for the field of geodesics $y = ax$, $y' = y/x$,

$$F_x^2 + F_y^2 = 1$$

and

$$F = (x^2 + y^2)^{\frac{1}{2}}.$$

14.17. The Problems of Mayer and Bolza

The simpler problems of the calculus of variations admit of a number of a generalizations, some of which are of considerable complexity. The most important of these extensions of the elementary theory are due to Lagrange, Mayer and Bolza.

Lagrange had formulated the variational problem for the integral

$$\int F(x,y,y')\ dx$$

where y is the vector with components $y_i(x)(1 \leqslant i \leqslant n)$ and y' is the vector $y_i'(x) = dy_i/dx$ while the components of y satisfy a number of differential equations,

$$g_k(x,y,y') = 0, \quad 1 \leqslant k \leqslant p < n.$$

This Lagrangian problem includes the variational problem for the integral

$$\int G(x,\eta,\eta',\ \ldots,\eta^{(m)})$$

in which the integrand G depends on the first m derivatives of a single function η, for we may consider the set $(\eta,\eta',\ \ldots,\eta^{(m)})$ to be the components of vector y which are subjected to the conditions

$$\eta' = d\eta/dx, \quad \eta'' = d\eta'/dx, \ \ldots.$$

A considerable generalization of Lagrange's theory was given by Mayer (1878, 1895) in the following unexpected form:

Let the functions $y_0, y_1, \ldots, y_n$ of a single variable x satisfy the $(p + 1)$ differential equations

$$g_k(x,y_0,y_1,y_1',\ \ldots,\ y_n,y_n') = 0, \quad 0 \leqslant k \leqslant p,$$

with the initial conditions

$$y_i = a_i \quad \text{at} \quad x = 0,$$
$$y_i = b_i \quad \text{at} \quad x = 1,$$

and $\qquad\qquad\qquad y_0 = c \quad \text{at} \quad x = 0.$

Then Mayer's problem is to minimize the value of y_0 at $x = 1$.

This problem is reducible to Lagrange's problem when the differential equations $g_k = 0$ are in the form

$$g_0 \equiv y_0' - f(x,y,y') = 0,$$
$$g_k(x,y,y') \equiv 0$$

when $\qquad\qquad\qquad I = c + \int_0^1 f(x,y,y')\ dx.$

But the most general form of this type of problem is due to Bolza (1913) in which the functions $y_1, y_2, \ldots, y_n$ satisfy the differential equations

$$g_k(x,y,y') = 0, \quad 1 \leqslant k \leqslant m < n, \quad \text{where} \quad y = (y_1, \ldots, y_n),$$

together with fixed end-conditions.

The problem then is to minimize the expression

$$I = g\{x,y(x_1),x_2,y(x_2)\} + \int_{x_1}^{x_2} f(x,y,y')\,\mathrm{d}x.$$

One important application of Bolza's problem is in the researches of Morse (Section 14.19) who uses the more general end-conditions

$$x_1 = a(t), \quad x_2 = b(t),$$

$$y_i(x_1) = \alpha_i(t), \quad y_i(x_2) = \beta_i(t),$$

where t is a set of parameters $(t_1, t_2, \ldots, t_n)$ and the function to be minimized is

$$I = g(t) + \int_{x_1}^{x_2} f(x,y,y')\,\mathrm{d}x.$$

Systematic studies of the Bolza problem languished until about 1930 when many memoirs were published by the school of mathematicians at Chicago.[1]

14.18. The Minimax Principle of George D. Birkhoff

The calculus of variations entered the new and wider field of 'global' problems with the work of George D. Birkhoff and Morse.

The study of geodesics on a two-dimensional surface presents problems in the *global* theory of the calculus of variations, and, following the researches of E. T. Whittaker (1904) and Signorini (1912), George D. Birkhoff gave a general treatment of this subject and enumerated the 'minimax' principle for an n-dimensional continuum possessing m-fold linear connectivity (G. D. Birkhoff 1917).

Hahn (1908) had already established the existence of at least one continuous non-constant function on an abstract topological space, and Birkhoff assumes the existence of an analytic function mapping the points of what we should now call a differential manifold (of n dimensions) onto the real numbers.

Suppose that this function $J(P)$ possesses l minimum points, P_1, P_2, $\ldots$, P_l and that any two points P_i,P_j can be joined by two paths which lie in the continuum and which can be continuously formed into one another within the continuum, while $J(P)$ remains bounded, then there exist at least $m + l - 1$ points of minimax within the continuum.

14.19. Morse and Variational Problems in the Large

The history of the calculus of variations from Weierstrass to Bolza is characterized by the attempts to liberate the theory from the limitations imposed by 'small variations'. In fact this theory, like so much classical analysis, is essentially a 'local' theory or a theory 'in the small'. From the

---

[1] *Contributions to the Calculus of Variations*, Chicago, 1931–2, 1933–7.

year 1930 onwards the most fruitful advances in variational theory have been the development of a 'global' theory or a theory 'in the large'.

These developments could not have taken place without the discoveries in group theory and topology and they owe much to the genius of Poincaré, who was one of the great pioneers in what we may call 'macroanalysis'.

In the field of variational problems the initial stimulus to the work of Morse seems to have been the investigations of George D. Birkhoff.

The work of Morse began with an analysis in the large of a function f of several real variables $(x_1, \ldots, x_n)$ defined on a Riemann manifold R which is itself a differential manifold equipped with the set of connectivities which completely specify its global topological character.

At a 'critical point' $x_i = a_i$ of f, the partial derivatives

$$\partial f / \partial x_i, \quad 1 \leqslant i \leqslant n,$$

are all zero, and unless the point is 'degenerate', a linear transformation of the variables $x_i - a_i$ expresses the leading terms of f in the form

$$-z_1^2 - z_2^2 - \ldots - z_k^2 + z_{k+1}^2 + \ldots + z_n^2.$$

The 'index' of the critical point is the number k.

Generalizing Birkhoff's minimax principle, Morse (1925) obtained the following relations between the number M_i of critical points of index i (of a non-degenerate function defined on R) and the connectivities R_j (mod 2) of R.

$$M_0 \geqslant R_0,$$

$$M_0 - M_1 + \ldots + (-1)^k M_k \geqslant R_0 - R_1 + \ldots + (-1)^k R_k, \quad 0 \leqslant k \leqslant n.$$

Here n is the dimension of R and when $k = n$ the last relation becomes an equality. Moreover, the same relation holds, not only for the whole of R but also for a suitably defined sub-region $\Sigma \subset R$. These are the famous 'Morse inequalities'.

Morse extended these relations to degenerate critical points (which may form closed sets) by defining equivalent sets of non-degenerate points in terms of singular chains and cycles on R.

Having laid this new and firm foundation Morse proceeded to the bold extension of the theory of the critical points of functions to a theory of variational problems in the large (Morse 1934).

Previous theories had attempted to study only the minimum and maximum values of an integral such as

$$\int_a^b F(x,y,y') \, dx.$$

In Morse's work such an integral is regarded as a *functional* and the curves $y_i = y_i(x)$ joining the end-points $x = a$, $x = b$ are regarded as 'points' in a space Ω. The curves determined by the Eulerian equations are the 'critical points' of the functional, and the Morse theory is not restricted to an examination of maxima and minima, but studies also the totality of minimaxes.

This extension of the theory presents new analytical and topological difficulties. It is necessary to replace the Fréchet definition of the distance between two curves in a Riemannian manifold R by a fresh definition. To deal with the infinitely many non-zero connectivity numbers of Ω a theory of deformations has to be created. The Morse theory triumphantly surmounts all these problems and issues in a complete and detailed generalization of the Morse inequalities given above.

14.20. *Tonelli and Semi-continuity*

The object of Tonelli's investigations (Tonelli 1921, 1923) was to construct an abstract existence theory which would play the same part in the calculus of variations as the classical minimum principle does in the theory of functions of real variables. In that theory a continuous function $f(x)$ defined on a bounded, closed set of points, say $a \leqslant x \leqslant b$, assumes its minimum value at some point of the set.

In the calculus of variations we consider a functional $I(f)$ defined on a certain class of functions $f(x)$ and we require its minimum. A new order of difficulty arises in this problem, which is illustrated by the following example.

The simplest problem of the calculus of variations is to determine the curve of minimum length joining two prescribed points A and B. If we restrict ourselves to curves of the form $y = f(x)$ where f is a differentiable function, then the curve length is given by an integral of the form

$$I(y) = \int_0^1 (1 + y'^2)^{\frac{1}{2}} \, dx; \quad y(0) = 0, \quad y(1) = 1.$$

Now the length of a curve is defined as the supremum of the total length of the sides of the inscribed polygons, and it therefore seems an artificial restriction to consider only those curves, $y = f(x)$, which are represented by a function $f(x)$ differentiable at all points which is continuous, monotone increasing with derivative equal to zero almost everywhere. Then the minimum length is no longer $\sqrt{2}$ but exactly 1.

There is clearly a great difference between the problem of minimizing a continuous function and minimizing the functional which represents the length of a curve. A continuous function is defined on a closed set of points $0 \leqslant x \leqslant 1$, but if the functional is defined only for differentiable functions, then these functions do not form a closed set. And, if we take $I(y)$ to be the length of any curve $y = f(x)$ (or polygon) joining the same end-points A and B, then $I(y)$ is not a continuous functional of $f(x)$.

The continuity of functionals is therefore not a satisfactory basis for existence theorems in the calculus of variations. Tonelli's researches were undertaken to discover and exploit an alternative foundation. A possible alternative was suggested to Tonelli by W. H. Young, viz. the concept of 'semi-continuity'.

The idea of 'semi-continuity' was first introduced by Baire (1904). A function $f(x)$ defined for a closed set of points x is lower semi-continuous at x_0 if

$$-\infty < f(x_0) \leqslant \underline{\lim}\, f(x) \quad \text{as} \quad x \to x_0,$$

(and there is a similar definition for upper semi-continuity). To extend this concept to a functional $I(f)$ defined for a curve $y = f(x)$ Tonelli considers all curves

$$x_k = x_k(t) \qquad 0 \leqslant t \leqslant 1, \quad k = 1, 2, \ldots, n,$$

in n-dimensional space, lying in the a fixed 'cube',

$$|x_k| \leqslant L,$$

with uniformly bounded lengths, each less than K, so that

$$|x_k(t_2) - x_k(t_1)| \leqslant K|t_2 - t_1|.$$

The crucial condition introduced by Tonelli is to consider only those functionals $I(C)$ which are semi-continuous in curves C of uniformly bounded length in a given cube. With this new basis Tonelli was able to prove that $I(C)$ actually attains its minimum for all such curves C.

14.21. L. C. Young and Generalized Curves

This brief account of the development of the calculus of variations may fittingly close with an indication of the great advances made by L. C. Young in his magisterial treatise (L. C. Young 1969). This immensely readable book provides an admirable account of Young's researches in a witty, cultured and sophisticated exposition. (The preceding account of Tonelli's work is based on the summary given by Young.)

L. C. Young's researches in this field began in 1933 when he began to introduce the concept of generalized curves in a Euclidean space of n dimensions (L. C. Young 1933). These 'curves', given by parametric vector equations

$$x = x(t), \quad t_0 \leqslant t \leqslant t_1$$

resemble a polygon in as much as they have a number (finite or enumerable) of directions specified by unit vectors θ_i. The attachment of these vectors to the curve is specified by functions $l_i(t)$ which give the Lebesgue measure of the set of points in the interval $[t_0, t]$ for which $dx/dt = \theta_i$. Then

$$\sum_i l_i(t)\, \theta_i = x(t) - x_0.$$

This concept was greatly extended and exploited in a series of papers published from 1938 to 1965 and expounded in the treatise referred to above (L. C. Young 1969).

In each problem of the calculus of variations we have to consider not just one 'Lagrangian' function f of a vector function $x(t)$ and its derivative $\dot{x}(t)$, but a multiplicity of functions, $x(t)$, $\dot{x}(t)$, usually considered as specifying smooth curves C. The revolutionary concept introduced by L. C. Young is to consider the curves C as functions g acting on the Lagrangian f to give

$$g(f) = \int_{t_1}^{t_2} f[x(t), \dot{x}(t)] \, dt.$$

This considerably simplifies the theory and dispenses with Tonelli's consideration of semi-continuity when $x(t)$ satisfies a Lipschitzian condition. It is also of great value in applications to Optimal Control Theory which forms the subject of the second part of L. C. Young's lectures.

15. Potential Theory

15.1. Extremals of Multiple Integrals

The basic problem of the variational calculus, to determine the minimum of an integral

$$\int_a^b f(x,y,y')\, \mathrm{d}x, \qquad y' = \mathrm{d}y/\mathrm{d}x,$$

is easily generalized to integrands $f(x, y_1, y_2, \ldots, y_1', y_2', \ldots)$ which depend upon several known functions $y_1, y_2 \ldots$ and the solution is found to depend upon the solution of the Eulerian equations. The necessary existence theorems have been available since the researches of Picard, and the variational problem is therefore, in essence, soluble.

But the situation is quite different when we consider multiple integrals such as

$$J = \iint f(x,y,u,u_x,u_y)\, \mathrm{d}x\, \mathrm{d}y$$

where u is a function of x and y, and the domain of integration is a region of the x,y plane. The Eulerian equation which provides necessary conditions for a minimum of J, is easily found in the form

$$\frac{\partial f}{\partial u} = \frac{\partial}{\partial x}\frac{\partial f}{\partial u_x} + \frac{\partial}{\partial y}\frac{\partial f}{\partial u_y},$$

where $u_x = \partial u/\partial x$ and $u_y = \partial u/\partial y$. But there remains the major problem of solving this partial differential equation, subject to prescribed values of u on the boundary of the domain of integration.

Even in the simplest case of the potential function u where

$$f = u_x^2 + u_y^2$$

and the Eulerian equation is

$$u_{xx} + u_{yy} = 0,$$

the question of the existence of solutions remained a problem which was not completely resolved until the researches of Poincaré (1887b).

In fact the researches on 'potential theory' have extended into a large and increasing domain which has penetrated and influenced the whole theory of partial differential equations and has greatly enriched our knowledge of classical analysis in general and of variational problems of multiple integrals in particular. We shall therefore devote a whole chapter to the history of potential theory.

226

Potential theory has developed out of the vector analysis created by Gauss, Green and Kelvin for the mathematical theories of gravitational attraction, of electrostatics and of the hydrodynamics of perfect fluids (i.e. incompressible and inviscid fluids). The first stage of abstraction was the study of harmonic functions, i.e. potential functions in space free from masses, charges, sources or sinks. This led to the inspired intuition of Dirichlet and the early attempts to justify his 'principle'.

The second stage in potential theory saw the introduction of measure theory and the powerful instrument of Lebesgue integration. The main topics of investigation were the concepts of 'energy' and 'capacity'.

In the third stage the distribution theory of Laurent Schwartz enabled potential theory to be developed with much greater precision, generality and simplicity.

Finally the introduction of topological abelian groups reduced potential theory to the study of the convolutions of measures.

All these investigations refer to potential theory in Euclidean space, but it is also appropriate to include in this chapter the generalization of harmonic functions to Riemann spaces of any number of dimensions and connectivity.

There is an extensive literature on potential theory, and Kellogg (1953), Brelot (1952, 1969) and Landkof (1972) give excellent historical summaries.

15.2. *Vector Analysis of Gauss, Green and Kelvin*

We must go back beyond the limits of our hundred years to uncover the roots of potential theory in the early, classical period of vector analysis, which is characterized by great physical intuition and bold, informal arguments, whose lack of rigour was an inevitable consequence of the lack of such essential techniques as a theory of integration and of convergence.

The starting point is Newton's law of gravitation which gave an expression for the intensity E of the gravitational force per unit mass exerted at a 'field point' P due to a given distribution of attracting matter. For a continuous distribution of attracting matter with density $\varrho(Q)$ at a 'charge' point Q, the gravitational intensity was expressed as the volume integral

$$E = -\int \varrho(Q) R^{-3} \boldsymbol{R} \, d\omega_Q$$

where $\boldsymbol{R}$ is the vector $\overrightarrow{PQ}$ and $d\omega_Q$ the element of volume at Q (suitable non-dimensional units being employed). But physicists have never hesitated to write this as a Stieltjes integral in the form

$$E = -\int R^{-3} \boldsymbol{R} \, dm_Q,$$

where dm_Q is an 'element' of mass at Q.

In the case of an electrostatic field ϱ is the density of electric charge (and may be positive or negative) and in hydrodynamics ϱ is a fictitious density of sources or sinks.

The pictorial representation of a field of force is due to Faraday, who introduced the concept of 'lines of force', i.e. curves in space, whose parametric equations

$$x = f(t), \quad y = g(t), \quad z = h(t),$$

are the integrals of the equations

$$\frac{\mathrm{d}x}{X} = \frac{\mathrm{d}y}{Y} = \frac{\mathrm{d}z}{Z} = \mathrm{d}t$$

where X, Y, Z are the components of the intensity $\boldsymbol{E}$. Thus the direction of $\boldsymbol{E}$ at a point P is along the tangent at P to the line of force passing through P.

In hydrodynamics the concept of a 'field of force' is replaced by the concept of a 'field of flow' and the intensity $\boldsymbol{E}$ is replaced by the local velocity of flow $\boldsymbol{v}$. This leads to the definition of the total 'flux' through a surface σ as the integral of the normal component of $\boldsymbol{v}$ over σ in the form $\int \boldsymbol{v} . \mathrm{d}\sigma$.

In gravitational theory the flux is similarly defined as $\int \boldsymbol{E} . \mathrm{d}\sigma$ and the fundamental theorem (Gauss 1813) takes the definitive form

$$\int \boldsymbol{E} . \mathrm{d}\sigma = \int \operatorname{div} \boldsymbol{E} . \mathrm{d}\omega$$

where σ is a closed surface bounding the volume ω and the 'divergence' of $\boldsymbol{E}$ is given by

$$\operatorname{div} \boldsymbol{E} = \frac{\partial X}{\partial x} + \frac{\partial Y}{\partial y} + \frac{\partial Z}{\partial z},$$

in a form invariant for all Cartesian coordinates x, y, z.

Now, by Stokes's theorem (Stokes, 1854)

$$\oint \boldsymbol{E} . \mathrm{d}\gamma = \int \operatorname{curl} \boldsymbol{E} . \mathrm{d}\sigma$$

where curl $\boldsymbol{E}$, the vector with Cartesian co-ordinates,

$$\frac{\partial Z}{\partial y} - \frac{\partial Y}{\partial z}, \quad \frac{\partial X}{\partial z} - \frac{\partial Z}{\partial x}, \quad \frac{\partial Y}{\partial x} - \frac{\partial X}{\partial y},$$

is integrated over any surface σ bounded by γ.

The presumed impossibility of withdrawing useful energy from a gravitational field without the expenditure of any mechanical work leads to the conclusion that the integral

$$\oint \boldsymbol{E} . \mathrm{d}\gamma$$

taken around any closed curve γ in free space is zero. Thus curl $\boldsymbol{E} = 0$, and therefore there exists a scalar potential function φ, such that $\boldsymbol{E} = \operatorname{grad} \varphi$, or, in Cartesian co-ordinates,

$$X = \partial\varphi/\partial x, \qquad Y = \partial\varphi/\partial y, \qquad Z = \partial\varphi/\partial z.$$

The electrostatics of charged conductors led to the concepts of 'capacity' and 'energy', which have dominated much recent work in potential theory. A single conductor in empty space is a closed surface carrying a total charge Q which produces a potential field φ, such that on the conductor φ is a constant, V. The linear relation between ϱ and $\boldsymbol{E}$ implies that Q/V is a constant C, the 'capacity' of the conductor. Its energy is given by physicists as $\frac{1}{2}QV$ or $\frac{1}{2}CV^2$ and by mathematicians as QV or CV^2.

The potential function φ is given by the integral

$$-4\pi\varphi = \int R^{-1}\varrho\,\mathrm{d}\omega$$

and it was inferred that φ must satisfy the equation

$$\varDelta\varphi = \operatorname{div}\operatorname{grad}\varphi = \varrho$$

where $\varDelta$ is the 'Laplacian' operator.

Green (1828, 1871) employed in his electrical researches a corollary to Gauss' divergence theorem in the form

$$\int(\varphi\operatorname{grad}\psi - \psi\operatorname{grad}\varphi)\,.\,\mathrm{d}\sigma = \int(\varphi\varDelta\psi - \psi\varDelta\varphi)\,\mathrm{d}\omega$$

where φ and ψ are two scalar functions.

A general solution of the potential equation can be obtained from Green's integral formula by taking ψ to be the fundamental potential function $1/r$, and by integrating over a region ω from which the point $r = 0$ is excluded by a small sphere $r = \varepsilon$.

In order to extend the theory of electrostatics to magnetostatics it was necessary to contemplate the existence, not only of point sources with charge m and potential $\varphi = -m/4\pi R$, but also of doublets with moment $\boldsymbol{m}$ and potential

$$\varphi = -\boldsymbol{m}\,.\operatorname{grad}(1/4\pi R)$$

A 'magnetic shell' was introduced as an ideal distribution of doublets over a surface σ such that the moment per unit area at any point P of σ was a vector $m\boldsymbol{n}$ where $\boldsymbol{n}$ is the normal to σ at P. The potential of such a surface distribution of doublets is

$$\varphi = -\int m\operatorname{grad}(1/4\pi R)\,.\,\mathrm{d}\sigma$$

The achievements of this first period of potential theory can perhaps be best summarized in the formula which shows that the potential φ in a region ω, bounded by a closed surface σ, can be expressed as the effect of a volume distribution in ω of density ϱ, a surface distribution of sources over σ of density s and a surface distribution of doublets over σ with density m, in the form

$$\varphi = \int\varrho R^{-1}\,.\,\mathrm{d}\omega + \int sR^{-1}\,.\,\mathrm{d}\sigma + \int m\operatorname{grad}R^{-1}\,.\,\mathrm{d}\sigma$$

where $\varrho = \varDelta\varphi$, $s = \operatorname{grad}\varphi$ and $m = -\varphi$, the local values of $\operatorname{grad}\varphi$ and φ at points on σ.

The unit point charge which produces a potential

$$\varphi = -1/(4\pi R)$$

has a volume density $\delta(\mathbf{r}) = {}^*\varDelta\varphi$ which is zero everywhere except at the origin where it is so infinite that (by a forgivable abuse of language) the convolution of any continuous function $f(\mathbf{r})$ and $\delta(\mathbf{r})$ is

$$\int f(\mathbf{r})\delta(\mathbf{r} - \mathbf{a})\mathrm{d}\omega = f(\mathbf{a}).$$

---

[1] The main contribution of Heaviside was not the development of potential theory, but the liberation of vector analysis from quaternionic shackles and the familiarization of electrical engineers with vector notation and method (Heaviside 1893b).

The 'generalized' function $\delta(\mathbf{r})$ is the famous Dirac function in three-dimensional space and one of the forerunners of Laurent Schwartz's 'distributions' (Section 12.10).

15.3. The Three Species of Potential Problems

There are three (or four?) species of mathematical problems associated with the potential equation,

$$\operatorname{div} \operatorname{grad} \varphi = 0$$

which are traditionally associated with the names of Dirichlet, Neumann and Robin, although many others had studied these equations before the mathematicians just mentioned. These problems are generalizations of certain problems of heat.

(a) Electrostatic problems. In electrostatics one of the earliest problems (Sir William Thomson 1848) was to calculate the potential V_0 of a closed conducting surface S carrying a total electric charge Q. The mathematical problem was to determine the potential function V at all points, outside S, which took the value V_0 at each point of S. The charge on S could then be found from Gauss's equation

$$4\pi Q = \int \operatorname{grad} V \cdot dS.$$

In the so-called Dirichlet's problem the potential V is required to reduce to a given non-constant function $f(P)$ at point P on the surface S, and there arises at once the question—for what functions $f(P)$ and for what surfaces S is the problem soluble?

(b) Hydrodynamic problems. In the hydrodynamics of a perfect fluid devoid of viscosity, compressibility and thermal conductivity, the fluid velocity $\mathbf{v}$ is the gradient of a potential function, $\mathbf{v} = \operatorname{grad} \varphi$. The fundamental problem is to determine the flow due to a solid body moving through a fluid, which may extend to infinity or be limited by various impervious boundaries, as in a wind-tunnel. The boundary conditions are that the velocity component along the normal $\mathbf{n}$ to the surface of the body (at each point P of S) is equal to the normal velocity of the point P of the body. In the so-called Neumann problem (C. Neumann 1887) the potential φ is required to satisfy a condition of the form

$$\mathbf{n} \cdot \operatorname{grad} \varphi = f(P)$$

where f is a prescribed function of P, an arbitrary point on S.

(c) Problems of thermal conduction. In the problem of the cooling of a conducting body by thermal reduction, the temperature T in the body satisfies the potential equation inside the body, and the flux of heat from the surface S satisfies an equation of the form

$$\mathbf{n} \cdot \operatorname{grad} T + kT = 0,$$

at each point of S (assuming that the rate of loss of heat by reduction is proportional to the temperature T, the temperature of the surroundings being taken to be zero).

Here k is a constant, but in the problem called after Robin (Robin 1887) k is a function of position on the surface S.

(d) Problems of elasticity. Long before this history opens Saint-Vincent (Love 1927) had solved the problem of the torsion of a cylinder of arbitrary cross-section

$$f(x,y) = \text{const.}$$

by expressing the displacements in the form

$$u = -\tau yz$$

$$v = \tau zx$$

$$w = \tau\varphi(x,y)$$

where τ is the twist and φ satisfies the potential equation

$$\partial^2\varphi/\partial x^2 + \partial^2\varphi/\partial y^2 = 0,$$

inside the cylinder, and the condition

$$\partial\varphi/\partial v = ly - mx$$

on the boundary, l and m being the direction cosines of the normal v.

The necessity of employing this awkward boundary condition can be avoided by introducing $\psi(x,y)$ the conjugate complex function of $\varphi(x,y)$ so that $\varphi + i\psi$ is an analytic function of $x + iy$. Then ψ is also a potential function of x and y, while the boundary condition assumes the simpler form

$$\psi - \tfrac{1}{2}(x^2 + y^2) = \text{const.}$$

15.4. *The Maximum Minimum Principle*

Gauss was familiar with a property of harmonic functions which later became of fundamental importance, viz. the mean value of any harmonic function φ averaged over a sphere of centre O is equal to the value of φ at O (Gauss 1840; p. 222).

In fact this 'theorem of the arithmetic mean' serves to characterize harmonic functions and Koebe (1906) proved that, if a function f is such that its spherical mean over any sphere in a region R is equal to its value at the centre of the sphere, then f must be harmonic in the interior of R.

A direct deduction from the theorem of the arithmetic mean is that a harmonic function cannot attain a maximum or minimum in any open region, but must attain these values at points on the boundary. A similar result, attributed by Maxwell (1892) to Earnshaw is that 'a charged body placed in a field of electric force cannot be in stable equilbrium.'

15.5. *'Dirichlet's Principle'*

If the physical facts of gravitation and electrostatics are correctly represented by the potential theory we have just summarized, then we can infer the existence of harmonic functions in a number of interesting conditions.

Thus Green (1828) (Section 15.2) argued that, if a unit charge is placed at a point Q inside a closed conducting surface σ maintained at zero potential, there must exist a function φ such that

$$\varphi - (4\pi R)^{-1}$$

is harmonic in the region ω bounded by σ, while $\varphi = 0$ on σ, R being the distance from the charge point Q to the field point P. This function $\varphi(P)$ is the famous 'Green's function'.

But one of Green's most remarkable researches (Green 1835) refers to the integral

$$V = \int \frac{d\xi_1 d\xi_2 \ldots d\xi_s \varrho(\xi_1, \xi_2, \ldots, \xi_n)}{[(x_1 - \xi_1)^2 + (x_2 - \xi_2)^2 + \ldots + (x_s - \xi_s)^2 + u^2]^{\frac{1}{2}(n-1)}},$$

taken over the interior of the degenerate ellipsoid E with $n \geqslant s$

$$\sum_{k=1}^{s} x_k^2/a_k^2 < 1,$$

the 'density function' ϱ being arbitrary. The inclusion of the variable u enabled Green to avoid the difficulties which would otherwise arise at the singularity at the point $\xi_k = x_k$ $(k = 1, 2, \ldots, s)$. Green proved that the function V must satisfy the generalized potential equation

$$\sum_{k=1}^{s} \frac{\partial^2 V}{\partial x_k^2} + \frac{\partial^2 V}{\partial u^2} + \frac{n-s}{u} \frac{\partial V}{\partial u} = 0.$$

Moreover he asserted that there must exist a function V_0 which minimizes the integral

$$\int \left\{ \left(\frac{\partial V}{\partial x_1}\right)^2 + \left(\frac{\partial V}{\partial x_2}\right)^2 + \ldots + \left(\frac{\partial V}{\partial x_s}\right)^2 + \left(\frac{\partial V}{\partial u}\right)^2 \right\} d\xi_1 d\xi_2 \ldots d\xi_s du \cdot u^{n-s}$$

(subject to certain conditions on the surface of the ellipsoid E), and that this function must satisfy the generalized potential equation.

This appears to be the first mention of the famous minimum principle subsequently used by Kelvin and later named after Dirichlet (1876) who had used similar considerations in his lectures in 1856–7.

Gauss (1840) considered the potential φ due to a distribution of positive density s on a closed surface σ with an assigned total mass M, and asserted that the integral $\int(\varphi - 2f)s\, d\sigma$ must attain a minimum when $\varphi - f$ is constant on σ, f being any prescribed continuous function on σ.

In 1847 Kelvin (Sir William Thomson 1847) was freely using a minimum principle for the integral

$$\int (\mathrm{grad}\ v)^2\, dx_1\, dx_2\, dx_3$$

and stating that Q attains its minimum value when

$$\operatorname{div} \operatorname{grad} v = 0.$$

It was Riemann who gave its name to the 'Principle of Dirichlet' in his paper on Abelian functions (Riemann 1857b), in which he avoided the introduction of Jacobi's theta functions by the consideration of the integrals of the form

$$\Omega = \int \left\{ \left(\frac{\partial \alpha}{\partial x} - \frac{\partial \beta}{\partial y} \right)^2 + \left(\frac{\partial \alpha}{\partial y} + \frac{\partial \beta}{\partial x} \right)^2 \right\} dv$$

taken over a 'Riemann surface' covering the x,y plane n times. Since $\Omega \geqslant 0$, Riemann inferred that for one form at least of the function α, Ω attains its minimum value (the function β being fixed).

There is no doubt that the energy integral

$$I = \int (\operatorname{grad} f)^2 \, d\omega$$

extended over a bounded region ω is bounded below, and that, if the values of f are prescribed over the boundary σ of ω, then I must possess a non-negative minimum. Moreover, if this minimum is actually attained when f is a function φ, then φ must be harmonic in ω.

The analogy with the theory of the minimum of continuous functions is incomplete, as Weierstrass demonstrated in his lectures (Weierstrass 1870). A continuous function defined in a closed interval necessarily attains its minimum in that interval. But no corresponding theorem was then available for the energy function I, which is a *functional*, even granted that, with suitable conventions, I is continuous in f.

The example given by Weierstrass is to minimize the integral

$$J = \int_{-1}^{1} x^4 \varphi'^2(x) \, dx$$

subject to the conditions $\varphi(-1) = a$, $\varphi(+1) = b$. Ignoring the Eulerian equations, Weierstrass noted that if

$$\varphi(x) = \tfrac{1}{2}(a + b) + \tfrac{1}{2}(b - a)(\tan^{-1}x/\varepsilon)/(\tan^{-1}1/\varepsilon),$$

then the boundary conditions are satisfied and

$$J < \tfrac{1}{2}\varepsilon^2(b - a)^2[\tan^{-1}(1/\varepsilon)]^2.$$

Hence the minimum value of J is zero. But this value can be attained only if $x^2\varphi'(x) = 0$ or $\varphi(x) = \text{const.}$ which is inconsistent with the boundary conditions.

15.6. Poisson's Integral

There is, fortunately, one problem where Dirichlet's problem can be completely solved, viz. when the values of the potential function φ are prescribed on a sphere ω.

Consider a sphere of centre O and radius a and two inverse points P, P'

on a radius OPP' at distances p and p' from O, so that $pp' = a^2$. Then the Green's function for a unit charge at P is

$$G(Q,P) = 1/r - a/(pr')$$

where $\qquad\qquad r = |PQ| \quad$ and $\quad r' = |P'Q|.$

Green's identity then allows us to express the potential φ at P in the form

$$4\pi\varphi(P) = -\iint\varphi(Q)\,\mathrm{grad}_Q\,G(Q,P)\,.\,\mathrm{d}\omega_Q$$

where $\mathrm{d}\omega_Q$ is a vector element of area and this is easily reduced to the form

$$\varphi(P) = -(p^2 - a^2)(4\pi a)^{-1}\!\int\varphi(Q)r^{-3}\,\mathrm{d}\omega.$$

This integral is usually attributed to Poisson but it does not appear in the papers (Poisson 1820, 1823) quoted by most writers on potential theory. Poisson's integral refers to the special case when the point P is on the surface of the sphere. The earliest form of 'Poisson's integral' (for the interior of a circle) that I have found is in the researches of H. A. Schwartz (1872).

15.7. The Construction of Harmonic Functions from Sequences

Under the fire of Weierstrass' criticism, Dirichlet's principle was temporarily abandoned as evidence of the existence of harmonic functions in favour of the construction of harmonic functions as limits of certain sequences of continuous functions.

Following H. A. Schwartz and C. Neumann, we may construct a sequence of harmonic functions with boundary values converging to a prescribed function, or following H. Poincaré we may construct a sequence of functions each of which has the assigned boundary values and which converge to a harmonic function.

The difficulty is that the sequences of functions most readily constructed in the hope of solving the Dirichlet problem may not be convergent, and, even if they are convergent, they require special investigation to establish that these limits are harmonic.

Stimulated by the desire to solve Dirichlet's problem, the nineteenth-century mathematicians triumphed over all these difficulties.

One of the most remarkable and general techniques was the discovery of Ascoli (1883) and Arzela (1895). It relates to the extraction of a convergent sequence from any enumerable collection of bounded, continuous functions $\{f_n(x)\}$ defined in a closed and bounded domain R. Such a collection is said to be 'equicontinuous' if each function $f_n(x)$ satisfies the same condition of continuity, viz. that, for any tolerance $\varepsilon > 0$ there exists a 'modulus' of continuity $\delta(\varepsilon)$ (independent of n), such that, if the distance from x to y is less than $\delta(\varepsilon)$ then,

$$|f_n(x) - f_n(y)| < \varepsilon.$$

The theorem states that, from any enumerable collection of functions, uniformly bounded and equicontinuous in R, we can extract a sequence $\{f_{\varphi(n)}(x)\}$, such that

$$n \leq \varphi(n) < \varphi(n+1)$$

and $f_{\varphi(n)}(x)$ converges uniformly in R to a continuous limit.

There are other theorems of convergence restricted to sequences of harmonic functions and these we shall now briefly describe.

15.8. Harnack's Convergence Theorems

As a preliminary to a study of convergent sequences of harmonic functions, Harnack (1887) showed that Poisson's integral (Section 15.5) leads immediately to the following inequality for a harmonic function φ which is positive in the interior of the sphere, viz.

$$\frac{a(a-p)}{(a+p)^2}\varphi(O) \leq \varphi(P) \leq \frac{a(a-p)}{(a-p)^2}\varphi(O)$$

where $p = |OP|$.

which relates the values of φ at the centre O and at a point P inside the sphere.

A similar inequality holds for any harmonic function which is positive in a closed region R, viz.

$$c\varphi(O) \leq \varphi(P) \leq k\varphi(O)$$

where the positive constants c and k depend only on R.

For sequences of harmonic functions Harnack established two convergence theorems:

(1) From the maximum–minimum theorem he deduced that if $\{\varphi_n\}$ is a sequence of functions harmonic in ω and converging uniformly on the boundary σ of ω, then φ_n converges uniformly throughout ω to a function φ harmonic in ω, and each partial derivative of φ_n converges to the same partial derivative of φ.

(2) From the inequality mentioned above, he inferred that an increasing sequence of functions $\{\varphi_n\}$ harmonic in ω, which is bounded at a single point O must converge uniformly in ω to a harmonic function.

15.9. The Alternating Procedure of H. A. Schwartz

The first serious attempt to dispense with Dirichlet's principle was made by H. A. Scwartz (1870) for the problem of constructing a harmonic function φ in a region ω of the x,y plane, such that φ takes assigned continuous values on the boundary σ of ω.

Schwartz's method depends upon an ingenious alternating method for solving this boundary value problem for a region ω_3 which is the intersection of two regions ω_1 and ω_2 for which the boundary value problem has already been solved (Courant and Hilbert 1953, vol. II, pp. 293–8).

By repeating this procedure a finite number of times we can solve the boundary value problem for any admissible region G which is the union of a finite number of overlapping domains, each of which is a disc, semi-disc, or half-plane—for which cases elementary solutions are readily constructed by the use of Poisson's integral.

Finally we can solve the problem for any domain G which can be covered by an enumerable set of overlapping admissible domains G_1, G_2, ... provided that the boundary of G is suitably smooth.

It is tempting to infer that a knowledge of Schwartz's alternating procedure inspired Poincaré to construct his own method of 'balayage' or 'sweeping'. But, since we have been assured by Freudenthal (Section 13.9) that Poincaré seldom troubled to study the writings of his predecessors, such an inference would be hazardous.

15.10. *C. Neumann and the 'Method of the Arithmetic Mean'*

C. Neumann (1870) invented the 'method of the arithmetic mean' to establish the existence of a harmonic function φ which takes prescribed values m on a *convex* surface σ.

His method depends upon a property of the potential φ due to a distribution of doublets over a closed surface σ. If the doublets are normal to the surface and have surface density m, then in the convenient units employed by Neumann,

$$2\pi\varphi = \int m\partial r^{-1}/\partial n \, d\sigma.$$

$\partial/\partial n$ denoting the derivative of r^{-1} along the normal to σ. Hence if the surface σ is convex, the value of φ at a point P on the surface is given by

$$2\pi\varphi(P) = -\int m \, d\Omega$$

where $d\Omega$ is the solid angle subtended at P by the element $d\sigma$ of the surface, i.e. $-\varphi(P)$ is the arithmetic mean of the values of m, and

$$\min m \leqslant -\varphi(P) \leqslant \max m.$$

Neumann's procedure (C. Neumann 1877, 1901) is to construct a sequence of harmonic functions $\{\varphi_n\}$ such that

$$2\pi\varphi_n = -\int \varphi_{n-1}^0 \partial r^{-1}/\partial n \, d\sigma,$$

φ_n^0 being the value of φ_n on the surface, and to use the relations

$$\varphi_n^- = \varphi_{n-1}^0 + \varphi_n^0$$
$$\varphi_n^+ = -\varphi_{n-1}^0 + \varphi_n^0$$

which connect the limiting values of $\varphi_n(P)$ as the point P approaches the surface from the interior of or from the exterior.

Neumann proved that, if σ is convex, the sequence $\{\varphi_n^0\}$ converges uniformly to the same constant c at all points of σ and the series

$$\sum_n (\varphi_{2n-1} - \varphi_{2n})$$

converges uniformly to a harmonic function U in ω, and takes the boundary values $m - c$.

Thus the harmonic function

$$c + \sum (\varphi_{2n-1} - \varphi_{2n})$$

solves the interior Dirichlet problem and a slight modification of the process solves the exterior problem.

Poincaré (1895b, 1896–7) has simplified Neumann's process and has proved that the restriction to convex surfaces is unnecessary.

G. Robin (1887) has given a method similar to Neumann's but using a surface distribution of sources instead of doublets, and Poincaré has proved that Robin's method is not restricted to convex surfaces σ.

15.11. *Poincaré and the Method of Balayage*

The methods of Schwartz and Neumann provide not only existence theorems for solutions of Dirichlet's problem, but also methods of calculating the potential functions which solve the problem. These methods are unfortunately hopelessly impractical, and subsequent researches are sharply divided into two classes—those which provide a pure existence theorem with no pretensions to numerical solution, and those which provide methods for the effective calculation of potential functions in terms of known and tabulated functions.

The first pure existence theorem is due to Poincaré (1887, 1890) and depends upon the method of 'balayage' or sweeping, which has retained a permanent place in the more abstract developments of potential theory.

Poincaré's method is applicable to the boundary values which are taken on a surface σ by any polynomial f in the Cartesian coordinates x, y, z, or to the boundary values taken by a uniformly convergent sequence of such polynomials.

The region ω is covered by an enumerable sequence of spheres $\{S_n\}$ which are 'swept' in the order

$$S_1, S_2, S_1, S_2, S_3, S_1, S_2, S_3, S_4, \ldots$$

Poincaré begins by replacing the polynomial f in S_1 by the harmonic function f_1 which takes on the boundary of S_1 the same values as f. This process, which 'sweeps' the charge of density $\varrho = -(4\pi)^{-1}\Delta f$ from the interior of S_1 onto the surface of S_1, is specified by Poisson's integral (Section 15.5).

The continuous function, equal to f_1 in S_1 and to f elsewhere in ω, is denoted by W_1, and is then replaced in S_2 by the harmonic function with the same boundary values as W_1 on S_2.

The continual repetition of this process yields a sequence of functions $\{W_n\}$, each of which satisfies the prescribed boundary conditions on σ and which converges to the required harmonic function φ.

We shall terminate this account of attempts to solve Dirichlet's problem with Hilbert's contribution, which is often described as a 'reestablishment' of Dirichlet's principle, but which is really a hopeful suggestion, later developed by Courant.

15.12. *Hilbert and the Direct Solution of Dirichlet's Problem*

Hilbert (1900) claimed that every problem of the calculus of variations has a solution (or at least a 'generalized' solution) provided the boundary conditions are suitably restricted in character, and he sketched the solution of two famous problems, viz. the problem of geodesics on a given surface, and Dirichlet's problem for a plane region ω.

The idea is to start from a sequence of functions $F_1(x,y)$, $F_2(x,y)$, ... satisfying the boundary conditions and such that the Dirichlet integrals

$$D(F_n) = \int \operatorname{grad} F_n \cdot \operatorname{grad} F_n \, d\omega$$

tend to a limit k. Each function F_n is then replaced by another function G_n, also satisfying the boundary conditions, but 'smoothed' so that

$$|\operatorname{grad} G_n| < \tan \varphi \qquad (n = 1, 2, \ldots)$$

φ being some fixed angle $(0 < \varphi < \tfrac{1}{2}\pi)$. Then the Dirichlet integrals $D(G_n)$ also tend to k.

Hilbert claimed that there exists a subsequence of functions $\{f_n\}$, selected from the sequence $\{G_n\}$ such that $f_n(x,y)$ converges uniformly to the desired harmonic function.

15.13. *Insoluble Dirichlet Problems*

In the happy, carefree days of Poincaré, no excessive attention was paid to the nature of the surface σ bounding a region ω, nor to the nature of the boundary function f defined on σ, in terms of which Dirichlet's problem was stated.

However, while Liapunov was still a young student of P. L. Chebyshev, he ventured to challenge an *obiter dictum* of Poincaré that a problem (such as the equilibrium configuration of a rotating mass of fluid) which has an obvious physical meaning does not require a rigorous mathematical solution.

In his researches on potential theory Liapunov (1898) introduced new standards of rigour. In establishing the existence of Green's function, he was careful to restrict his investigations to a special type of surface, now called 'Liapunov surfaces' and to boundary functions satisfying a special type of continuity (Hölder continuity) more restrictive than uniform continuity.

The absolute necessity of some restrictions on the boundary surface σ and the boundary function f was made startlingly clear by the examples of insoluble Dirichlet problems given by Zaremba (1911) and H. Lebesgue (1924).

If the region ω is bounded externally by a sphere S and internally by a point sphere at the centre O of S, and if f is zero on S and unity at O, then, as Zaremba showed, there is no function harmonic in ω and taking these boundary values.

Lebesgue established the insolubility of the Dirichlet problem for a much more regular type of boundary, viz. the region inside the sphere, $r = 1$, and outside the 'Lebesgue spine',

$$(x^2 + y^2)^{\frac{1}{2}} = \exp(-1/z), \qquad z > 0.$$

Finally Hadamard (1906) showed that even if there exists a harmonic function

$$u = \tfrac{1}{2}a_0 + \sum a_n r^n \cos n\theta$$

represented by a series convergent in the unit disc, $r \leqslant 1$, the Dirichlet integral

$$\int (\operatorname{grad} u)^2 r \, dr \, d\theta$$

may be infinite. In fact it suffices to take

$$a_n = 0 \quad \text{if} \quad n \neq 2^{2p}, \quad a_n = 2^{-p} \quad \text{if} \quad n = 2^{2p}.$$

A similar example, also due to Hadamard, is provided by the harmonic function

$$u = \sum r^\nu n^{-2} \sin \nu\theta$$

where
$$\nu = n!$$

15.14. *Capacity*

A further advance in the study of boundary values is due to Wiener (1924) who examined this problem in the light of the concept of 'capacity'.

In electrostatics if σ is a closed surface (with regular points only) and u is the harmonic function which attains the uniform boundary value $u = V$ on σ then the charge on σ is

$$Q = -(4\pi)^{-1} \int \operatorname{grad} u \cdot d\sigma$$

and its capacity is
$$c = Q/V.$$

Wiener extended this concept from a smooth surface to the boundary σ of a domain ω consisting of an open set of points Ω together with its limit points Ω^*. If a method of sequences gives a function u, harmonic outside σ, constructed from the boundary function $f = 1$ on σ, then u is called the 'conductor' function of Ω and the capacity of Ω is defined by the Gauss integral as for a smooth surface σ.

Now a generalized solution of Dirichlet's problem attains the prescribed boundary values at the 'regular' points of σ. Wiener showed that a regular point q can be characterized in terms of the capacity γ_n of the set of points in ω lying between two spheres of radii λ^n and λ^{n+1} of centres q. Then q is regular or exceptional accordingly as the series

$$\sum q_n \lambda^{-n}$$

diverges or converges.

15.15. *Superharmonic and Subharmonic Functions*

The first reported appearance of superharmonic and subharmonic functions is in a paper by Hartogs (1906) which is mainly concerned with power series of the form

$$\sum_{n=0}^{\infty} y^n f_n(x)$$

where x and y are complex variables and $f_n(x)$ a holomorphic function of x in some region Ω. The series is convergent in a region $|y| < R(u,v)$ where

$$u + iv = x,$$

and Hartogs proved that $F(u,v) = \log R(u,v)$ is a subharmonic function of u and v, i.e.

$$\frac{\partial^2 F}{\partial u^2} + \frac{\partial^2 F}{\partial v^2} \leqslant 0.$$

The simplest examples are provided by the series

$$\sum y^n x^n$$

for which $|R| = |x|^{-1}$ and $F = -\tfrac{1}{2}\log(u^2 + v^2)$,

and $$\sum y^n \cos nx$$

for which $R = \exp -|u|$ and $F = -(u^2 + v^2)^{\tfrac{1}{2}}$.

But a formal and fully developed theory was not given until the researches of Perron (1923) and Riesz (1925, 1926, 1930).

In one dimension the harmonic functions satisfy the equation $d^2u/dx^2 = 0$ (and hence are represented by straight lines), and also satisfy the trivial form of Gauss' mean-value theorem,

$$2u(x) = u(x - h) + u(x + h).$$

Perron, whose definition of major and minor functions (Section 11.20) had so greatly simplified the theory of the Denjoy integral, applied a similar technique to the theory of potential. Gauss' theorem expresses any function h, harmonic in a closed sphere S as the mean value $M(h)$ of h over ω, the interior of S, or over σ, the surface of S. Perron defined a subharmonic function by the inequality

$$v \leqslant M(v)$$

and a superharmonic function by the inequality

$$v \geqslant M(v)$$

these relations being enforced for every sphere S in the domain of definition of v.

Subharmonic functions have many of the properties of harmonic functions, and the negative of a subharmonic function is superharmonic. In particular the potential u of a non-negative distribution f which satisfies the equation,

$$\Delta u = f,$$

is a subharmonic function.

Perron obtained a concise solution of Dirichlet's problem by considering the functions v which are subharmonic in a region ω and which satisfy the inequality $v \leqslant \varphi$ on σ the boundary of ω. Then Perron's solution is the function h, which is the least upper bound of all such subharmonic functions (which are bounded above since $v \leqslant \max \varphi$). In fact h is harmonic in ω and $h = \varphi$ on σ.

15.16. Barriers and the Attainment of Boundary Values

In order to examine the approach of a function u harmonic in ω to its limiting values on σ, the boundary of ω, Poincaré (1890; especially pp. 226–7) and H. Lebesgue (1912) introduced the concept of 'barrier' functions. If p is a field point in ω and q a point on the boundary σ, a function $w(p,q)$ is said to be a barrier for ω at the boundary point q if it is continuous and superharmonic in ω, if $w(p,q) \to 0$ as $p \to q$, and if it has a positive lower bound outside any sphere centre q.

Poincaré (1890) and H. Lebesgue (1912) showed that a necessary and sufficient condition for the existence of a solution u to the Dirichlet problem (u harmonic in ω, and u a given function f continuous on σ) is that a barrier $w(p,q)$ exists at every point of σ. It follows that the harmonic function u, constructed by the methods of sequences, due to Schwartz or Poincaré, approaches the prescribed boundary value at each "regular" point of σ, i.e. each point q for which a barrier $w(p,q)$ exists.

Even if the boundary σ is not sufficiently 'regular' to ensure the existence of a 'strict' solution of the Dirichlet problem, there may exist a 'generalized' solution in the sense that the harmonic function assumes the boundary value f 'on the average'. Thus in Weyl's method of orthogonal projection (Section 15.17), a function f prescribed in ω is to be split into the sum of a harmonic function u and an anharmonic function ω, which vanishes on σ, the boundary of ω. Then, if k is any function, harmonic in ω,

$$\int \operatorname{grad} k \,.\, \operatorname{grad} w \, d\omega = 0$$

where

$$\int w \operatorname{grad} k \,.\, d\sigma = 0.$$

Thus w is zero 'on the average' on σ.

15.17. Weyl and the Method of Orthogonal Projection

Weyl (1940) gave a completely new and powerful method of solving Dirichlet's problem by employing the theory of projection operators in a Hilbert space.

We know from Hadamard's researches that a necessary condition for the solution of the interior Dirichlet problem is that the boundary function on the surface σ must be the restriction to σ of a function f, defined in the region ω bounded by σ, and such that the Dirichlet integral

$$D(f) = \int (\operatorname{grad} f)^2 \, d\omega$$

is finite.

Now if the Dirichlet problem has a solution φ, and if

$$\psi = f - \varphi$$

then $D(\varphi)$ and $D(\psi)$ must be finite and ψ must be zero on σ.

Moreover

$$D(\varphi,\psi) = \int \operatorname{grad} \varphi \cdot \operatorname{grad} \psi \, d\omega$$
$$= \int \psi \operatorname{grad} \varphi \, d\sigma$$
$$= 0 \quad \text{since} \quad \psi = 0 \text{ on } \sigma.$$

Hence, if we regard f, φ and ψ as elements of a Hilbert space with the metric or norm specified by

$$\|f\| = [D(f)]^{\frac{1}{2}},$$

then the Dirichlet problem consists in splitting the given function f into two parts, the harmonic part φ and the anharmonic part ψ, such that

$$f = \varphi + \psi,$$
$$\Delta\varphi = 0 \quad \text{in} \quad \omega,$$
$$\psi = 0 \quad \text{on } \sigma,$$

and $$D(\varphi,\psi) = 0.$$

The operations by which φ and ψ are obtained from f are orthogonal projections onto the subspaces of harmonic functions and of anharmonic functions respectively. The existence of these projection operations is guaranteed by standard theorems of functional analysis.

15.18. *Fredholm's Integral Equations*

The first method which had any serious pretensions to furnish a numerical solution of Dirichlet's problem was due to Fredholm (1900, 1903). Poincaré (1895b) had already noticed that the Dirichlet problem and the Neumann problem for regions interior or exterior to a surface ω can be represented by integral equations. The Dirichlet problem is solved by expressing the potential φ at a point p in terms of a doublet distribution on σ, in the form

$$\varphi(p) = (2\pi)^{-1} \int m \operatorname{grad} r^{-1} \, d\sigma.$$

Then m satisfies the equation

$$\varepsilon f(p) = m(p) - \lambda \int m(q) K(p,q) \, d\sigma,$$

where in the interior problem, $\varepsilon = -1$ and $\lambda = 1$, while in the exterior problem $\varepsilon = 1$ and $\lambda = -1$, while

$$K(p,q) = (2\pi)^{-1} \partial r^{-1} / \partial v,$$

the differentiation being along the outward normal v to σ.

Similar relations hold in the Neumann problem when the potential is expressed in terms of a source distribution on σ.

Poincaré (1895b) had expressed the solutions of these equations as power series in λ, convergent when $|\lambda| \leqslant 1$, but Fredholm obtained a general solution, by considering the integral equations as limiting forms of linear algebraic equations such as

$$\varepsilon f(p_j) = m(p_j) - \lambda \sum m(p_k) K(p_j, p_k).$$

Fredholm's researches not only provide an existence theorem but also a method by which we can obtain approximate numerical values of the potential function.

15.19. *The Complex Newtonian Potential of Marcel Riesz*

The concept of fractional integration dates back to Liouville (1834) and to Riemann, who considered the integral

$$I(\alpha) = \frac{1}{\Gamma(\alpha)} \int_0^x (x - s)^{\alpha-1} f(s) \, ds$$

as a repeated integral of order α. Thus for $f(x) = 1$

$$I(\tfrac{1}{2}) = 2(x/\pi)^{\frac{1}{2}}.$$

But these authors considered only positive, real values of α. M. Riesz (1936, 1938, 1940) considered non-Newtonian potentials of the form

$$\int r^{-\alpha} \, d\mu$$

where r is the distance from the source $d\mu$ to the field point, but the parameter α is not equal to unity, but is a complex number.

Many properties of the Newtonian potential survive, and indeed, it is possible to go much further and to study 'potentials' of similar form in which R^α is replaced by some symmetric function of the charge point and the source point.

However, such generalizations are useless unless the energy principle is verified, and the complex Newtonian potentials and these generalizations are now only of interest in so far as they developed in two very different directions—in wave theory and in H. Cartan's topological group theory of potential. (§15.25)

15.20. *Evans and the Use of Stieltjes Integrals*

The introduction of superharmonic and subharmonic functions represents the greatest and final achievement of classical potential theory. The theory was transformed into a much more abstract and general theory by the use of the powerful methods of integration due to Stieltjes, Radon and Schwartz.

In the classical theory it was necessary to treat separately the potentials due to distributions over volumes, surfaces, lines and points. Evans (1920, 1927) seems to have been the first to use the Stieltjes integral as a uniform method of synthesizing these different representations of potentials. Thus Evans expresses the Poisson integral for the logarithmic potential of a distribution over the unit circle, $r = 1$, in the form

$$u(r,\theta) = \frac{1}{2\pi} \int_0^{2\pi} \frac{1 - r^2}{1 + r^2 - 2r \cos(\varphi - \theta)} \, dF(\varphi)$$

where $F(\varphi)$ is of bounded variation.

15.21. *Frostman and the Use of Radon Integrals*

By the use of a still more powerful method of integration due to Radon, Frostman (1935) gave a rigorous and concise treatment of a problem enunciated by Gauss.

The problem of equilibrium, i.e. to find the sources on a surface σ which make the potential constant over σ, was solved by Frostman by proving that, for the potentials $U(\mu)$ due to sources μ of positive measure on σ, the minimum energy $\int U(\mu)\, d\mu$ is actually attained. The three main concepts utilized by Frostman are:

(1) The principle of energy, viz. $\int U(\mu)\, d\mu$ is essentially non-negative, and vanishes only if $\mu = 0$.

(2) The maximum principle, viz. if $0 \leqslant U(\mu) \leqslant k$ on a closed set of points occupied by sources, then $U(\mu) \leqslant k$ everywhere.

(3) A physical concept of Lebesgue's idea of 'almost everywhere', in the form of 'everywhere except on a set S such that every compact set in S has zero capacity'.

15.22. *Orthonormal Systems of Harmonic Functions*

For the numerical solution of problems of potential theory, physicists have been accustomed to rely upon the well-known systems of harmonic functions, such as spherical harmonics, which figure so largely in the treatises on electromagnetic theory by Maxwell, Jeans, Smythe and their coworkers. With the advent of electronic computers these methods are no longer of such importance, but they have at least one permanent claim on our attention inasmuch as they provide proofs of the 'completeness' of the classical systems of harmonic functions.

A satisfactory theory of series of orthonormal functions became possible only after the invention of the Lebesgue integral, and is largely due to the researches of F. Riesz.

An enumerable sequence of functions $\{\varphi_n(x)\}$, $n = 1, 2, \ldots$ summable over a finite or infinite domain R, is orthonormal if

$$(\varphi_m, \varphi_n) = \int_R \varphi_m(x)\varphi_n(x)\, dx = \delta_{mn},$$

where the Kronecker symbol δ_{mn} is equal to 0 if m and n are distinct, or unity if $m = n$.

The sequence $\{\varphi_n\}$ is said to be complete if the only functions $f(x)$ of summable square, such that

$$(f, \varphi_n) = \int_R f(x)\varphi_n(x)\, dx = 0$$

for all n, are zero almost everywhere in R.

The most familiar example is the sequence of trigonometric functions

$$(2\pi)^{-\frac{1}{2}},\ \pi^{-\frac{1}{2}}\cos x,\ \pi^{-\frac{1}{2}}\sin x,\ \ldots,\ \pi^{-\frac{1}{2}}\cos nx,\ \pi^{-\frac{1}{2}}\sin nx,\ \ldots,$$

for the interval $-\pi \leqslant x \leqslant \pi$.

In general the 'Fourier constants' of a summable function $f(x)$ for an orthonormal sequence $\{\varphi_n\}$ are given by

$$c_n = (f, \varphi_n)$$

and the corresponding truncated 'Fourier series' is

$$f_n = \sum_{k=1}^{n} c_k \varphi_k.$$

The central theorem, due to F. Riesz (Section 12.4) is that $f_n(x)$ converges 'on the average' to the generating function $f(x)$, i.e.

$$\int_R [f(x) - f_n(x)]^2 \, dx \to 0 \quad \text{as} \quad n \to \infty.$$

In potential theory we consider the expression of a harmonic function in terms of an enumerable set of harmonic functions of certain special types, such as spherical harmonics. The crucial property of these special systems is that they are complete.

This result can be deduced from the classical theory of Legendre functions but it is more rapidly inferred from E. T. Whittaker's integral representation of harmonic functions (E. T. Whittaker 1903). Whittaker showed that *any* harmonic function which is analytic in the neighbourhood of the origin can be expressed in the form

$$\sum A_{mn} S_{mn}, \quad -n \leqslant m \leqslant n, \; n = 0, 1, 2, \ldots$$

where S_{mn} is the coefficient of e^{imu} in the expansion of

$$(z + ix \cos u + iy \sin u)^n.$$

This representation provides an enumerable set of linearly independent harmonic functions, S_{mn}, and shows that this set of harmonic functions is complete. To obtain an orthonormal set of harmonic functions we have only to employ Schmidt's process of orthogonalization. This is a recursive process which starts with any enumerable, linearly independent sequence of functions $\{f_n\}$ and constructs an orthonormal sequence $\{\varphi_n\}$ by the formulae

$$\varphi_1 = c_{1,1} f_1, \; \varphi_{n+1} = \sum_{k=1}^{n} c_{n+1,k} \varphi_k + c_{n+1,n+1} f_{n+1},$$

$$(\varphi_m, \varphi_n) = \delta_{mn} \, (m \leqslant n) \quad n = 0, 1, 2, \ldots.$$

Whittaker's integral representation of a harmonic function (E. T. Whittaker 1903) has the form

$$\int_{-\pi}^{\pi} f(z + ix \cos u + iy \sin u) \, du$$

in three dimensions, f having partial derivatives of the second order, and is mainly of interest for the way in which it can be adapted to provide representations of Bessel functions, Mathieu functions and ellipsoidal harmonics.

15.23. *The Kernel Functions of Bergman and Schiffer*

A considerable unification of potential theory, especially in two dimensions, is effected in the researches of Bergman and Schiffer (1953), which were developed for elliptic equations of the type

$$\Delta u = q(x,y)u(x,y), \quad q > 0$$

$$\Delta = \partial^2/\partial x^2 + \partial^2/\partial y^2.$$

For these equations the Dirichlet or energy integral is

$$E(u) = \iint (u_x^2 + u_y^2 + qu^2) \, dx dy,$$

extended over a domain ω, and the scalar product of u and v is

$$E(u,v) = \iint_\omega (u_x v_x + u_y v_y + quv) \, dx dy.$$

From any closed enumerable set of solutions of the equation $\Delta u = qu$ we construct an orthonormal set $\{u_m\}$ such that $E(u_m, u_n) = \delta_{mn}$. The Kernel function $K(P,Q)$ is then defined by the uniformly convergent series

$$K(P,Q) = \sum_{n=1}^{\infty} u_n(P) u_n(Q).$$

A 'fundamental singularity' is a solution $S(P,Q)$ of the given equation, regarded as a function of P with a logarithmic singularity at Q, such that

$$S(P,Q) = -(2\pi)^{-1} \log |PQ| + s(P,Q),$$

where $s(P,Q)$ is twice differentiable in ω, and

$$\lim_{\varrho \to 0} \int \partial S(P,Q)/\partial \nu_\varrho \cdot d\sigma_\varrho = -1$$

where ν_ϱ is the outward drawn normal to the circle of centre Q and radius ϱ.

Then the Green's function is given by

$$G(R,Q) = S(R,Q) - E\{S(P,Q), K(P,R)\}$$

and the Neumann function by

$$N(R,Q) = S(R,Q) + K(R,Q) - E\{S(P,Q), K(P,R)\}$$

so that

$$K(P,Q) = N(P,Q) - G(P,Q).$$

The Green's function vanishes on the boundary σ of ω, and the Neumann function has a vanishing normal derivative on this boundary. Hence

$$u(Q) = \int_\sigma u(P)\partial G(P,Q)/\partial v_P \, d\sigma = -\int N(P,Q)\partial u(P)/\partial v \, d\sigma.$$

There is a slightly more complicated result for Robin's function $R_\lambda(P,Q)$ which enables us to satisfy the boundary condition

$$\partial u/\partial v - \lambda u = f(s),$$

if s is on the boundary.

15.24. Deny and Potentials as Distributions

H. Cartan (1945) had remarked that the pre-Hilbertian space of potentials was not complete in the classical case of Newtonian potentials, and Deny (1950a, b) therefore was led to complete this space by the introduction of Schwartz's distributions. For Deny the charges are specified by a distribution T of compact support and the elementary function R^{-1} by the corresponding distribution. The Newtonian potential U is then the convolution

$$U^T = R^{-1} * T$$

and
$$\Delta U^T = -4\pi T$$

using generalized derivatives in the Laplacian Δ.

The theory can be enlarged by replacing the Newtonian nucleus R^{-1} by any 'tempered' distribution N which necessarily possesses a generalized Fourier transform

$$\mathcal{F}(N) = \mathcal{N}.$$

If $\mathcal{T}$ is the corresponding Fourier transform of the distribution T which specifies the charges, then Deny defines the energy to be

$$\int \mathcal{N}|\mathcal{T}|^2 \, dv.$$

The space of charge distributions T with finite energy is then endowed with the scalar product

$$\int \mathcal{N}\mathcal{T}_1\bar{\mathcal{T}}_2 \, dv$$

and the norm

$$\{\int \mathcal{N}|\mathcal{T}|^2 \, dv\}^{\frac{1}{2}}.$$

This is a complete Hilbert space.

The transformation of potential theory can be carried still further. In the more abstract work of Deny there need be no reference to charges or to potentials. Instead we have a measure N replacing R^{-1} and called a 'nucleus', and another measure μ of compact support replacing the charges. The potential is replaced by the convolution $N*\mu$.

The method of balayage for a bounded open set ω with adherance $\bar{\omega}$ is to determine a positive measure μ' such that

$$N^*\mu' \leqslant N^*\mu, \text{ everywhere,}$$

and $$N^*\mu' = N^*\mu \text{ in } \omega.$$

15.25. H. Cartan and Functional Analysis

Radon integrals and measures played only a modest part in the researches of Evans and Frostman and it was left to H. Cartan (1941) to take full advantage of these techniques and to liberate potential theory entirely from the confines of Euclidean space and of any metrical concept of potential.

Following a suggestion due to Marcel Brelot (H. Cartan 1941) Cartan transports potential theory into the space of a topological group, and bases the new theory on the formula for the composition of measures due to M. Riesz.

The points of Euclidean space are replaced by the elements x, y, ... of an abstract topological group G and the masses or charges are represented by Radon measures μ, which are defined as additive functionals operating on continuous functions f of compact support.

The harmonic functions in Cartan's theory are generalizations of the potentials of M. Riesz and, like them, depend on a parameter α. Cartan's potentials are constructed from Radon measures $\varepsilon_\alpha(x)$ and the measure μ of a charge distribution. The composition of these two measures $[\varepsilon_\alpha,\mu]$ has the form $U(x)\,\mathrm{d}x$ where $U(x) = U^\mu_\alpha(x)$ corresponds to the classical potential of the charges μ.

If μ_1 and μ_2 are two distributions of positive charge their mutual energy

$$\int U^{\mu_1}_\alpha(x)\,\mathrm{d}\mu_2(x) = \int U^{\mu_2}_\alpha(x)\,\mathrm{d}\mu_1(x)$$

defines a scalar product and a norm

$$\|\mu\| = \left\{\int U^\mu\,\mathrm{d}\mu\right\}^{\frac{1}{2}}$$

in a 'pre-Hilbertian' space of distributions which is complete and which provides an abstract approach to the method of 'balayage' and to the 'maximum principle' for potentials.

15.26. Kelvin and Circulation

The foundations of the theory of hydrodynamics were laid by Kelvin (Sir William Thomson 1869) and Stokes before the centennial period of our history, but an account of their contributions to potential theory throws considerable light on the modern topological theory of harmonic integrals.

We owe to Kelvin the concept of the perfect fluid—uniform, incompressible, inviscid and non-conducting—and, although this was regarded as a first approximation to a real fluid, Kelvin always had in view the possibility of a theory of atoms as vortex rings in the one fluid, both real and perfect, viz. the aether.

Kelvin's great contribution to the theory of the perfect fluid is the introduction of the 'circulation', i.e. the line integral Γ of the velocity v along a closed curve C

$$\Gamma = \oint_C v \cdot d\mathbf{r}.$$

The equations of motion of the fluid imply that the circulation around any closed curve C which is convected by the fluid is always constant, and hence that, if the fluid is initially at rest, and the velocity is a continuously differentiable function, then the circulation Γ is always zero for all closed curves. By Stokes' theorem (Stokes 1854) this implies that

$$\text{curl } v = 0.$$

Now the condition that the fluid should be incompressible is div $v = 0$. Thus the velocity vector must satisfy the two equations

$$\text{curl } v = 0 \quad \text{and} \quad \text{div } v = 0.$$

If the region of flow is simply connected, or 'acyclic' then there must exist a single-valued function φ, the 'velocity potential', such that

$$v = -\text{grad } \varphi \quad \text{and} \quad \Delta\varphi = 0,$$

so that φ is a harmonic function.

If, however, the region of flow is multiply connected, the situation, however artificial it may appear to the physicist, is much more interesting to the mathematician.

15.27. Maxwell and Cyclosis

Riemann's researches on multiply-connected spaces were restricted to two-dimensional regions (the famous 'Riemann surfaces'), but the systematic study of multiply-connected three-dimensional spaces was taken up by Listing (1861) and brought to the notice of English physicists by Maxwell (1873; Vol. I, pp. 16–20).

A simply connected region is homotopic to a single point, but Maxwell noted that a multiply-connected region is homotopic to a *graph* in which a number, say p, points are connected by, say, l lines, and in which the number of independent loops, or closed paths, or 'cycles' is $k = l - p + 1$ (the 'cyclomatic' number).

In such a region, although the flow may be 'irrotational' i.e. the vorticity curl v is everywhere zero, it does not follow that the circulation Γ around a closed curve is necessarily zero.

15.28. Kelvin and Barriers

To study the flow in a multiply-connected region R, Kelvin, following Riemann, introduced the concept of 'barriers', i.e. surfaces which divide R into a number of simply-connected regions. Any closed curve Γ will cross a specified barrier B_j, p_j times in the positive direction and p_j' in the negative direction, and in irrotational flow the circulation around Γ can be proved to have the form

$$\Gamma = n_1 k_1 + n_2 k_2 + \ldots + n_m k_m,$$

where $n_j = p_j - p_j'$ and k_j is a constant independent of where the closed curve crosses the barrier B_j.

Under these circumstances the velocity potential φ must be a multiply-valued function. The constants $k_1, k_2, \ldots, k_m$ are called its 'periods'.

Kelvin noted that Green's Theorem needs to be extended when such 'cyclic' functions are considered and he established the general formula

$$\iint \varphi \operatorname{grad} \psi \, dS + \sum_j k_j \iint \operatorname{grad} \psi \, dB_j$$

$$= - \iint (\operatorname{grad} \varphi \operatorname{grad} \psi) \, dV - \int \varphi \Delta \psi \, dV.$$

15.29. *Hodge and Harmonic Integrals*

The generalization of classical potential theory from Euclidean space of three dimensions to Euclidean space of n dimensions is a trivial exercise. However, the generalization to a Riemannian space of n dimensions leads to an entirely new and fertile field of mathematical research. This great advance was made by W. V. D. Hodge (1932, 1933, 1936) and is described in his book, *The Theory and Applications of Harmonic Integrals* (Hodge 1941).

The kind of generalizations which are involved may be roughly indicated in the following stages:

(1) Harmonic vectors such as the fluid velocity are replaced by harmonic tensors in Riemannian space R.
(2) The circulation $\oint v . dr$ is replaced by an integral of harmonic tensors over closed p-dimensional manifolds in R.
(3) This restricts us to harmonic tensors which are completely anti-symmetric or 'skew'.
(4) We are then led to clarify a distinction, which is obscured in classical fluid dynamics, viz. the distinction between a vector and a skew tensor.

The classical equations

$$\operatorname{curl} v = 0 \quad \text{and} \quad \operatorname{div} v = 0$$

may be written in integral form as

$$\int v . dr = 0 \quad \text{and} \quad \oint v . dS = 0$$

as integrals along a closed curve C and a closed surface S (each homotopic to a point). On C the velocity is a vector, but on S it is a skew tensor, i.e. the dual of the vector.

In Hodge's generalizations we have two equations, one involving a skew tensor $\mathbf{P}$ of rank p and the other the dual skew tensor $\mathbf{P}^*$ of rank $n - p$, where n is the dimension of the Riemann space R.

(5) Finally the classical expression for the circulation T is replaced by the integral of the skew tensor $\mathbf{P}^*$ over a closed manifold of dimension p which has the form

$$\sum \lambda_i w^i$$

where the coefficients λ_i are integral and the constants w_i are the periods of the integral.

15.30. De Rham and Currents

The researches of de Rham (1931) on the differential geometry of chains suggested to him that 'there is a deep analogy, in an n-dimensional manifold, between the p-dimensional chains and the forms of degree $n - p$. It suggests that they can be considered as particular cases of a more general concept which will be called a *current*. From this point of view, the operation d (differential) and B (boundary) will be particular cases of the same operation, and the exterior product of differential forms will correspond to the topological intersection of chains, (de Rham and Kodaira, 1950). This analogy, already adumbrated in the papers of de Rham, could be clearly and explicitly expressed only be utilizing the theory of distributions of Laurent Schwartz. In fact de Rham attributes the exact definition of a current to Schwartz, but priority of publication resides with de Rham, for Schwartz's contribution did not appear in print until 1950 (L. Schwartz 1950, Vol. 1, p. 12; 1966).

A distribution is a 'continuous' linear functional defined on the linear space of all test functions φ, i.e. infinitely differentiable point functions with compact support in an n-dimensional Euclidean space R. To define a *current* de Rham makes the following generalizations:

(1) R is replaced by a topological n-dimensional manifold M with an infintely differentiable structure.

(2) The test functions, φ, are replaced by infinitely differentiable skew symmetric tensors of rank p with compact carriers (or supports).

(3) The linear functional $T(\varphi)$ is replaced by the integral of such a tensor over an $(n - p)$-dimensional manifold in M.

A current is then a 'continuous' linear functional, defined on the linear space of all infinitely differentiable forms of degree p with a compact carrier.

De Rham expresses his theory in terms of the exterior differential forms introduced by E. Cartan (Section 15.33) but extends Cartan's theory by associating tensors of *odd* kind with the classical covariant tensors of *even* kind. The former are defined by introducing the factor $J/|J|$ into the transformation equations where J is the Jacobian of the transformation from one coordinate system to another.

15.31. Differential Forms on a Manifold

The concept of a differential form may be roughly indicated by considering the integral of a tensor T with components $T_{\alpha_1 \alpha_2 \ldots \alpha_p}$ of rank p over a manifold M of dimension p in a Riemannian space with coordinates u^1, u^2, $\ldots$, u^n. The infinitesimal $du^{\alpha_1} du^{\alpha_2} \ldots du^{\alpha_p}$ may be regarded as one of the components of the tensor which represents an infinitesimal element of M. The integral of T over M is then given by the expression

$$\int T_{\alpha_1 \alpha_2 \ldots \alpha_p} du^{\alpha_1} du^{\alpha_2} \ldots du^{\alpha_p} \int \omega$$

summed in accordance with Einstein's convention. The integrand ω is a differential p-form.

Differential forms first appeared in analysis in the simple case when $p = 1$ as proposing the problem of discovering the integral equivalent to the differential equation

$$X_1 dx_1 + X_2 dx_2 + \ldots + X_p dx_p = 0.$$

The equation is 'exact' if $X_q = \mu \partial f / \partial x_q$ and the integral equivalent is $f = \text{const}$. But the equation

$$y dx + dz = 0$$

is not exact and its integral equivalent requires two equations

$$z = f(x), \quad y = -f'(x).$$

The early history of the subject is given by Forsyth (1890). In the period with which we are concerned the main contributions are due to Grassmann (1844,1862), Lie (1877), Frobenius (1877), E. Cartan (1899), who revitalized the theory, and to de Rham (1936, 1955), who combined it with L. Schwartz's theory of distributions to obtain a topological theory of chains, currents and harmonic integrals.

15.32. *Grassmann's Calculus of Extension*

The infinitesimal element of the manifold M has an orientation depending upon the order of the suffixes α_1, α_2, $\ldots$ α_p. Thus in three dimensions the area is readily recognized to be the vector product of du^1 and du^2. It follows that the integral of the tensor T over M involves only the skew-symmetric elements of T. In three dimensions the integral reduces to

$$\int (T_{23} dx^2 dx^3 + T_{31} dx^3 dx^1 + T_{12} dx^1 dx^2),$$

but in the usual notation the fact that T is a skew-symmetric tensor of rank two is obscured by the equivalence of T to a vector, or tensor of rank one.

In general the coefficients $T_{\alpha_1, \alpha_2 \ldots \alpha_p}$ and the elements of M, $dx^{\alpha_1} dx^{\alpha_2} \ldots dx^{\alpha_p}$ are skew-symmetric, and the algebraic manipulation of differential forms therefore involves long and cumbrous expressions and the daunting appearance of the generalized Kronecker symbols.

These technical complexities can be avoided by the explicit or implicit use of Grassmann's calculus of extension. For operations in an n-dimensional space Grassmann employs an algebra generated by n units e_1, e_2, $\ldots$, e_n with the combinatorial law of multiplication

$$e_p e_q + e_q e_p = 0$$

for *all* p and q, so that $e_p^2 = 0$. Multiplication is associative and the product $e_1 e_2 \dots e_n$ is identified with the numerical unity 1. The 'complementary' units are given by

$$E_p = (-1)^{p-1} e_1 e_2 \dots e_{p-1} e_{p+1} \dots e_n,$$

with

$$E_1 = e_2 e_3 \dots e_n$$

and

$$E_n = (-1)^{n-1} e_1 e_2 \dots e_{n-1},$$

whence $e_p E_p = e_1 e_2 \dots e_n = 1$ (no summation over p!).

To employ Grassmann's calculus we replace the element of M by the algebraic element

$$du = \varepsilon_{\alpha_1} \, du^{\alpha_1} \varepsilon_{\alpha_2} \, du^{\alpha_2} \dots \varepsilon_{\alpha_p} \, du^{\alpha_p}$$

and the tensor T by the algebraic element

$$T = (p!)^{-1} E_{\alpha_1} E_{\alpha_2} \dots E_{\alpha_p} T_{\alpha_1 \alpha_2 \dots \alpha_p}.$$

Then the differential form ω, which is the integrand in the integral of T over M is

$$\omega = (-1)^{n-1} t \, du.$$

To evaluate the product of two algebraic elements, such as du and dv, or t and s, it is only necessary to use the definition of E_p and the anticommutative rule of multiplication of the units e_p.

15.33. *E. Cartan and the Exterior Derivative*

The first and most important application of the theory of differential forms is to the generalized form of Stoke's theorem. This theorem requires the introduction of the first derived form ω' of a given linear differential form

$$\omega = u_1 \, dx_1 + \dots + u_n \, dx_n$$

and the obvious definition is

$$\omega' = \sum du_p \, dx_p = \tfrac{1}{2} \sum \{\partial u_p / \partial x_q - \partial u_q / \partial x_p\} \, dx_q \, dx_p.$$

E. Cartan gave definitions of the higher derived forms $\omega^{(p)}$ which are by no means intuitively evident. In order to discuss the Pfaffian equation, $\omega = 0$, or systems of such equations, he writes

$$\omega^{(2m-1)} = (m!)^{-1} (\omega')^m$$

and

$$\omega^{(2m)} = \omega \omega^{(2m-1)}.$$

15.34. *Harmonic Differentials*

In the convenient but illogical notation of differential forms such as

$$\alpha = \sum a(i_1, i_2, \dots i_p) \, dx^{i_1} \wedge \dots \wedge dx^{i_p}$$

where a is a skew-symmetric tensor, the differential of α is given by

$$d\alpha = \sum da(i_1, i_2, \dots i_p) \wedge dx^{i_1} \wedge \dots \wedge dx^{i_p}$$

where $da(i_1, i_2, \ldots, i_p)$ is the completely skew form derived from

$$\partial a(i_1, i_2, \ldots i_p)/\partial x^{i_q}.$$

In general for any current $T(\varphi)$ acting on a p-form φ, the differential is defined by

$$dT(\varphi) = (-1)^{p+1} T[d\varphi].$$

Now the adjoint form to α (given above) is

$$*\alpha = \sum \alpha^*(j_1, j_2, \ldots, j_{n-p}) \, dx^{j_1} \wedge \ldots \wedge dx^{j_{n-p}},$$

where the values $j_1, j_2, \ldots, j_{n-p}$ are different from $i_1, i_2, \ldots i_p$. We can now construct the adjoint of the current $T(Q)$ by the formula

$$*T[\varphi] = T.$$

De Rham then introduces the 'codifferential' δT defined by the formula

$$\delta T = (-1)^{np+n+1} d*T$$

for a form T of degree p in space of n dimensions.

Now $d^2 T = 0$ and $\delta^2 T = 0$ and the harmonic operator Δ is defined by Rham and Kodaira (1950) as

$$\Delta = d\delta + \delta d.$$

A current α is said to be harmonic if $\Delta \alpha = 0$ from which it follows that

$$\delta \alpha = 0 \quad \text{and} \quad d\alpha = 0.$$

This is the pair of equations used by Hodge to define a harmonic tensor. The scalar product of a current T and a form φ is defined as

$$(T, \varphi) = T[*\varphi] = \int T \wedge *\varphi.$$

Harmonic forms can now be introduced by means of an operator H which is defined by the formula

$$(T, h) = (HT, h)$$

for each harmonic form h. Then Hodge's theorem is that if $c_1, \ldots, c_r$ are r closed chains of the same dimension and kind, no linear combination of which bounds a current, then $Hc_1, \ldots, Hc_r$ are r linearly independent harmonic forms, and, if $p_1, \ldots, p_r$ are r arbitrarily given numbers, we can always find a harmonic form h such that $(Hc_i, h) = p_i$. The form $*h$ is harmonic, and, since

$$\int_{c_i} *h = (c_i, h) = p_i$$

it has the periods $p_1, \ldots p_r$.

16. Mathematical Logic

16.1. Introduction

There is an enormous literature on the subject of mathematical logic and, in order to make the concise account in the present chapter intelligible, I have taken the bold, if not hazardous, step of organizing the historical material in terms of a personal interpretation of the significance of the interactions between mathematics and logic. I propose therefore to disregard the customary divisions of this subject and to discuss it under the following titles, which will then be shortly explained, before we proceed to a more detailed account in historical order.

For mathematicians, the fundamental problem is the relation of mathematics to logic. Are these subjects independent or related? And, if they are related, is one subordinate to the other, or are they mutually dependent? The possible answers to these questions suggest that this chapter of our history may be divided into the following sections:

   (a) 'Mathematical logic', i.e. literally understood as logic expressed in mathematical language.
   (b) 'Logical mathematics', i.e. mathematics developed as a part of formal logic.
   (c) 'A-logical mathematics', i.e. mathematics independent of formal logic.
   (d) Mathematics as essentially incomplete.
   (e) Mathematics repudiating classical logic.

(a) 'Mathematical logic'. Aristotle and the mediaeval logicians had used letters of the alphabet to denote the subject and predicate in a proposition, but Boole (1847, 1848, 1854) was the first to express a proposition as an algebraic equation. He had no doubts of the complete autonomy of the 'laws of thought' and he discovered with surprise and some perplexity that the symbolic expression of these laws involved the invention of a new algebra.

Boole's algebra can be regarded as the mathematical expression of the calculus of classes (i.e. set theory) or of the calculus of propositions. The proper appreciation of Boole's achievement had to wait upon the work of Schröder (1890–5) and the development of lattice theory by Garrett Birkhoff (1948), of the theory of filters by Henri Cartan (1937a, b) and of the theory of ultra-products by Łoś (1955).

Frege rejected the classical analysis of a proposition into a subject and predicate united by a copula (the little word 'is') and proposed to express

255

a proposition in the language of mathematical analysis as a 'propositional function', i.e. a mapping φ of the relevant subjects, say, p, q, ..., into appropriate predicates, $\varphi(p)$, $\varphi(q)$, ...

(b) 'Logical mathematics'. For Boole mathematics was merely an appropriate language for logic, but for Peano it is almost true to say that logic furnished the appropriate language for mathematics. Frege, however, transcended both of these exercises in an ambitious programme which sought to exhibit the fundamental concepts of mathematics, especially arithmetic, as purely logical concepts, such as the concepts of class and of implication.

This programme was realized with considerable success in the monumental work of Russell and Whitehead (Section 16.3), which left a precious legacy to philosophers, logicians and mathematicians in the form of the famous Russell paradox, which seems to vitiate the whole work of Frege, Peano and Russell himself.

(c) A-logical mathematics. If the 'logicism' of Russell is abandoned, little seems to be left of classical mathematics, which it was thought could be firmly based on logic, and the question naturally arises: 'Can mathematics be developed independently of logic?

The first attempt to answer this question must be credited to Hilbert, who, having set in order the science of geometry by an appeal to arithmetic, turned his attention to the logical problems posed by the latter discipline. Rejecting Russell's appeal to the external world to provide a realistic justification for concepts of number, Hilbert was forced to abandon any attempt to establish the objective truth or falsity of arithmetic, and to be content with a proof of its internal consistency. Such a proof he never achieved, and indeed the subsequent work of Gödel showed that it can never be reached.

However, certain fragments of arithmetic can be constructed as independent consistent systems without any appeal to logic. Such a system was first constructed by Skolem, and later developments are due to Bernays and to Goodstein.

(d) Mathematics as a 'model' for logic. Although Hilbert's programme was not and could not be realized in arithmetic, it had already been realized in a number of algebraic systems, in which all possible consequences of the axioms could be constructed and proved to be free of any contradiction, as for example the 'dual' algebra of the elements 0 and 1, with addition and multiplication defined modulo 2. Unfortunately (perhaps!) Hilbert concentrated on the foundations of arithmetic—a much more difficult and intractable subject than algebra—and Gödel's investigations proved that his programme was unrealizable.

The truth-tables of Post and Wittgenstein implicitly appeal to the Boolean dual algebra. Bernay's proof of the consistency of the propositional calculus uses a simple matrix algebra. The many-valued logic of Post and of Lukasiewicz requires a similar algebraic support. In all these investigations some rather simple algebras are employed as 'models' validating logical axioms.

In the work of Skolem on axiom-free arithmetic, mathematics is no longer employed to validate logic but is boldly expounded as an independent system, which is constructed step by step by recursive definitions. This work was further developed by Bernays and Goodstein so as to include a substantial part of the theory of functions of a real variable.

The method of recursive construction initiated by Skolem had two consequences of the greatest importance. On the one hand it led to the profound study of recursive systems, for their intrinsic interest, and on the other to their application to mechanical and electronic methods of computing—a work initiated by Turing (1937).

(e) Mathematics repudiating classical logic. A new era in the development of mathematical logic opened with the complete repudiation of classical two-valued logic by Brouwer. Hitherto logicians had accepted that a meaningful proposition must be either 'true' or 'false', but Brouwer maintained that in mathematics the only significant characterization of propositions is either 'proved to be true', 'proved to be false' or '(as yet) unproved and undecided'. He therefore rejected the Hilbertian test of consistency based on the classical principle of contradiction and would allow only purely constructive proofs and definitions.

As an unquestioned basis for this new and revolutionary revision of mathematics Brouwer accepted the (signless) integers as a datum given by intuition and not to be subjected to any critical examination.

The three-valued logic preached by Brouwer divided the mathematical world into two camps. One party, led by Hilbert, angrily repudiated the gospel according to Brouwer, precisely because it was destructive of almost all existing mathematics. The other party, led by Weyl, accepted the new teaching with sorrow and repentance for earlier misdeeds.

Time heals many wounds, and at the present time mathematicians are inclined to tolerate both two-valued and three-valued logic as valid alternatives, comparable perhaps to Euclidean and non-Euclidean geometries. It must also be admitted that one of the attractions of Brouwer's intuitionism is that it presents the mathematician with the challenging problem of reconstructing mathematics without the aid of the principle of contradiction, i.e. it is no longer allowable to assume that if the contradictory of a proposition is proved to be false, then the proposition must be true.

This synopsis of the development of mathematical logic, incomplete, partial and prejudiced as it may be, will perhaps help to guide the reader through the luxuriant forest of literature of this most fertile and fundamental discipline. We can now proceed to discuss in more detail the work which has been so briefly summarized above.

Mathematical Logic

16.2. Frege and Propositional Functions

Frege (1879, 1884) was dissatisfied with the classical analysis of a judgement into a subject and predicate united by the copula, because of the existence of equivalent propositions in which the roles of subject and predicate are interchanged, as in the following example which he cites: 'At Plataea the Greeks defeated the Persians' and 'At Plataea the Persians were defeated by the Greeks'. These judgements have a common conceptual context, and Frege therefore proposed that, in his formula language, this common meaning should be expressed in the form: 'The defeat of the Persians by the Greeks at Plataea is a fact', and that all judgements should be expressed in this form with a single predicate, viz. 'is a fact'.

Next Frege noticed that this judgement still remains a judgement (either true or false) if we replace the words 'Persians' or 'Greeks' by the names of any nationalities, and the word 'Plataea' by the name of any place and that all the judgements so obtained are of the form: 'The defeat of x by y at z is a fact', which he thought of as a function $\varphi(x,y,z)$ with a domain consisting of the words x,y and z which give φ a meaning, and with a value, which was either 'truth' or 'falsehood'. The concept of the 'truth-value' of a proposition was not new—it can be traced back to the logicians of Megara and scholastic philosophers—but it had been forgotten or ignored. By resurrecting this important and fertile idea Frege initiated a new development in mathematical logic and had a profound influence on the work of Wittgenstein.

16.3. The Propositional Calculus

Having settled that the subject-matter of logic is the theory of propositional functions, Frege proceeds to define material implication in terms of truth values and uses without analysis the negation and conjunction of propositions ('p' *and* 'q'). Thus he explicitly introduces 'conditionality', (i.e. 'p' implies 'q'), as meaning

p is true and q is true

or p is false and q is false

or p is false and q is true,

excluding only the possibility that p is true and q is false, just as in Philonian logic.

This definition is, of course, completely vacuous since according to this analysis of implication we can infer any proposition, true or false. It is therefore necessary for Frege to restrict material implication by imposing a number of axioms, or rules of inference, such as

$$[p \to (q \to r)] \to [(p \to q) \to (p \to r)]$$

and $$(p')' = p$$

where p' denotes the negation of p.

The relation which Frege takes as fundamental in the propositional calculus—viz. Philonian implication—is effectively the same as that called 'implication' by Russell (1903), which is the relation between propositions p and q which holds whenever q is true or p is false. This relation is equally vacuous and needs to be supplemented by ten axioms, such as, if $p \rightarrow q$ and $q \rightarrow r$, then $p \rightarrow r$. In fact the Philonian definition states that a false proposition implies any proposition, true or false, and that a true proposition is implied by any proposition, true or false; it is therefore ineffective as a basis for a theory of deduction.

The propositional calculus was given a systematic form by Whitehead and Russell (1910, Vol. 1) who treated propositions as undefinables, p, q, ... which can be converted into other propositions by two undefined operations:

(i) 'negation', denoted there by $\sim p$ and here by p' (and read as 'not p');
(ii) 'disjunction', denoted there by $p \vee q$ (and read as 'p or q'). Implication $(p \supset q)$ (read as 'if p, then q') is defined as $p' \vee q$ and conjunction (pq) (read as p and q), is defined as $(p' \vee q')$.

The five axioms are:

(Tautology)	$p \vee p \supset p$
(Addition)	$q \supset p \vee q$
(Permutation)	$p \vee q \supset q \vee p$
(Association)	$p \vee (q \vee r) \supset q \vee (p \vee r)$
(Summation)	$(q \supset r) \supset [(p \vee q) \supset (p \vee r)]$.

When these axioms are divested of the naïve and natural interpretations suggested by the verbal equivalents (such as 'p or q implies q or p') three questions arise about their logical structure:

(i) Are they consistent?
(ii) Are they independent?
(iii) Are they complete?

The first two questions were answered in the affirmative by Bernays (1926) in his *Habilitationsschrift* delivered at Göttingen in 1918 and published in 1926. Meanwhile Post (1921) had published independently another examination of the same questions.

The method of proof of consistency adopted by Bernays is to eliminate the relation of implication by means of definition in terms of negation and disjunction, and to express any proposition A in a certain 'conjunctive normal form' B, i.e. as the conjunction of a number of propositions, each of which is the disjunction of a number of the basic propositions and their negations. To express A in terms of B means simply to show that

$$A \supset B \quad \text{and} \quad B \supset A.$$

Logical Mathematics

16.4. Russell and Logicism

The publication of *The Principles of Mathematics* (Russell 1903) is commonly regarded as marking the beginning of a new epoch in the study of the foundations of mathematics, but the significance of Russell's work is often misunderstood.

Russell's programme as announced in the first sentence of the preface of this book is (*i*) the proof that all mathematics deals exclusively with concepts definable in terms of a very small number of fundamental logical concepts, (*ii*) that all its propositions are deducible from a very small number of fundamental logical principles, and (*iii*) 'the explanation of the fundamental concepts which mathematics accepts as undefinable'.

Russell planned to write a second volume in collaboration with Whitehead in which the demonstration of the first thesis should be established by strict symbolical reasoning, but this plan was abandoned and instead these two authors collaborated in the three massive volumes of the *Principia Mathematica* (Whitehead and Russell 1910).

Russell was not the first to apply a minute logical analysis to the principles of mathematics and to envisage a detailed examination of the methods of proof employed, and he himself pays generous tribute to the work of Cantor and Peano. Russell had also been anticipated by Frege, whom he repaid by communicating the famous 'Russell paradox' which invalidated the whole of the Frege–Russell theory of classes (Section 16.5).

But Russell's work had a much greater influence than that of Frege or of Peano, primarily because of the high literary quality and quiet persuasiveness of Russell's style, which contrasts so markedly with the cumbrous symbolism of Frege's *Begriffsschrift* (Frege 1879) and the awkward notation invented by Peano (1889) (Section 13.7).

The significance of Russell's *Principles of Mathematics* is much clearer now that it can be compared with the work of Hilbert and Brouwer. The essential point of the comparison is the concept of 'existence' in mathematics.

In Hilbert's 'formalist' interpretation of mathematics the criterion of existence is the self-consistency of the fundamental axioms, without any reference to their possible reference to the physical world. Indeed Hilbert's system provides no justification for the process of counting as employed, for example, by shepherds or soldiers. Russell's system, by contrast, is refreshingly realistic and convincingly identifies the natural numbers with the common arithmetical properties naïvely attributed to collections or 'multitudes' of entities.

In Brouwer's 'intuitionist' system the criterion of existence is the exhibition of a logical construction in a finite number of steps and the consequent repudiation of the *tertium non datur* for infinite sets. Russell, who, in spite of a sad disclaimer, always retained the prejudices of Platonic idealism, did not attempt any refutation of the finitist doctrine but evaded its disastrous consequences by asserting the 'axiom of infinity'.

The realist theory of the natural numbers (confusingly called the nominalist theory by Russell) depends upon his definition of number and the corresponding existence theorem that numbers exist as logical entities.

Numbers are defined in terms of the 'similarity' of two classes of entities, i.e. the existence of a 'one–one' correlation between these terms—a concept which is independent of and logically prior to the concept of the number 'one'. The number of a class C is then defined as the class of classes which are similar to the given class C. Russell then proceeds to give a constructive theory of the natural numbers.

The necessary definitions are as follows:

  (i) 0 is the empty set, or null class;
  (ii) 1 is the class of classes which are not empty and which reduce to the null class when an element is removed;
  (iii) $n + 1$ is the number of the class resulting from adding an element to a class of n elements.

In Russell's theory Peano's axioms are proved as theorems and Russell's claim to have reduced arithmetic to logic appears to be completely justified. Unfortunately the robust realism of Russell's theory leads to the famous paradox relating to predicates which are predicable of themselves. This is far from implying that the whole theory must be abandoned, but is a salutary indication that the concept of 'class' needs most careful examination and that 'no subtlety in distinguishing is likely to be excessive'.

16.5. *The Russell Paradox*

In the initial development of set theory by Cantor there did not seem to be any necessity to give anything more than a general description of the concept of a 'set', or to give any justification for asserting such 'obvious' and 'commonsense' properties as the three following:

  (*i*) 'extensionality', i.e. if two sets have the same members, then they are identical;
  (*ii*) 'abstraction', i.e. if P is any 'property', there exists a set whose elements are those with the property P;
  (*iii*) 'the axiom of choice', which has been discussed in Section 5.6.

The account which we have given of the controversy concerning the axiom of choice suffices to show that this axiom certainly requires a more profound examination before it can be admitted into the canons of set theory, and we shall give later a summary of the decisive investigations of Gödel (Section 16.20).

The axiom of abstraction seems most natural and indeed self-evident, but when Russell commenced his searching investigation (June 1901) into the 'Principles of mathematics' (Russell 1903) he found at once that the paradox of Burali-Forti (1897) was reducible to a contradiction inherent in this very axiom. For this axiom of 'abstraction' guarantees the existence of a class C of sets x such that x is not an element of x, and it is quickly inferred that the class C both is and is not an element of itself.

Russell communicated a brief account of this paradox to Frege (16 June 1902) who immediately replied (22 June 1902) with surprise and consternation but with a hope that the difficulty might be surmounted. Indeed Frege soon satisfied himself that he could solve this problem. Meanwhile Russell himself came to think that it would be sufficient to distinguish 'the class as many' (elements) from 'the class as one' (collection).

However, Russell soon became dissatisfied with this facile solution and added Appendix B to the first edition of *Principles of Mathematics* (Russell 1903) in which he gave a first sketch of theory of types. The essence of this theory is best stated in terms of 'propositional functions' $\varphi(x)$ i.e. propositions concerning a class of entities x. Russell pointed out that a propositional function, like a mathematical function $f(x)$ of an argument x, has a definite 'domain' (unfortunately called a 'range' by Russell), i.e. the propositional function $\varphi(x)$ is significant only if x is restricted to this domain. Then and only then can arise the question whether $\varphi(x)$ is true or false.

In order to apply this principle Russell distinguished a hierarchy of domains, beginning with classes of individuals and proceeding to collections of classes, and so on. Mathematical logic was to be governed by the self-denying ordinance that a propositional function was significant only in its proper domain. This should rule out any confusion between elements, sets of elements and classes of sets, and restrain us from speaking of classes which are not elements of themselves.

However, Russell had become dissatisfied with this theory and proposed no less than three different theories to overcome the paradoxes of set theory. These theories scarcely need discussion for they were quickly abandoned in favour of a more sophisticated theory of types (Russell 1908).

The two essentially novel features in this paper are the 'vicious circle principle' and the 'axiom of reducibility'. The vicious circle principle is a reformulation of Poincaré's notion of impredictability in the form that the domain of a propositional function is not an element of the domain (but belongs to a higher type), whence the elements of a class must not be defined in terms of the class. Thus we must not define the natural numbers to be all the successors, $n + 1$, of the elements n beginning with 0, for this involves the concept of *all* natural numbers.

But to enforce this principle seemed to render much of elementary mathematics impossible and some method had to be found to evade its consequences. Russell's method was to employ the axiom of reducibility according to which any propositional function, φ, with a domain of any type, is equivalent to a first-order propositional function of the form 'x belongs to the class α', provided x belongs to the domain of φ. Perhaps the best comment on the use of the axiom of reducibility to solve the paradoxes is Quine's verdict that Russell's solution is 'enigmatic'. This and the original papers and Quine's introduction are given in Heijenoort (1967).

16.6. Zermelo and the Construction of Sets by 'Separation'

Almost at the same time as Russell proposed the theory of types as the final and 'commonsense' solution of the paradox which he himself had discovered, Zermelo (1908b) gave a completely different approach to set theory by rigorously restricting the methods by which sets may be constructed for use by the working mathematician. This paper presents the first explicitly axiomatic account of set theory in which the only notion which still required clarification was the idea of a 'definite property' or definite propositional function which should determine without any doubt whether or not each individual x of a given class possesses the said property.

Zermelo retains the 'axiom of extensionality', but rigorously restricts the 'axiom of abstraction' to the 'axiom of separation' (*Aussonderung*), according to which, if φ is any propositional function, definite for any actually existing class M, then M possesses a subset N containing as elements those elements x of M for which $\varphi(x)$ is true. Unfortunately the separation axiom is far from definite, even if the symbol for the property specifying a set is printed in Gothic type!

This axiom effectively excludes such concepts as 'the set of all sets', or 'the set of all sets S such that S is not a member of S', for such imaginary sets cannot be constructed by 'separation' from other sets known to exist.

Indeed it seems at first sight that the new axiom has made a clean sweep of all sets, but fortunately one set survives, viz. the null set, and this can be made to yield a hierarchy of sets if we add two further axioms, viz.

(*i*) if T is any set, then there exists the 'power set' of T which contains as elements all the subsets of T;

(*ii*) if T is any set, then there exists a set which contains as elements all elements of the elements of T.

Thus from the null set '0' which has no elements, we construct the set '1' having '0' as its only element. Then we construct the set {1} having as elements '0' and '1'.

To make further progress and to make set theory applicable to classical analysis Zermelo added further axioms, of which the most significant are the 'axiom of infinity', which guarantees the existence of infinite ordinals, and the 'axiom of choice', which Zermelo states in the form of the 'multiplicative axiom', viz. that a 'distinguished' element can be chosen from each of an infinite collection of *disjoint* sets.

16.7. Fraenkel and Ur-elements

Fraenkel (1922) endeavoured to complete Zermelo's theory by clarifying the concept of a 'definitive property' and by proving that the axiom of choice is independent of the other axioms adopted by Zermelo. The proof is carried out by a construction, according to the prescriptions of Zermelo's axioms (excluding of course the axiom of choice), which starts from the null set and enumerable distinct pairs of objects $(a_n, \bar{a}_n)$, none of which is a set, and which have subsequently been called 'Ur-elements'. It is shown that the construction leads to a set for which no 'choice set' exists. It follows that not all sets obey the multiplicative axiom, which must therefore be independent of the other axioms postulated by Zermelo.

The introduction of the 'Ur-elements' may have seemed a gratuitous fantasy to pure mathematicians in 1922 but it became a familiar feature to quantum theorists after the invention of Pauli's exclusion principle (Pauli

1925). According to this principle electrons are indistinguishable entities and the properties of an assembly of electrons are unchanged by the interchange of any two electrons. Thus for a collection of atoms, each containing many electrons, the only significant propositional functions are those unchanged by any electronic interchanges. Therefore no selection principle can exist for such a collection of (electronic sets), i.e. there are collections of sets for which the multiplicative axiom is false.

16.8. *J. von Neumann and the Segment Theory and Function Theory of Ordinal Numbers*

In 1923 J. v. Neumann offered an alternative method of introducing transfinite numbers which seemed to give greater precision to the somewhat vague idea of Cantor that the order type of an ordered set M is obtained by abstracting from the specific nature of the elements of M and considering only their ordering. The guiding concept in this new method is that an ordinal number should be simply the set of preceding ordinals (i.e. a segment or *Abschnitt* of a well-ordered set). The difficulty is that such a definition of transfinite ordinals requires the use of transfinite induction and J. von Neumann's first paper is mainly concerned with the provision of apparatus for the justification of transfinite induction.

A considerable simplification of this theory was given later by R. M. Robinson (1937) who directly defined an ordinal u as a set such that

(*i*) if $x \in u$ and $y \in u$, then either $x \in y$, $x = y$ or $y \in x$;
(*ii*) if $x \in y$ and $y \in u$, then $x \in u$.

The aim of von Neumann's first paper (J. v. Neumann 1923) was to give 'unequivocal and concrete form to Cantor's notion of ordinal number'. The second paper (J. v. Neumann 1925) was to give 'a logically unobjectionable axiomatic presentation of set theory'. This aim he achieved, with considerable success, by basing the theory not on an axiomatical concept of 'set' but on an axiomatical concept of 'function'. Thus J. v. Neumann has two kinds of objects to consider, viz. 'arguments' and 'functions', and a two-variable operation $[x,y]$—the value of the function x for the argument y.

J. v. Neumann unconsciously adopted Boole's theory of classes as selections by an operator p from some universe of discourse, but instead of using the axiom $p^2 = p$ he defined a class to be a function f such that $[f,x]$ is always equal, either to an argument A or an argument B, for all relevant arguments x. (We may think of A as 1 and of B as 0). A set was then defined to be restricted kind of class, whose 'size' was limited by the condition that we exclude any function f such that the totality of arguments x for which $[f,x] \neq A$ can be mapped onto the totality of all arguments by some 'function'.

This sharp distinction between classes and sets enables J. v. Neumann to establish Zermelo's axiom of separation and Fraenkel's axiom of replacement, and to prove the well-ordering theorem, thus eliminating any use of the axiom of choice.

In fact there are classes, which Quine conveniently calls 'ultimate' classes, which are not sets and are not elements of any other class. The class of all sets and the class of all ordinals are both ultimate classes so that the paradoxes of Russell and of Burali-Forti cannot arise in this system.

16.9. *Skolem and the Enumerability of Domains*

The first attempt to clarify Zermelo's concept of a 'definite proposition' was made by Weyl (1921), but a far more searching criticism was made by Skolem (1922)

Skolem clarified Zermelo's concept of a 'definite proposition' by restricting this name to finite expressions constructed from elementary propositions of the form $a \in b$ or $a = b$ by means of the fundamental operators of conjunction (a and b), disjunction (a or b), negation, universal quantification ('for all a') and existential quantification ('for some a').

However, the propositions in question, namely the propositions of set theory, are asserted about the elements of a 'domain' B, which Skolem showed was not uniquely determined by Zermelo's axioms; whence there arises a relativity of set-theoretic notions, accordingly as we operate in different Zermelo domains.

Lowenheim (1915) had already proved that if any definite proposition, as defined above, is satisfied in any domain, then it is satisfied in an enumerably infinite domain. Skolem (1920) had simplified this proof by introducing a certain 'normal form' of a proposition, and using the axiom of choice. In the 1922 paper he was able to make the proof independent of this axiom.

It thus appeared that if Zermelo's axioms are consistent then it must be possible to introduce an infinite sequence of symbols, 0, 1, 2, ... in such a way that they form a Zermelo domain B.

Hitherto set-theoreticians had held that the notion of integer should be defined and that the principle of mathematical induction should be proved, but the Skolem–Lowenheim theorem suggests that we can take the notions of integer and induction as more fundamental than the set-theoretic axioms, and thus obtain a clear, natural and unquestionable basis.

The active and critical mind of Skolem did not rest content with this suggestion but proceeded to a more radical reformulation of arithmetic, which was not only independent of the logical principles of set theory, but also dispensed with the concept of 'existence'.

A-logical Mathematics

Finally we come to what is perhaps the most surprising development of mathematical logic—the deliberate invention of mathematical structures which are independent of set theory and, to a greater or less extent, independent of the logic of implications.

This ideal is not completely realized in Peano's theory of the natural numbers, but that theory undoubtedly provided the stimulus to later work in 'a-logical mathematics' by showing what a small amount of set theory and logic was necessary.

16.10. *Grassmann's Arithmetic*

According to Wang (1957) 'the first serious and rather successful attempt to put numbers on a more or less axiomatic basis' was made by Grassmann (1861). There the arithmetic of all integers, positive, negative and zero, is presented as what we should now describe as the abstract algebra of an ordered, integral domain in which each set of positive integers has a least element. Thus if s and t are integers, it is postulated that $-s$, $s+t$ and st are integers.

It is remarkable that Grassmann did not state that different numbers have different successors, nor that 1 is not the successor of any positive integer. Nor did he give any formal analysis of such logical terms as the connectives 'and', 'or', equality and inclusion.

16.11. *Dedekind and Mathematical Induction*

Dedekind's theory of chains and quasi-chains has been described in Sections 5.4 and 5.5, but it seems appropriate to include here a summary of his theory of mathematical induction—an indispensable method of proof, but one which seems itself to demand the status of an axiom.

In the terminology of Dedekind the natural numbers are based on an infinite system N with an ordering relation, $\varphi : N \to N$ and a base element a such that a does not belong to $\varphi(a)$.

From his definition Dedekind was able to give a proof of the validity of the principle of mathematical induction within his system.

In common parlance, this principle states that if the element α possesses a property P, and if P is a 'hereditary' property, then all elements of the chain K_α possess the property P. (A property is hereditary if 'ξ possesses property P' implies that '$\varphi(\xi)$ also possesses property P'.)

16.12. *Peano's Postulates*

Peano is justly famous for the invention of a symbolic notation which gives an awkward but concise and precise expression to the logical ideas required in a rigorous development of arithmetic, but his lasting contribution to mathematical logic was a simplification of Dedekind's theory of chains, to which he frankly acknowledged his indebtedness.

Peano's work begins with his *Arithmeticae Principia* (Peano 1889) and is continued in a series of papers (Peano 1891) and in the various editions of the *Formulaire de Mathématiques* (Peano 1895–1903). It had a great influence on Bertrand Russell who summarizes the rather diffuse and incomplete exposition of Peano (Russell 1903; Vol. 1, Chap XIV).

Peano starts with two undefined terms, viz. 0, natural number (or finite, positive integer), an undefined relation of 'successorship' and five primitive propositions, viz.

(*i*) 0 is a number;

(*ii*) if *a* is a number, then *a* has a unique successor, *a* + 1, which is also a number;

(*iii*) if *a* and *b* have the same successor, then they are identical;

(*iv*) 0 is not the successor of any number;

(*v*) if 0 belongs to a class *K*, and if '*a* ∈ *K* 'implies that' *a* + 1 ∈ *K*', then every number belongs to *K*.

Thus by the last postulate Peano makes the principle of mathematical induction an axiom instead of a theorem as in Dedekind's treatment.

16.13. Consistency and Categoricity of Peano's Postulates

As his letter to Keferstein in 1890 shows (Heijenoort 1967, pp. 98–103), Dedekind recognized that one of the most difficult parts of his system was to establish the categoricity of his system, i.e. to prove that, if *S* and *S** are any two systems which satisfy his definitions, then *S* and *S** are isomorphic in the sense that to any number *n* in *S* there corresponds a number *n** in *S** in such a way that the base element *O** in *S** corresponds to the base element *O* in *S* and the successor of *n** in *S** corresponds to the successor of *n* in *S*. Moreover, the correspondence should be one–one. Well-known examples of such isomorphic systems are (1, 2, 3, . . .) and (2, 3, 4, . . .) or (1, 2, 3, . . .) and (1, 4, 9, . . .).

It is easy to prove that any two 'Dedekindian' systems *S* and *S** are isomorphic provided we assume that any number is obtainable by a *finite* number of applications of the successor operation, viz. $n \to n + 1$. Dedekind realized the inadequacy of this naïve argument and sought to provide another proof which should exclude all 'alien intruders' from the set of natural numbers.

The impossibility of such a proof is shown by the simple counter-example of the Veronesean system (Section 3.2) of elements consisting of ordered pairs of non-negative integers (p,q) with the transformation law

$$\varphi(p,q) = (p + 1, q)$$

and the base element (0,0).

Of course, similar considerations apply to the system based on Peano's postulates.

In Russell's account of Peano's system (Section 16.12) he makes no reference to the possible existence of non-isomorphic realizations of the system, but makes stringent criticisms of the undoubted existence of many isomorphic realizations such as (1, 2, 3, . . .) and (101, 102, 103, . . .) with different base elements. According to Russell the failure to identify the base element with the unit class (or, better still, the null class) renders Peano's postulates entirely useless in applied mathematics, physics and any science which involves counting.

Peano has been criticized for the complete absence of any formal rules of inference, and, indeed, for the absence of any logical connectives between the successive statements which constitute his exposition of arithmetic. But the historian may be pardoned if he fancies that this cavalier treatment of logic, both in Peano and in Grassmann, is a prophetic sign of the conscious and explicit attempt to liberate arithmetic from logic which is found in Skolem, Curry and Goodstein.

16.14. *Skolem and Recursive Definitions*

Skolem's study of Whitehead and Russell's *Principia Mathematica* had suggested to him that there was a great difference between the logical concepts of conjunction, disjunction and negation on the one hand, and the idea of existence, as expressed in either of the forms

(*i*) a given proposition $\varphi(x)$ is true always, i.e. for all x; or

(*ii*) a given proposition $\varphi(x)$ is true sometimes, i.e. there exists some x for which $\varphi(x)$ is true.

Skolem's discovery (Skolem 1923) was that the general theorems of arithmetic can be founded rigorously without the use of the notions 'always' and 'sometimes'.

This discovery he made by the use of recursive definitions which he expressed in terms of 'variables' x, y, ... (which were undefined but which are later recognized to be the natural numbers), and of the 'successor' $x + 1$ of x. Thus, for example, the 'sum' of a and $(b + 1)$ is defined in terms of the sum of a and b as $a + (b + 1) = (a + b) + 1$.

Skolem does not refer to Peano, but there is a remarkable similarity in their arithmetic theories. Nor does he refer to Brouwer in this paper, although later (Skolem 1929) he recognized that his own ideas were similar to the intuitionist concepts put forward by Brouwer.

The great merits of Skolem's 'primitive recursive arithmetic' are that each definition and theorem is stated and proved by a *finite* number of steps, and that it is impossible to state the negation of universal propositions. Thus the principle of the excluded middle is only used when a finite number of instances are concerned and the paradoxes of the infinite are avoided.

Moreover, by the methods of recursive arithmetic, both logic and arithmetic can be constructed simultaneously with full rigour and in complete detail in a calculus in which the only statements are equations of the form $a = b$.

The methods of recursive arithmetic are strictly 'elementary' in character, but nevertheless they lead to some surprising conclusions.

16.15. *Curry and Recursive Arithmetic*

The treatment of arithmetic by Dedekind and Peano was considerably advanced by Skolem (1923) and greatly developed by Hilbert and Bernays (1934, 1939) on the basis of an underlying logical calculus. Curry (1941) however, carried out a formalization of recursive arithmetic as an entirely independent logical system from which the propositional calculus can be obtained as an 'interpretation' of arithmetic in which

$\sim a$ becomes 1 if $a = 0$ or   0 if $a > 0$,

$a \vee b$ becomes the product ab,

$a \wedge b$ becomes the sum $a + b$,

$a \rightarrow b$ becomes b if $a = 0$ or   0 if $a > 0$.

16.16. Goodstein and the Equation Calculus

Independently of Curry, Goodstein (1945) greatly advanced the theory of a-logical mathematics by his invention of the axiom-free equation calculus in which logic and arithmetic are constructed simultaneously in a free variable system in which a complete elimination of the logical signs leads to an axiom-free, self-sustained pure equation calculus.

As in the work of Skolem, the familiar elementary functions, the sum function $x + y$, the product function xy and the exponential function x^y (for the natural numbers x, y) are introduced by recursive equations of the type

$$F(x,0) = a(x), \quad F(x,y') = b[x,y,F(x,y)],$$

where y' is the 'successor' of y, and $a(x)$, $b(x,y,z)$ are previously defined.

In Goodstein's calculus the principle of mathematical induction is superfluous, since the equivalence of two functions $f(x)$ and $g(x)$ follows from the equation $f(0) = g(0)$ and the expression of $f(x + 1)$ as a function of $f(x)$, together with the expression of $g(x + 1)$ as the same function of $g(x)$. Similarly the universal and existential operators of classical logic are equally superfluous.

Fractions, i.e. rational numbers, are expressed in the usual way as ordered pairs of natural numbers. But the remarkable feature of Goodstein's system is that it incorporates the concepts of convergence, of continuity, and of the fundamental 'kinematic' property of continuous functions (Section 10.2).

The most complete account of the early researches on recursive functions is given by Péter (1951) and an elementary (but complete) introduction is given by Goodstein (1964).

16.17. Post and Valuation Theory

It is a remarkable fact that while so much attention was directed to the establishment of an autonomous arithmetic, independent as far as possible, even from the propositional calculus, mathematicians were slow to realize the existence of *algebraic* systems, completely free from formal logic. The characteristic of these algebras is that they involve only a finite number of terms and a finite number of relations, and that it can be explicitly proved that they are completely consistent.

The proof of consistency is possible because we can display all the consequences which can be deduced from the axioms of the algebra and note that they contain no contradictions. Thus in the 'two-valued algebra' of Boole and Schröder there are two terms, say α and β, and two relations, addition and multiplication, such that

$$\alpha + \alpha = \alpha, \quad \alpha + \beta = \beta + \alpha, \quad \beta + \beta = \beta,$$

$$\alpha\alpha = \alpha, \quad \alpha\beta = \alpha = \beta\alpha, \quad \beta\beta = \beta.$$

It is an easy exercise to verify that addition and multiplication are distributive and associative, and to verify that the complete tale of theorems deducible in this algebra contains no contradiction.

The self-consistency of this two-valued algebra is sufficient to establish the self-consistency of the classical propositional calculus. This result was first stated explicitly by Post (1921) who also gave the generalization to the so-called 'n-valued logic'.

The appropriate algebra for this generalization is most easily exhibited by using the numerals, 1, 2, 3, ..., merely as an ordered set of marks to differentiate the terms t_1, t_2, ..., t_m. Then addition is defined as

$$t_j + t_k = t_k + t_j = t_j \quad \text{if} \quad j \leqslant k,$$

$$\text{or} \quad t_k \quad \text{if} \quad k \leqslant j,$$

and the generalization of negation or complementation by the formula

$$t'_j = t_{j+1} \quad \text{if} \quad j < n,$$

and $\qquad t'_n = t_1.$

Multiplication is then defined as

$$t_j t_k = (t'_j + t'_k)'.$$

16.18. Bernays and 'Der Methode der Aufweisung'

In his *Habilitationsschrift* at Göttingen in 1918 (Bernays 1926) Bernays examined the independence of the axioms of inference of the propositional calculus by constructing seven different algebras, the elements of which could be taken to be propositions. The possibility of such an investigation is crucially dependent on the self-consistency of each algebra and its independence of any logical system.

The simplest of the algebras consists of three elements, α, β, γ with the addition defined by the relations

$$\alpha + \alpha = \alpha + \beta = \alpha + \gamma = \beta + \alpha = \gamma + \alpha = \alpha,$$

$$\beta + \beta = \beta, \quad \beta + \gamma = \gamma + \beta = \gamma, \quad \gamma + \gamma = \alpha,$$

and complementation by the relations

$$\alpha' = \beta, \quad \beta' = \alpha, \quad \text{and} \quad \gamma' = \gamma.$$

Multiplication is then defined as usual by the relation

$$pq = (p' + q')'.$$

The significance of Bernay's investigations is that each of his algebras can be shown to be self-consistent by actually displaying all the (finite) number of theorems which can be deduced from the axioms of addition, multiplication and complementation.

Mathematics as Essentially Incomplete

16.19. Hilbert and the Problem of Consistency

During the years 1922 to 1931 Hilbert (Heijenoort 1967; p. 367) devoted a series of papers and lectures to problems of the foundations of mathematics, including the axiomatization of the real number system, the continuum hypothesis and the well-ordering hypothesis, but his most impressive contribution was the launching of a programme (Hilbert 1926) to provide a formalization of both logic and arithmetic, which would prove rigorously the complete consistency of the fundamental axioms.

   (*i*) The guiding principles of Hilbert's programme was the principle that every mathematical problem could be solved;

   (*ii*) the eradication of any intuitive meaning or physical significance from the fundamental terms and axioms;

   (*iii*) the reduction of the problem of consistency to a proof that the formula $1 \neq 1$ cannot be deduced from the axioms.

The first principle is indeed the tacit assumption of all mathematicians that some method or some axioms can and will be found to solve every problem, but in Hilbert's programme it meant that the ordinary axioms of arithmetic were sufficient to prove or to disprove all arithmetical questions, whether finite or transfinite. In this restricted form Hilbert's conviction was later conclusively refuted by Gödel (Section 16.20).

The second principle would probably be denied by Poincaré, Brouwer and the strict intuitionists, but it is acceptable to most other mathematicians as the only way of giving mathematics (and logic) an independent establishment. Hilbert strenuously repudiated any attempt to find any empirical justification for the mathematics of the infinite, whether discrete or continuous, on the undeniable ground that this lies outside our experience.

The third principle is almost self-evident as a necessary and sufficient condition for the self-consistency of the fundamental axioms, for those who accept the *tertium non datur*, but would be contested by the intuitionists.

Hilbert's axioms contain the Aristotelian principles of contradiction, of double negation and of the excluded middle, but their most original feature was the introduction of Hilbert's transfinite logical choice function ε. This was defined by the formula $A(a) \to A(\varepsilon(A))$, A being any proposition, which is interpreted to mean that either

   (*i*) $A(a)$ is false for all a, or
   (*ii*) there is some a, viz. $\varepsilon(A)$, for which A is true.

The introduction of this axiom allows Hilbert to *define* what is meant by the quantifiers 'all' and 'there exists', and also to eliminate transfinite induction from certain definitions by recursion.

After the publication of Gödel's famous investigation of formally undecidable propositions (Gödel 1931), it soon became accepted that Hilbert's programme could not be realized. Nevertheless his work has an enduring value as the first explicit recognition of the need for 'meta-mathematics', i.e. a system which, as it were, surveys mathematics from the outside as an inventory of formulae, which are the 'ideal elements' in the new higher theory, devoid of any meaning except such as may be implied by their axiomatic foundation.

16.20. Gödel and the Incompleteness Theorem

In 1930 Gödel announced his discovery of a proof that even within the confines of elementary mathematics there are theorems which cannot be proved or disproved (Gödel 1931a). This assault has been described as the most important contribution to logic since Aristotle and as an epoch-making discovery in mathematical logic. The main paper was published in 1931 (Gödel 1931b) and English translations have been provided by Meltzer (1962) and Heijenoort (1967). Detailed analyses of Gödel's arguments have been given by Goodstein (1964) and Mostowski (1964).

Gödel's work shows that the realization of the Hilbert programme is much more difficult than was anticipated, for it shows that finitary proofs of self-consistency cannot be established *within* the formalism of classical mathematics, or of set theory, or of the axioms of *Principia Mathematica* (supplemented by Peano's axioms). It seems to leave open the possibility of finitary proofs that cannot be expressed in those formal structures, but this has been poor consolation to Hilbert and his followers.

The motivation of Gödel's argument is the group of semantic paradoxes, especially the 'Liar paradox' ('I am lying') and Richard's paradox (1905) which Gödel remodelled in a form free from the subtle errors in Richard's original formulation.

Mostowski (1964) has presented Richard's paradox in the following concise form: Any definition of the properties of integers can be expressed by a finite number of words, and such definitions therefore form an enumerable sequence,

$$W_1, \ W_2, \ \ldots, \ W_n, \ \ldots.$$

Hence, according to the *tertium non datur*, if n and p are any integers, either

(i) n possesses the property W_p or,
(ii) n does not possess this property.

Denote these alternatives by the symbols

(i) $\vdash W_p(n)$ and
(ii) $\sim \vdash W_p(n)$.

Now consider the property denoted by $\sim W_n(n)$ which must be one of the properties $W_1, \ W_2, \ \ldots, \ W_n, \ \ldots$. Let it be $\vdash W_q(n)$. Now choose n to be equal to q. Then we obtain $\vdash W_q(q)$ in contradiction to our hypothesis $\sim W_q(q)$.

In Gödel's paper all that is vague in this argument (e.g. the concept of 'property', 'definition') is made rigorous and precise by his invention of the 'arithmetization of metamathematics'. This consists of a code in which the basic elements of arithmetic are given prime numbers according to such rules as $0 \rightarrow 1$, 'the successor of' $\rightarrow 3$, 'not' $\rightarrow 5$, 'or' $\rightarrow 7$ etc. Any formula is just a succession of these elements, and, if n_0, n_1, n_2, ... are the prime numbers assigned to the component elements of a formula, then the formula itself is given the number

$$2^{n_0} \cdot 3^{n_1} \cdot 5^{n_2} \ldots p_k^{n_k}$$

where p_k is the kth odd prime. Again, any theorem is a sequence of formulae which have been assigned numbers $f_0, f_1, \ldots, f_k, \ldots$ and to such a theorem is assigned the number

$$2^{f_0} \cdot 3^{f_1} \ldots p_k^{f_k}.$$

This new and original method assigns integral numbers (the 'Gödel numbers') to the basic signs of arithmetic, to propositions which are expressed by series of basic signs, and to theorems which are series of propositions. Thus Gödel does for metamathematics what Descartes did for geometry—i.e. he reduces it to arithmetic.

By this method Gödel is able to survey the whole province of arithmetic including all possible propositions and theorems. As in Hilbert's programme there is no concept of the 'truth' or 'falsity' of any proposition, but only the precise concept of its 'provability' in terms of the basic axioms.

The basic axioms are those of Peano's arithmetic together with the logic of *Principia Mathematica*, and the axiom of choice. From these axioms we obtain a formal system of arithmetic, and Gödel's main theorem is that, if the axioms are consistent, then the system is not complete, i.e. we can construct a proposition A, such that neither A nor its negation $\bar{A}$ is provable in the system.

Mathematics Repudiating Classical Logic

16.21. 'Existence' in Mathematics

Russell, Zermelo, Fraenkel and J. v. Neumann endeavoured to rehabilitate set theory by a meticulous examination of the basic concepts, embodied in explicit axioms which were framed so as to exclude the 'rogue' or 'exceptional' or 'paradoxical' sets of Cantor's naïve set theory. But a far more radical revision of the foundations of mathematics had been envisaged by Poincaré and Kronecker and was carried out ruthlessly by Brouwer and his school of 'intuitionist' mathematicians.

Poincaré had roundly asserted that the paradoxes of set theory arose from the basic error of assuming that infinite sets actually exist as completed wholes, and Krönecker had earlier protested against Cantor's methods of generating transfinite numbers. But the full consequences of these protests were not realized until Brouwer's inaugural address at the University of Amsterdam (1912) on 'Intuitionism and formalism' (Brouwer 1913b).

Brouwer's numerous writings and his seemingly interminable speeches were, and still are, notoriously difficult to understand. He undoubtedly maintained that so far from mathematics being a part of logic, logic itself should be dependent on mathematics, and that mathematics arose from fundamental and basic intuitions of the natural numbers, and of our experience of the continuity of the temporal process. Neither Brouwer nor any of his disciples has succeeded in making the philosophical foundations of 'intuitionism' completely intelligible, but fortunately it is possible to extract the mathematical essence of the new doctrine in terms of the concept of 'existence'.

No phrase occurs more frequently in mathematical literature than the apparently innocuous and inevitable phrase 'There exists . . .'. But what does 'existence' mean in mathematical context?

Many mathematicians, including Bertrand Russell in his youth, cheerfully accept the Platonic doctrine that their work does not consist in inventing or constructing mathematical entities such as arithmetical numbers and geometrical figures, but in discovering such entities, which are already pre-existing in an ideal world. But this assertion still leaves open the question, How do we know that our presumed mathematical discoveries are genuine revelations and not the products of self-delusion?

It is immediately obvious that a necessary condition for the existence of any mathematical entity is the absence of any logical contradiction in the proof offered for its existence. Thus, for example, the existence of non-Euclidean geometries was accepted once their axioms were thought to be self-consistent. But the self-consistency of the argument for a mathematical entity is only a necessary condition for its existence: it is not a sufficient condition. It establishes logical possibility but not existential reality.

However, it is often a difficult task to investigate, let alone to prove, the self-consistency of a mathematical scheme. The self-consistency of non-Euclidean geometries was proved only by assuming the self-consistency of Euclidean geometry, and the latter only by assuming the self-consistency of arithmetic. A thoroughgoing study of the self-consistency of the foundations of mathematics is obviously a necessary, challenging, and daunting task. This heroic task was undertaken by Hilbert and is described in another section (Section 16.19).

Brouwer, however, not only stressed the insufficiency of self-consistency as a criterion of mathematical existence, but even threw doubt on the possibility of establishing this self-consistency. For Brouwer a mathematical entity exists if and only if it has been *constructed* in a finite number of steps from our fundamental data—the collection of natural numbers.

16.22. *The Principle of the Excluded Middle*

The concept of 'constructability' as a necessary and sufficient condition for the 'existence' of a mathematical entity has won the approval of many eminent mathematicians, but, it must be confessed, they differ greatly in their criteria of constructability.

For example, Denjoy (1912) uses transfinite induction to 'construct' an integral by his method of totalization, but energetically rejects Perron's use of the infimum (or greatest lower bound or lower envelope) of a collection of 'majorantes'. Poincaré regarded the use of finite induction as the central method of analysis, but, for Brouwer, this method is unacceptable.

In Brouwer's theory the criterion of constructability is enforced in the strict form, that would have won the approval of Kronecker, that a mathematical entity must be positively defined by a finite number of operations starting with a finite set of natural numbers. Previously, most, if not all, mathematicians would have agreed that an entity x exists, if the assumption that x does not exist leads to a contradiction. This is the well-known principle of Aristotelian logic of the *tertium non datur*, i.e. either x exists or x does not exist, there is no medium between existence and non-existence. Brouwer is led to repudiate this principle because he interprets 'existence' to mean 'constructability' according to Brouwerian axioms.

There are three possibilities—the entity x may be proved to be constructable, or it may be proved not to be constructable, or the question of its constructability may not yet be decided. Thus:

(*i*) the imaginary unit i, such that $i^2 = -1$ is constructable in the form of an ordered pair of integers (0,1) with the proper definition of multiplication;

(*ii*) there is no rational number, p/q, such that $(p/q)^2 = 2$;

(*iii*) it is not known whether there is an integer n, greater than 2, such that $x^n + y^n = z^n$, x, y, z being positive integers.

The rejection of the principle of the excluded middle would invalidate most of classical analysis, or to be more precise, it would invalidate most of the theorems of classical analysis. In particular it would invalidate the usual proofs of the Bolzano–Weierstrass theorem that a bounded set of natural numbers has a least upper bound, and Rolle's theorem that if $f(x)$ is differentiable in the interval $[a,b]$ then there exists a number s such that $a \leqslant s \leqslant b$ and $(b-a)f'(s) = f(b) - f(a)$. In topology it would invalidate Novikov's fundamental theorem that the complement of a closed set of points is an open set (Heijenoort 1967, pp. 414–37, especially p. 436).

Brouwer's principles not only question the proofs of so many classical theorems, but also lead to a number of unexpected and unwelcome consequences. It *is* possible to give a constructive definition of a function of a real variable in strict accord with intuitionism, but it then appears that such a function must be continuous.

Nevertheless Brouwer must be given full credit for proposing a precise criterion of the 'existence' of mathematical entities.

16.23. Heyting's Intuitionist Calculus

Brouwer himself repudiated classical logic because its empirical justification was limited to finite classes and therefore inappropriate for infinite sets. He also maintained that mathematics was essentially an intellectual activity by means of which certain mathematical entities were intuitively seen to exist and certain principles to be true (but not the *tertium non datur!*). It therefore seems paradoxical that Brouwer's intuitionism should be formalized, but nevertheless this task has been attempted by Heyting (1934, 1956).

Heyting tentatively suggests that mathematical reasoning according to the canons of intuitionism can and should be based on a number of axioms for a propositional calculus which are expressed in terms of certain undefined logical relations, which are, however, represented by the symbols commonly used in classical logic for conjunction $(p \cdot q)$, disjunction $(p \vee q)$, negation $(\neg p)$ and implication $(p \supset q)$. He proved that from these relations it is not possible to derive either of the two principles rejected by Brouwer, viz.

$$p \vee \neg p \quad \text{and} \quad \neg \neg p \supset p.$$

16.24. Kolmogorov and Pseudo-truth

A most influential convert to intuitionism was undoubtedly Weyl who announced his conversion in several lectures delivered at Zurich in 1920 (Weyl 1921) and confirmed it in the presence of Hilbert at Hamburg in 1927. But Weyl's positive contribution to intuitionism seems to be limited to a defence of Poincaré's criticism of the circularity in Hilbert's justification of mathematical induction.

Another great mathematician who accepted Brouwer's criticism of traditional mathematics was Kolmogorov (Heijenoort 1967, pp. 414–6) who undertook to show why the illegitimate use of the principle of the excluded middle had not, in fact, led to any contradictions, and also why this illegitimate use had often been unnoticed.

The essence of Kolmogorov's contribution is to distinguish between p (the assertion that the proposition p is true) and $\neg \neg p$ (the double negation, which asserts the falsehood of the negation of p). In intuitionism it is not permitted to assert $p \vee \neg p$, but it is possible to show that $\neg(\neg p \text{ and } \neg \neg p)$ i.e. we must accept either the negation of p or the double negation $\neg \neg p$, which Kolmogorov calls its 'pseudo-truth'.

He proves that classical mathematics can be translated into intuitionist mathematics provided that each theorem p is replaced by the theorem of its pseudo-truth.

16.25. Gödel's Interpretation of Heyting's Formulae

Gödel has given two different interpretations of Heyting's axioms in order to permit a comparison with classical logic.

In the first interpretation (Gödel 1932a) he proposed to define disjunction and implication in the classical sense by the formulae

$$p \oplus q = \neg(\neg p \cdot \neg q),$$

$$p \rightarrow q = \neg(p \cdot \neg q)$$

using new symbols $\oplus$ and $\rightarrow$ for 'or' and 'implies'. He thence deduced the principle of the excluded middle in the form $p \oplus \neg p$ and the principle for the elimination of double negation in the form $\neg\neg p \rightarrow p$. It would thus appear that classical logic can be deduced from intuitionism, but the appearance is illusory since it depends upon the interpretation of Heyting's symbol $\neg p$, to mean 'p is false', whereas intuitionists maintain that it means 'p is absurd or impossible'.

In the second paper (Gödel 1932b) he proposed to interpret Heyting's symbols by introducing a new symbol Δ to mean 'it is demonstrable that'. We then have the interpretation in the following table

Heyting's symbols	Gödel's symbols
$\neg p$	$\Delta(\sim\Delta p)$
$p \cdot q$	$\Delta p \cdot \Delta q$
$p \vee q$	$\Delta p \vee \Delta q$
$p \rightarrow q$	$\Delta p \supset \Delta q$.

Here $\sim p$ is the classical symbol meaning 'p is false'. This interpretation certainly shows that, if classical logic is valid, then Heyting's axioms are self-consistent, and that Brouwer's repudiation of the excluded middle can be regarded simply as the rejection of the thesis that all mathematical problems are soluble.

Let us conclude with the following informal argument which indicates the motivation of the intuitionist theorem that any function is continuous.

Any number x such that $0 \leqslant x \leqslant 1$ can be 'constructed' as a monotone sequence of approximations,

$$x_0, x_2, x_2, \ldots, x_n, \ldots$$

such that

$$x_n = \sum_{m=0}^{n} a_m 2^{-m}$$

where $$a_m = 0 \quad \text{or} \quad 1.$$

Any function $f(x)$ can be 'constructed' as a sequence of approximations

$$f_0(x), f_1(x), \ldots, f_n(x), f_n(x), \ldots$$

where $f_n(x)$, depends only on $a_0, a_1, \ldots, a_n$ and where

$$|f_s(x) - f_t(x)| < 2^{-s}, \text{ if } s < t,$$

for all s and t. Hence, if $|x - y| < 2^{-\alpha}$,

$$f_s(x) - f_s(y) = 0, \quad \text{if} \quad s \leqslant \alpha,$$

and $f_s(x) - f_s(y) = [f_s(x) - f_\alpha(x)] - [f_s(y) - f_\alpha(y)] + [f_\alpha(x) - f_\alpha(y)],$

whence $$|f_s(x) - f_s(y)| < 2^{-\alpha+2}.$$

16.26. *Modal Logic*

Intuitionist logic must be distinguished from another variety of logic, 'modal' logic, which was known to the Megarian logicians and the scholastics, and was reintroduced by McColl (1878, 1879, 1880, 1906) and C. I. Lewis (1918). In addition to the two truth-values 'true' and 'false', these authors add the idea of 'impossibility' and its negation—'consistency'. 'Strict implication', symbolized by $p < q$, is defined as 'it is impossible that p is true and q is false'. Unfortunately strict implication, like the ordinary material 'implication', brings its own paradoxes, e.g. if it is impossible that p be false, then p is strictly implied by any proposition q.

Lewis endeavoured to eliminate these paradoxes, but in the latest edition of his treatise withdrew the chapter on strict implication, although the possibility of a satisfactory revisions seems to be still open (Lewis and Longford 1960).

17. Conclusion

The preceding chapters have described the development of a number of mathematical topics during the last hundred years with especial emphasis on their earlier stages, and they must inevitably lead the reader to ponder the fascinating and difficult question of the nature of mathematical discovery and invention. In surveying this problem historical facts are preferable to psychological speculation, and we are fortunate indeed to have the evidence of a number of outstanding practising mathematicians, among whom may be cited Poincaré (1908, 1913), Hadamard (1945) and Dame Mary Cartwright (1955).

There is almost complete unanimity that mathematical research is not achieved by a passive expectation of some great new concept or powerful method which will transform the subject, but by the hard and continuous study of a difficult and challenging problem. As evidence for this ineluctible truth we may cite the authoritative declarations of Weil and Hilbert.

In the introduction to his great treatise, *Foundations of Algebraic Geometry*, Weil made this penetrating comment on the subject's historical development (Weil 1946):

> Algebraic geometry, in spite of its beauty and importance, has long been held in disrepute by many mathematicians as lacking proper foundations. The mathematician who first explores a promising new field is privileged to take a good deal for granted that a critical investigator would feel bound to justify step by step; at times when vast territories are being opened up, nothing could be more harmful to the progress of mathematics than a literal observance of strict standards of rigour. Nor should one forget, when discussing such subjects as algebraic geometry, and in particular the work of the Italian school, that the so-called 'intuition' of earlier mathematicians, reckless as their use of it may sometimes appear to us, often rested on a most painstaking study of numerous special examples, from which they gained an insight not always found among modern exponents of the axiomatic creed. At the same time, it should always be remembered that it is the duty, as it is the business, of the mathematician to prove theorems, and that this duty can never be disregarded for long without fatal effects. The experience of many centuries has shown this to be a matter on which, whatever our tastes or tendencies, whether 'creative' or 'critical', we mathematicians dare not disagree. As in other kinds of war, so in this bloodless battle with an ever-retreating foe which it is our good luck to be waging, it is possible for the advancing army to

279

outrun its services of supply and incur disaster unless it waits for the quartermaster to perform his inglorious but indispensable tasks.

We must also cite the famous address given by Hilbert to the International Congress of Mathematics in Paris on 'Mathematical problems' (Hilbert 1900). This great catalogue of unsolved problems did much more than throw down a challenge to mathematicians and thereby stimulate much fruitful research; it also expressed Hilbert's conviction that real progress in mathematics must begin with the attack on limited, specific problems. He forcefully maintained that the necessary and sufficient condition for advance in mathematics is the solution of a new, concrete and difficult problem, or the proof that the problem has no solution. And he optimistically proclaimed his conviction (which he insisted that every mathematician shares although none has given a proof!) that every definite mathematical problem must necessarily be capable of rigorous settlement, either by an answer to the question at issue or the proof that no answer is possible.

In stressing the importance of great, concrete, fruitful problems, Hilbert himself maintained:

> As long as a branch of science affords an abundance of problems, it is full of life; want of problems means death or cessation of independent development. Just as every human enterprise prosecutes final aims, so mathematical research needs problems. Their solution steels the force of the investigator; thus he discovers new methods and view points and widens his horizon.

These quotations leave us with the really pressing question—how in fact are mathematical problems solved? There is no doubt about the necessity of sheer hard work, of continued application to a definite problem, of the use of hypotheses, of counter-examples, of all the apparatus of the method of trial and error, as brilliantly described by Lakatos (1976). But it is equally certain that this painful mental exertion is scarcely ever effective. In fact it is usually when the mind, exhausted by the fruitful search for the solution of a problem, relaxes and turns aside for a mental holiday that light begins to dawn. The answer to the problem then appears, without any conscious effort by the mathematician, suddenly, with clarity and certainty. Further hard work may be necessary to translate this sudden vision, to relate it to existing theories, and to complete the detailed exposition, but in principle the problem is solved.

This seems to be confirmed by Poincaré's discovery of Fuchsian functions which is related in some detail in the essay by Hadamard quoted earlier.

The nature of the psychological process by which the mind of the mathematician is suddenly enlightened has been compared with poetic, pictorial and musical inspiration in a remarkable series of thirty-eight confessions collected by Ghiselin (1952), and it is reminiscent of the methods employed in Zen Buddhism to prepare the novice to cease from discursive thought and to listen to the voice of the Buddha (Suzuki 1934).

But no satisfactory account of the psychology of mathematical illumination has yet been given.

There is one feature of the development of mathematics since 1879 which deserves especial mention, namely the influence of a number of splendid text-books and of magnificent encyclopaedic works. University teaching has been greatly indebted to the fine 'Cours d'analyse' of the French mathematicians, among whom we must mention Goursat (1902), Picard (1891–8) and Jordan (1882–7), and the modest text-book of Hardy (1908), written when 'analysis was neglected in Cambridge' and 'like a missionary talking to cannibals'.

From Germany we have had the great *Encyklopaedie der Mathematische Wissenschaften* (1898–1921) the earlier volumes of which were translated and expanded in French.

But in more recent years the greatest influence has been the magisterial treatise of the group of French mathematicians, led by André Weil writing under the name of Bourbaki (Bourbaki 1939–   ), which is still in progress and has as its sublime purpose a complete exposition of all mathematics from its most fundamental concepts to its most recent developments. To this must be added the successive volumes we owe to Dieudonné (1960–   ) which unite the architectonic excellence of Bourbaki with the pedagogic persuasiveness of Hardy.

It would be presumptious for an historian to judge what are the outstanding advances in mathematics during the last hundred years, but a study of a few undoubtedly great achievements may throw some light on the character of the greatest researches.

In the humble science of arithmetic we have the great theorem of Gödel, which seems to forbid any progress with Hilbert's programme, but, on the other hand, we have the invention of recursive methods by Skolem and Goodstein which liberates arithmetic from propositional logic.

In analysis the invention of distributions, or generalized functions, by Laurent Schwart has enormously increased the range of the subject and has greatly extended the domain of 'soluble' differential equations.

In geometry the introduction of algebraic concepts and methods, originally due to Emmy Noether, has completely transformed the subject into the theory of 'sheafs' and 'schemes' as expounded by Grothendieck.

Algebra itself has been transformed by the study of 'Universal algebra', initiated by Garrett Birkhoff's theory of lattices, and by the great unification produced by the theory of categories and functors, due to Eilenberg and MacLane.

Two characteristics of the development of mathematics during the period under survey call for especial remark:

(1) A hundred years ago mathematicians were beginning to realize that rigorous proofs of theorems (especially in analysis) depended utterly on clear and effective definitions (such as Dedekind's definition of real numbers). But the study of basic subjects, such as the foundations of geometry and mathematical logic, which require elementary and undefinable concepts, replaced definitions by axioms, as in Hilbert's reformulation of Euclid. A set of axioms,

however, is useless without a proof of consistency, and this is usually algebraic in character. The latest stage is thus the creation of mathematical structures almost entirely free from axioms. An admirable account of the changes in mathematical education and in the effect of the examination system in the United Kingdom is given by Griffiths (1971).

(2) In the 1870s mathematics consisted of a number of separate and diverse disciplines, such as arithmetic, algebra, geometry and infinitesimal calculus. In the 1970s this is no longer the case. There has been an enormous increase in the volume of material which a mathematician has to master, but there are no longer any boundaries between the different 'branches' of the subject, which are now so closely interdependent that no mathematician can afford to specialize. In the 1970s to be a good mathematician is to be a polymath.

Bibliography

At the time of writing there is no history of mathematics exclusively devoted to the work of the last hundred years, but the concluding chapters of the standard histories of mathematics do venture into the first half of the period which is the subject of this book. I owe much to their authors and gratefully record in particular the following works.

E. T. Bell, *The Development of Mathematics*, New York and London, 1st ed. 1940, 2nd ed., 1945.

C. B. Boyer, *A History of Mathematics*, New York, London and Sydney, 1968.

M. Kline, *Mathematical Thought from Ancient to Modern Times*, New York, 1972.

For the specialist there are:

J. Cavaillès, *Philosophie mathématique*, Paris, 1962.

N. Bourbaki, *Éléments de mathématiques: Éléments d'histoire des mathématiques*, 2nd ed., Paris, 1969.

Much valuable material is to be found in the mathematical 'source' books for which we are indebted to American writers and publishers:

J. R. Newman, *The World Of Mathematics*, Vol. 1–4, London, 1956.

G. Birkhoff, *A Source Book in Classical Analysis*, Cambridge, Mass., 1973.

J. van Heijenoort, *From Frege to Gödel*, Cambridge, Mass., 1967.

Finally the reader must be referred to the journals which specialize in the history of mathematics:

Zentralblatt fur Mathematik
Archive for history of exact sciences
Historia Mathematica

References

Abel, N. H. (1836). 'Untersuchungen über die Riehe $1 + (m/1)x + [m(m + 1)/2]x^2 + \ldots$', *Crelles Journal.*, **1**, 311–339 (especially footnote to p. 316).

Alexander, J. W. (1922). 'On transformations with invariant points', *Trans. Am. math. Soc.*, **23**, 89–95.

— (1938). 'A theory of connectivity in terms of gratings', *Ann. Math.*, **39**, 883–912.

Alexander, S. (1916–8). 'Space, time and deity', *Gifford lectures*, London.

Alexandroff, P. S. (1924). 'Über die Äquivalenz des Perronschen und des Denjoyschen Integralbegriffes', *Math. Z.*, **20**, 213–22.

— (1925). 'Paul Urysohn', *Fundam. Math.*, **7**, 138–40.

— (1928a). 'Untersuchungen über Gestalt und Lage abgeschlossener Mengen beliebiger Dimension', *Ann. Math (2)*, **30**, 101–87.

— (1928b). 'Über den allgemeinen Dimensionsbegriff', *Math. Annln*, **98**, 617–35.

— and Hopf, H. (1935). *Topologie I*, Berlin.

Andronov, A. and Witt, A. (1930). 'Zur Theorie des Mitnehmens von van der Pol', *Arch. Elektrotech.*, **24**, 99–110.

Antosik, P., Mikusiński, J. and Sikorski, R. (1973). *Theory of distributions*, Amsterdam–Warsaw.

Appleton, E. V. and Greaves, W. M. H. (1923). 'On the solution of the representative differential equation of the triode', *Phil. Mag.*, **45**, 401–14.

— and Pol, B. van der (1922). 'On a type of oscillation hysteresis . . .', *Phil. Mag.*, **43**, 177–93.

Aquino, S. Thomae de (ca. 1260). *In librum Boethii de Crinitate*, qu. V, art. III.

Arzela, C. (1895). 'Sulle funzioni di linee', *Memorie R. Accad. Sci. Ist. Bologna (Ser. V)*, **5**, 225–44.

Ascoli, G. (1883). 'Le curve limite di una varietà deta di curve', *Atti Accad. naz. Lincei Memorie (Serie Terza) (Classe di scienze fisiche e matematiche)*, **18**, 521–86.

Baire, R. (1899a). 'Sur la théorie des ensembles', *C. r. hebd. Séanc. Acad. Sci., Paris*, **129**, 946–9.

— (1899b). 'Sur les fonctions de variables réelles', *Annali Mat. pura appl. (IIIa)*, **3**, 1–123 (especially p. 10).

— (1904). *Leçons sur les fonctions discontinues*, Paris, p. 71.

— (1907). 'Sur la non-applicabilité de deux continus . . .', *Bull. Sci. math. (2)*, **31**, 94–9.

Banach, S. (1922). 'Sur les opérations dans les ensembles . . .', *Fundam. Math.*, **3**, 133–81 (especially p. 160).

— (1929). 'Sur les fonctionelles linéaires', *Studia math.*, **1**, 211–6, 223–39.

— (1932). *Théorie des opérations linéaires*, 1st ed., Lwów (reprinted by Chelsea Publishing Company, New York).

Barré, A. (1845). 'Mémoire sur les sommes et les différences géometriques', *C. r. hebd. Séanc. Acad. Sci., Paris*, **21**, 620–5.

Bauer, H. (1915). 'Der Perronsche Integralbegriff . . .', *M. Math. Phys.*, **26**, 153–98.

Beltrami, E. (1868a). 'Saggio di interpretazione della geometria non-Euclidea', *G. Mat.*, **6**, 284–312.

— (1868b). 'Sulla teorie delle linee geodetiche', *Rc. Ist. lomb. Sci. Lett. (2)*, **1**, 708–18.

Bendixson, I. (1883). 'Quelques theorèmes de la théorie des ensembles de points', *Acta math.*, **2**, 415–29.

— (1901). 'Sur les courbes définis par les équations différentielles', *Acta math.*, **24**, 1–88.

Bergman, S. and Schiffer, M. (1953). *Kernel functions*, New York.

Berkeley, Bishop G. (1734). *The analyst*, London.

Bernays, P. 1926). 'Axiomatische Untersuchung des Aussagen-Kalkuls der *Prinipia mathematica*', *Math. Z.*, **25**, 305–20.

Bernstein, F. (1905). 'Untersuchungen aus der Mengenlehre', *Math. Annln*, **61**, 117–55.

Berry, G. G. (1906). 'Les paradoxes de la logique', *Revue de métaphysique et de morale*, **14**, 627–50.

Bertrand, J. (1867). *Traité d'arithmetique*, 4th ed., Paris.

Betti, E. (1871). 'Sopra gli spazi di un numero qualunque di dimensioni', *Annali Mat. pura appl. (2)*, **4**, 140–58.

Birkhoff, G. (1948). *Lattice theory* (revised edition), New York.

Birkhoff, G. D. (1913). 'Proof of Poincaré's geometric theorem', *Trans. Am. math. Soc.*, **14**, 14–22.

— (1917). 'Dynamical systems with two degrees of freedom', *Trans. Am. math. Soc.*, **18**, 199–300.

— (1926). 'An extension of Poincaré's last geometric theorem', *Acta math.*, **47**, 297–311.

— (1927). *Dynamical systems*, New York.

— and Kellogg, O. D. (1922). 'Invariant points in function space', *Trans. Am. math. Soc.*, **23**, 96–115.

Bochenski, I. M. (1961). *A history of formal logic* (translated by I. Thomas), New York, Section 23.

Bochner, S. (1932). *Vorlesungen über Fouriersche Integrale*, Leipzig. (English translation: *Lectures on Fourier integrals*, Princeton, 1959).

— (1946). 'Linear partial differential equations with constant coefficients', *Ann. Math.*, **47**, 202–12.

— (1952). 'Théorie des distributions', *Bull. Am. math. Soc.*, **58**, 78–85.

— and Martin, W. T. (1948). *Several complex variables*, Princeton, p. 112.

Bois-Reymond, P. du (1879). 'Erlauterungen zu den Anfangsgründen der Variationsrechnung', *Math. Annln*, **15**, 283–314 (especially p. 297).

— (1882). *Die allegemeine Funktionentheorie I*, Darmstadt, p. 189. (French translation by G. Milheud and A. Girod, Nice, 1887, p. 153.)

Bolza, O. (1903). 'Some instructive examples in the calculus of variations', *Bull. Am. math. Soc. (2)*, **9**, 1–24.

— (1904). *Lectures on the calculus of variations*, Chicago.

— (1909). *Vorlesungen über Variationsrechnung*, Leipzig.

— (1913). "Über den "abnormalen Fall" beim Lagrangeschen und Mayerschen Problem', *Math. Annln*, **74**, 430–46.

Bolzano, B. (1817), 'Rein analytischen Beweis', *Abh. K. Böhm. Ges. Wiss. (3)*, Vol. 5, Prague. See also *Ostwald's Klassiker*, No. 154 (1905).

— (1834). 'Funktionenlehre', *Schriften*, Vol. 1, Prague, 1930.

— (1950). *Paradoxes of the infinite*, London (translated by Fr. Prihonsky with historical introduction by D. A. Steele).

Bonola, R. (1908). *Die nicht-Euklidische Geometrie*, Leipzig.

Boole, G. (1847). *The mathematical analysis of logic*, London and Cambridge.

— (1848). 'The calculus of logic', *Camb. Dublin Math. J.*, **3**, 183–98.

— (1854). *An investigation of the laws of thought*, London.

Borel, E. (1895). 'Sur quelques points de la théorie des fonctions', *Annls. scient. éc. norm. sup., Paris (3)*, **12**, 1–55 (especially p. 51).

— (1898). *Leçons sur la théorie des fonctions*, 1st ed., Paris, p. 46.

— (1905). 'Quelques remarques sur les principes de la théorie des ensembles', *Math. Annln.*, **60**, 194–5.

— (1914). *Leçons sur la théorie des fonctions*, 2nd ed., Paris. Note IV: 'Les polémiques sur le transfini et sur la démonstration de M. Zermelo', especially 'Cinq lettres sur la théorie des ensembles'. These letters appeared originally in *Bull. Soc. math. Fr.*, **33**, 261–73 (1904).

Bourbaki, N. (1939–   ). *Éléments de mathématiques*, Paris.

— (1969). *Éléments de mathématiques: Éléments d'historie des mathématiques*, 2nd ed., Paris.

Brauer, R. and Weyl, H. (1935). 'Spinors in *n* dimensions', *Am. J. Math.*, **57**, 425–49.

Brelot, M. (1952). 'La theorie moderne du potential', *Annls Inst. Fourier Univ. Grenoble*, **4**, 113–43.

— (1969). *Historical introduction to potential theory*, C.I.M.E. 1 Coclo, Stresa, p. 3–21.

Bremermann, H. J. and Durand, L. (1961). 'On analytic continuation, multiplication and Fourier transforms of Schwartz distributions' *J. Math. Phys.*, **2**, 240–58.

Brianchon, C.-J. (1817). *Mémoire sur les lignes du seconde ordre*, Paris.

Brill, A. and Noether, M. (1894). 'Bericht über die Theorie der algebraischen Funktionen', *Jber. dt. MatVerein.*, **3**, 353.

Brillouin, L. (1926). 'Remarques sur la mécanqiue ondulatoire', *J. Phys. Radium*, **7**, 353–68; 'La méchanique ondulatoire', *C. r. hebd. Séanc. Acad. Sci., Paris*, **183**, 24–6.

Briot, C. A. A. and Bouquet, J. C. (1856). 'Recherches sur les fonctions définis par les équations différentielles', *J. Éc. polytech.*, **21**, 133–98.

Brouwer, L. E. J. (1910). 'On the structure of perfect sets of points', *Proc. Sect. Sci. K. Ned. Akad. Wet.*, **12**, 785–94.

— (1911). 'Beweis der Invarianz der Dimensionenzahl', *Math. Annln*, **70**, 161–5.

— (1912a). 'Über Abbildung von Mannigfaltigkeiten', *Math. Annln*, 97–115.

— (1912b). 'Beweis der Invarianz des *n*-dimensionalen Gebeit', *Math. Annln*, **71**, 305–19.

— (1913a). 'Über den natürlichen Dimensionsbegriff', *Crelles Journal*, **142**, 146–52.

— (1913b). 'Intuitionism and formalism', *Bull. Am. Math. Soc.*, **20**, 81–96.

Browder, F. (1960). 'On the fixed-point index', *Summa Bras. math.*, **4**, 253–93.

Brown, R. E. (1971). *The Lefschetz fixed-point theorem*, Glenview, Illinois.

Burali-Forti, C. (1897). 'Una questione sui numeri transfiniti', *Rc. Circ. mat. Palermo*, **11**, 154–64.

Cantor, G. (1872). 'Über die Ausdehnung eines Satzes aus der Theorie der trigonometrischen Reihen', *Math. Annln*, **5**, 123–32.

— (1878). 'Ein Beitrag zur Mannigfaltigkeitslehre', *Crelles Journal*, **84**, 242–58.

— (1879). 'Über einen Satz aus der Theorie der stetigen Mannigfaltigkeiten', *Nachr. Ges. Wiss., Göttingen*, 127–35.

— (1882). 'Über ein neues und allgemeines Condensationsprinzip der Singularitäten von Funktionen', *Math. Annln*, **19**, 588–94.

— (1883). 'Über unendlichen, linearen Punktmannigfaltigkeiten', *Math. Annln*, **21**, 545–91.

— (1884a). 'Über unendlichen, linearen Punktmannigfaltigkeiten: No. 6', *Math. Annln*, **23**, 453–88.

— (1884b). 'De la puissance des ensembles parfaits de points', *Acta math.*, **4**, 381–92.

— (1889a). 'Über unendlichen linearen Punktmannigfaltigkeiten', *Math. Annln*, **33**, 545–86.

— (1889b). 'Bermerkung mit Bezug auf den Aufsatz: Zur Weierstrass' Cantor's-chen Theorie der Irrationalzahlen in *Math. Annln*, 33, 154', *Math. Annln*, **33**, 476.

— (1895). 'Beiträge zur Begründung der transfiniten Mengenlehre, I', *Math. Annln*, **46**, 481–512.

— (1897). 'Beiträge zur Begründung der transfiniten Mengenlehre, II', *Math. Annln*, **49**, 343–7.

Cartan, E. (1899). 'Sur certains expressions différentielles et le problème de Pfaff', *Annls scient. Éc. norm. sup., Paris*, **16**, 239–332.

— (1913). 'Les groupes projectifs qui ne laissent invariante aucune multiplicité plane', *Bull. soc. math. Fr.*, **41**, 53–96.

— (1914). 'Les groupes projectifs qui ne laissent invariante aucune multiplicité plane', *J. Math. pures appl. (6)*, **10**, 149–56.

— (1937). *Leçons sur la théorie de spineurs*, Paris. (English translation, 1966.)

— (1951). 'La méthode du repère mobile', *Leçons sur la géométrie des espaces de Riemann*, 2nd ed., Paris, Chap. IX.

Cartan, H. (1937a). 'Théorie des filtres', *C. r. hebd. Séanc. Acad. Sci., Paris*, **205**, 595–8.

— (1937b). 'Filtres et ultrafiltres', *C. r. hebd. Séanc. Acad. Sci., Paris*, **205**, 777–9.

— (1941). 'Sur les fondements de la théorie du potentiel', *Bull. Soc. math. Fr.*, **69**, 71–96.

— (1945). 'Théorie du potentiel Newtonien ...', *Bull. Soc. math. Fr.*, **73**, 74–106.

Cartwright, M. L. (1955). *The mathematical mind* (James Bryce Memorial Lecture), London.

— and Littlewood, J. E. (1945). 'On non-linear differential equations of the second order', *J. Lond. math. Soc.*, **20**, 180–9.

Cauchy, A.-L. (1821). 'Cours d'analyse', *Oeuvres*, Vol. 2, Paris, Part III.

Cavaillès, J. (1932). 'Sur la 2e définition des ensembles finis donnée par Dede-kind', *Fundam. Math.*, **19**, 143.

— (1962). *Philosophie mathématique*, Paris.

Cayley, A. (1859). 'A sixth memoir upon quantics', *Phil. Trans. R. Soc.*, **149**, 61–90. (Also appears in *Collected Papers*, Vol. II, No. 58, Cambridge, 1889.)

Čech, E. (1932). 'Théorie générale de l'homologie dans un espace quelconque', *Fundam. Math.*, **19**, 149–83.

Chasles, M. (1852). *Traité de géométrie supérieure*, Paris, p. 446, paragraph 623.

Chesterton, G. K. (1925). *The victorian age in literature*, London, pp. 7–8.

Chevalley, C. (1954). *The algebraic theory of spinors*, New York.

Chittendon, E. W. (1917). 'On the equivalence of écart and voisinage', *Trans. Am. math. Soc.*, **18**, 161–6.

Christoffel, E. B. (1869). 'Über die Transformation der homogenen Differential-calculusdrücke zweiten Grades', *Crelles Journal*, **70**, 46–70, 241–5.

Clifford, W. K. (1878). 'Applications of Grassmann's extensive algebra', *Am. J. Math.*, **1**, 350–8.

Cohen, P. J. (1966). *Set theory and the continuum hypothesis*, New York –Amsterdam.

Conway, A. W. (1911). 'On the application of quaternions to some recent developments', *Proc. R. Ir. Acad.*, **29**, 1–9.

Courant, R. and Hilbert, D. (1953). *Methods of mathematical physics*, Vol. II, London, pp. 293–8.

Cousin, P. (1895). 'Sur les fonctions de n-variables complexes', *Acta math.*, **19**, 1–61 (especially Section 10).

Cronin, J. (1964). *Fixed points and topological degree in nonlinear analysis*, Providence, Rhode Island.

Crowe, M. J. (1967). *A history of vector analysis*, Notre Dame.

Curry, H. B. (1941). 'A formalization of recursive arithmetic', *Am. J. Math.*, **63**, 263–82.

Darboux, G. (1875). 'Mémoire sur les fonctions discontinus', *Annls scient. Éc. norm. sup.*, *Paris (2)*, **4**, 72–112.

— (1889). *Théorie des surfaces*, Part 2, Book V, Paris, pp. 438–511.

Darwin, C. G. (1927). 'The electron as a vector wave', *Proc. R. Soc. (A)*, **116**, 227–53.

— (1928). 'The wave equation of the electron', *Proc. R. Soc. (A)*, **118**, 654–80.

Dedekind, R. (1872). *Stetigkeit und irrationalen Zahlen*, Brunswick.

— (1877). 'Über die Anzahl der Ideal-Klassen in den verschiedener Ordnungen eines endlichlen Korpus', *Gesammelte Werke*, Vol. 1, Berlin, pp. 104–58 (reprinted by Chelsea Publishing Company, New York, 1969).

— (1888). *Was sind und was sollen die Zahlen*, Brunswick.

— (1932). 'Zweite Definition (1889) des Endlichen und Unendlichen', *Gesammelte Werke*, Vol. 3, Brunswick, pp. 450–60.

— and Weber, H. (1882). 'Theorie der algebraischen Funktionen einer Veränd-lichen', *Crelles Journal*, **92**, 181–290.

Dehn, M. (1926). *Die Grundlagung der Geometrie in historische Entwicklung*, Berlin.

Delauncy, C. (1841). 'Thèse sur la distinction des maxima et des minima', *J. Math. pures appl. (1)*, **6**, 209–37.

Denjoy, A. (1912). 'Une extension de l'intégrale de M. Lebesgue', *C. r. hebd. Séanc. Acad. Sci.*, *Paris*, **154**, 859–62.

— (1941). *Leçons sur le calcul des coefficients . . .*, Paris, pp. 677–83.

Deny, J. (1950a). 'Les potentiels d'énergie finie', *Acta math.*, **82**, 107–83.

— (1950b). 'Sur la définition de l'énergie en théorie du potentiel', *Ann. Inst. Fourier Univ. Grenoble*, **2**, 83–99.

Dieudonné, J. (1960–). *Foundations of modern analysis* (English translation by I. G. MacDonald), New York–London.

— (1966). *Fondements de la géométrie algébrique moderne*, Montreal.

— (1972). 'The historical development of algebraic geometry', *Am. math. Mon.*, **79**, 827–66.

Dini, U. (1878). *Fondamenti per la teorica delle funzioni di variabili reale*, Pisa.

Dirac, P. A. M. (1928). 'The quantum theory of the electron', *Proc. R. Soc. (A)*, **117**, 610–24; **118**, 351–61.

— (1930). *The principles of quantum mechanics*, Cambridge.

Dirichlet, P. G. L. (1829). 'Sur la convergence des séries trigonométriques', *Crelles Journal*, **4**, 157–69.

— (1871). *Vorlesungen über Zahlentheories*, 2nd ed., Brunswick (1st ed., 1863).

— (1876). *Vorlesungen über die im ungekehrten Verhältnis des Quadrates des Entfernung wirkenden Kräfte*, Leipzig.

Duffing, G. (1918). *Erzwungene Schwingungen bei veranderlichen Eigenfrequenz und ihre technische Bedentung*, Brunswick.

Dunford, N. and Schwartz, J. T. (1957). *Linear operators*, Part 1, New York, Chap. 2.

Eddington, A. (1946). *Fundamental theory*, Cambridge.

Eilenberg, S. and MacLane, S. (1945). 'General theory of natural equivalences', *Trans. Am. math. Soc.*, **58**, 231–94.

— and Steerod, N. (1952). *Foundations of algebraic topology*, Princeton.

Einstein, A. (1916). 'Die Grundlagen der allgemeinen Relativitätstheorie', *Annln Phys.*, **49**, 769.

Enriques, F. (1898). *Lezioni di geometria proiettiva*, Bologna.

Erdmann, G. (1877). 'Über unstetige. Lösungen in der Variationsrechnung', Crelles Journal., **82**, 21–30.

— (1878). 'Zur Untersuchung der zweiten Variation einfacher Intregrale', *Z. Math. Phys.*, **23**, 362–79.

Euler, L. (1748). 'Sur la vibration des cordes', *Mém. Acad. Sci. Berl.*, 69–85.

— (1749). 'De vibratione chordarum exercitatio', *Nova acta eruditorum*, 512–27.

Evans, G. C. (1920). 'Fundamental points of potential theory', *Rice Inst. Pamp.*, No. VII, 252–329.

— (1927). 'The logarithmic potential', *American Mathematicial Society Colloquium Publications*, Vol. VI.

Fatou, P. (1906). 'Séries trigonométriques et séries de Taylor', *Acta math.*, **30**, 325–400.

Fermi, E. (1922). 'Supra i fenomeni che avvengono in vicinanza di una linea oraria', *Atti Accad. naz. Lincei Rc. Sed. solen.*, **31**, 21–3, 51–2, 101–3.

Fischer, E. (1907). 'Sur la convergence en moyenne', *C. r. hebd. Séanc. Acad. Sci., Paris*, **144**, 1022–4.

Floquet, G. (1883). 'Sur les équations différentielles linéaires à coefficients périodiques', *Annls scient. Éc. norm. sup., Paris*, **12**, (2), 47–82.

Forder, H. G. (1941). *Calculus of extension*, Cambridge.

Forsyth, A. R. (1890). *Theory of differential equations*, Vol. I: *Exact equations and Pfaff's problem*, London.

— (1890–1906). *Theory of differential equations*, Vols. I–VI, London.

— (1893). *Theory of functions of a complex variable*, Cambridge, p. 6.

— (1927). *Calculus of variations*, Cambridge, p. vii.

— (1935). 'Our tripos days at Cambridge', *Math. Gaz.*, **19**, 162–79 (especially p. 170).

Fowler, R. H., Gallop, E. G., Lock, C. W. H. and Richmond, H. W. (1920). 'The aerodynamics of a spinning shell', *Phil. Trans. R. Soc.*, **221A**, 295–387 (especially pp. 337–9).

Fraenkel, A. A. (1922). 'Der Begriff "definit" ...', *Sber. preuss. Akad. Wiss. Phys.-math. Klasse*, 253–7.

—, Bar-Hillel, Y. and Levy, A. (1958). *Foundations of set theory*, 1st ed., Amsterdam–London, Chap. 1 (2nd ed., 1958).

Fréchet, M. (1906). 'Sur quelques points du calcul fonctionnel', *Rc. Circ. mat. Palermo*, **22**, 1–74.

— (1907). 'Sur les ensembles de fonctions et les opérations linéaires', *C. r. hebd. Séanc. Acad. Sci., Paris*, **144**, 1414–6.

— (1910). 'Les dimensions d'un ensemble abstrait', *Math. Annln*, **68**, 145–68.

— (1914). 'Sur la notion de différentielle d'une function de ligne', *Trans. Am. math. Soc.*, **15**, 135–61.

— (1915). 'Sur l'intégrale d'une fonctionelle étendue à un ensemble abstrait', *Bull. Soc. math. Fr.*, **43**, 249–67.

— (1926). *Les espaces abstraits*, Paris (2nd ed., 1951).

Fredholm, I. (1900). 'Sur une nouvelle méthode pour la résolution du problème de Dirichlet', *Förh. svenska tek. VetenskAkad. Finl.*, **54**, 39–46.

— (1903). 'Sur une classe d'équations fonctionnelles', *Acta math.*, **27**, 365–90.

Frege, G. (1879). *Begriffsschrift, eine der arithmetischen nachgebildete Formalspreche des reinen Denkens*, Halle.

— (1884). *Die Grundlagen der Arithmetik*, Breslau.

Freudenthal, H. (1955). 'Poincaré et les fonctions automorphes', *Le livre du centenaire de la naissance de Henri Poincaré, Paris*, pp. 212–9.

Friedrichs, K. O. (1939). 'On differential operators in Hilbert space', *Am. J. Math.*, **61**, 523–44.

— (1944). 'The identity of weak and strong extensions of differential operators', *Trans. Am. math. Soc.*, **55**, 132–51.

—, Corbeiller, Ph. le, Levinson, N. and Stoker, J. J. (1942–3). *Non-linear mechanics* (mimeographed notes).

Frobenius, G. (1887). 'Über des Pfaff'sche Problem', *Crelles Journal*, **82**, 230–315.

Frost, P. (1863). *Newton's Principia, with notes*, 2nd ed., London, Sections I–III.

Frostman, O. (1935). 'Potentiel d'équilibre et capacité des ensembles', *Meddn Lunds Univ. mat. Semin.*, **3**.

Gauss, C. F. (1800). 'Grundbegriffe der Lehre von der Reihe', *Gesammelte Werke*, Vol. 10 (1), Leipzig, 1917, pp. 390–2.

— (1801). 'Disquisitiones arithmeticae', *Gesammelte Werke*, Vol. 1, Göttingen, 1863.

— (1813). 'Theoriea attractionis corporum . . .', *Gesammelte Werke*, Vol. 5, Göttingen, 1867, pp. 1–22.

— (1827). 'Disquisitiones generales circa superficies curvas', *Gesammelte Werke*, Vol. 4, Göttingen, 1873, pp. 217–58.

— (1840). 'Allgemeine Lehrsatze in Bezeihung auf die im Verkehrten Verhältnisse des Quadrets der Enternungwirkenden Anziehungs und Abstossungskrafte', *Gesammelte Werke*, Vol. 5, Göttingen, 1867, pp. 197–242.

Ghiselin, B. (1952). *The creative process*, Berkley (2nd ed., 1964, New York).

Gibbs, J. W. (1902). *Elementary principles in statistical mechanics*, Yale.

Gödel, K. (1931a). 'Einige metamathematische Resultate über Entscheidungsdefinitheit und Widerspruchsfreiheit', *Anz. Akad. Wiss. Wien, Math.-nat. Klasse*, **67**, 214–5.

— (1931b). 'Über formal unentscheidbare sätze . . .', *Mh. Math. Phys.*, **38**, 173–98.

— (1932a). 'Eine Interpretation des intuitionischen Aussagenkalküls', *Ergebnisse eines mathematischen Kolloquiums*, Vol. 4, Vienna, pp. 39–40.

— (1932b). 'Zur intuitionistischen Arithmetick und Zahlentheorie', *Ergebnisse eines mathematischen Kolloquiums*, Vol. 4, Vienna, pp. 34–8.

— (1940). *The consistency of the axiom of choice and of the generalized continuum hypothesis with the axioms of set theory*, Princeton, N. J.

Goodstein, R. L. (1945). 'Function theory in an axiom-free equation calculus', *Proc. Lond. math. Soc. (2)*, **48**, 401–34 (presented in 1941).

— (1964). *Recursive number theory*, Amsterdam.

Gordon, P. (1868). 'Beweis, dass jede Covariante und Invariante einer binären Form eine ganze Funktion einer endlichen Anzahl solche Formen ist', *Crelles Journal*, **69**, 323–54.

Goursat, E. (1902). *Cours d'analyse mathematique*, 1st ed., Paris.

Grassmann, H. (1844). *Die lineale Ausdehnungslehre, ein neuer Zweig der Mathematik*, Leipzig; *Gesammelte mathematische und physike Werke*, Vol. 1, Part 1, Leipzig, 1894, pp. 1–319.

— (1861). *Lehrbuch der Arithmetik für höhere Lehrenstalten*, Berlin.

— (1862). *Die Ausdehnungslehre vollständig und in strenger Form bearbeitet*, Berlin.

Greaves, W. M. H. (1923a). 'On a certain family of periodic solutions . . .', *Proc. R. Soc. (A)*, **103**, 516–24.

— (1923b). 'On the stability of the periodic solutions of the triode oscillator', *Proc. Camb. phil. Soc., math. phys. Sci.*, **22**, 16–23.

Green, G. (1828). *An essay on the application of mathematical analysis to the theories of electricity and magnetism* (published by subscription), Nottingham.

— (1835). 'On the determination of the exterior and interior attractions of ellipsoids of variable densities', *Trans. Camb. phil Soc.*, **5**, 395–430.

— (1838). 'On the motion of waves in a variable canal . . .', *Trans Camb. phil. Soc.*, **6**, 457–62.

— (1871). In *Mathematical papers* (Ed. N. M. Ferrers), London, pp. 3–115.

Greene, G. (1940). 'The austere art', *The Spectator*, 20 December, p. 682.

Griffiths, H. B. (1971). '1871: Our state of mathematical ignorance', *Am. math. Monthly*, **78**, 1067–85.

Hadamard, J. S. (1903a). 'Sur les opérations fonctionnelles', *C. r. hebd. Séanc. Acad. Sci., Paris*, **136**, 351–4.

— (1903b). *Leçons sur le propagation des ondes*, Paris.

— (1906). 'Sur le problème de Dirichlet', *Bull. Soc. math. Fr.*, **34**, 135–8.

— (1910). *Leçons sur le calcul des variations*, Paris.

— (1932). *Le problème de Cauchy*, Paris.

— (1945). *An essay on the psychology of invention in the mathematical field*, Princeton.

— (1952). *Lectures on Cauchy's problem*, New York.

— *et al.* (1905). 'Cinq lettres sur la théorie des ensembles', *Bull. Soc. math. Fr.*, **33**, 261–73.

Hahn, H. (1907). 'Über die nicht archimedischen Grössensystem', *Sber. Akad. Wiss. Wien*, **116**, 601–55.

— (1908). 'Bemerkungen zur den Untersuchungen des Herrn. M. Fréchet, "Sur quelques points du calcul fonctionnel"', *Mh. Math. Phys.*, **19**, 247–57.

— (1927). 'Über lineare Gleichungssysteme ...', *Crelles Journal*, **157**, 214–29 (especially p. 217).

Hamilton, W. R. (1837). 'Theory of conjugate functions or algebraic couples ...', *Trans. R. Ir. Acad.*, **17**, 293–422.

— (1843). *Royal Irish Academy Council Book*, October 16.

— (1862). 'On the existence of a symbolic biquadratic equation', *Phil. Mag.*, **24**, 127–8.

— (1864). 'On the existence of a symbolic biquadratic equation', *Proc. R. Ir. Acad.*, **8**, 190–1.

Hammer, P. C. (1961). 'Extended topology', *Nieuw Archf. Wisk.*, **9**, 16–24, 25–33, 74–86.

— (1962). 'Extended topology', *Nieuw Archf. Wisk.*, **10**, 55–77.

Hankel, H. (1822). 'Untersuchung über die unendlichen oft oszillierenden und unstetigen Funktionen', *Math. Annln*, **20**, 63–121.

Hardy, G. H. (1908). *A course of pure mathematics*, 1st ed., Cambridge.

— (1910). *Orders of infinity*, Cambridge.

— (1927). *Srinivasa Ramanajan* (Ed.), Cambridge.

— (1969). *A mathematician's apology*, 2nd ed., Cambridge.

Harnack, A. (1881). *Die Elemente der Differential- und Integralrechnung*, Leipzig.

— (1885). 'Über den Inhalt von Punktmengen', *Math. Annln*, **25**, 241–50.

— (1887). *Grundlagen der Theorie des logarithmischen Potentials*, Leipzig.

Hartogs, F. (1906). 'Zur Theorie der analytischen Funktionen ...', *Math. Annln*, **62**, 1–88.

— (1915). 'Über das Problem der Wohlordnung', *Math. Annln*, **76**, 438–43.

Hausdorff, F. (1914). *Grundzuge der Mengenlehre*, 1st ed. (reprinted by Chelsea Publishing Company, New York, 1949).

— (1957). *Set theory* (English translation of *Grundzuge der Mengenlehre*), New York (2nd ed., 1962).

Heaslet, M. A. and Lomax, H. (1948). 'The calculation of downwash behind supersonic wings ...', *N.A.C.A. Technical Report*, No. 1620.

—, — and Jones, A. L. (1947). 'Volterra's solution of the wave equation', *N.A.C.A. Technical Report* No. 889.

Heaviside, O. (1892). 'On operations in physical mathematics', *Proc. R. Soc. (A)*, **52**, 504–29.

— (1893a). 'On operations in physical mathematics', *Proc. R. Soc. (A)*, **54**, 105–43.

— (1893b). 'The elements of vectorial algebra and analysis', *Electromagnetic theory*, London, Chap. 3.

Hedrick, E. R. (1902). 'On the sufficient conditions in the calculus of variations', *Bull. Am. math. Soc. (2)*, **9**, 11–24.

Heegaard, P. (1898). *Forstudier til en topologisk teori*, Copenhagen.

Heijenoort, J. van (1967). *From Frege to Gödel* (Ed.), Cambridge, Mass.

Heine, E. (1872). 'De Elemente der Funktionenlehre', *Crelles Journal*, **74**, 172–188.

Heisenberg, W. (1926). 'Zur Quantenmechanik II', *Z. Phys.*, **37**, 557–615 (especially pp. 596–605).

Helmholtz, H. von (1862). *Die Lehre von der Tonempfindungen*, 1st ed., Braunschweig. (English translation: 'On the sensations of tone', *Theory of combinational tones*, 1875, London, Appendix XII, pp. 621–3).

— (1868). 'Über die Thatsachen, die der Geometrie zu Grunde liegen', *Nachr. Ges. Wiss. Göttingen, Math.-physik, Kl.*, **15**, 193–221.

Hermite, C. (1873). 'Sur la fonction exponentielle', *C. r. hebd. Séanc. Acad. Sci., Paris*, **77**, 18–24, 74–9, 226–33, 285–93; *Oeuvres*, Vol. 2, Paris, pp. 150–81.

Hesse, O. (1857). 'Über die Kriterion des Maximums und Minimums der einfachen Integral', *Crelles Journal.*, **54**, 227–73.

Hessenberg, G. (1909). 'Kettentheorie und Wohlordnung', *Crelles Journal*, **135**, 81–133.

Heyting, A. (1934). *Mathematische Grundlagenforschung*, Berlin.

— (1956). *Intuitionism*, Amsterdam–London. (This gives a bibliography of Brouwer's papers from 1907–1955.)

Hilbert, D. (1889). 'Über die Enlichkeit der Invariantensystems für binäre Grundformen', *Math. Annln*, **33**, 323–6.

— (1890). 'Über die Theorie der algebraischen Formen', *Math. Annln*, **36**, 473–534.

— (1897). 'Die Theorie der algebraischen Zahlkörper', *Jber. dt. MatVerein*, **4**, 175–546.

— (1899). Les fondements de la géométrie (best edition of *Grundlagen der Geometrie*) (Ed. P. Rossier), Paris, 1971.

— (1900). 'Mathematische Problem', *Gesammelte Abhandlungen*, Vol. 3, Berlin, pp. 290–329; *Comptes Rendus du IIe Congrès International des Mathematiques, Paris*, 1900; *Arch. Math. Phys. (3)*, **1**, 44–63, 213–37 (1901).

— (1902). 'Über die Grundlagen der Geometrie', *Math. Annln*, **56**, 381–422.

— (1903). 'Über der Satz von der Basiswinkel in gleichschenkliger Dreieik', *Proc. Lond. math. Soc.*, **35**, 51–68.

— (1926). 'Über das unendliche', *Math. Annln*, **95**, 161–90.

— and Bernays, P. (1934). *Grundlagen der Arithmetik*, Vol. 1, Berlin.

— — (1939). *Grundlagen der Arithmetik*, Vol. 2, Berlin.

Hill, G. W. (1886). 'On the part of the motion of the lunar Perigree', *Acta math.*, **8**, 1–36 (first published in 1877).

Hobson, E. W. (1931). *The theory of spherical and ellipsoidal harmonics* (reprinted by Chelsea Publishing Company, New York, 1955).

Hodge, W. V. D. (1932). 'A Dirichlet problem for harmonic integrals . . .', *Proc. Lond. math. Soc. (2)*, **36**, 257–303.

— (1933). 'Harmonic funtionals in a Riemann space', *Proc. Lond. math. Soc. (2)*, **38**, 72–95.

— (1936). 'The existence theorem for harmonic integrals', *Proc. Lond. math. Soc. (2)*, **41**, 483–96.

— (1941). *The theory and applications of harmonic integrals*, Cambridge.

Hölder, O. (1901). 'Die Axiome der Quantität und die Lehre von Mass', *Ber. Verh. sächs. Akad. Wiss., Leipzig*, **53**, 1–64.

Horn, J. (1899). 'Uber eine lineare Differentialgleichung zweiten Ordnung . . .', *Math. Annln*, **52**, 271–92.

294 *References*

Houel, G. J. (1869).). *Sur le calcul des equipollences*, Paris.

Huntingdon, E. V. (1913). 'A set of postulates for abstract geometry', *Math. Annln*, **73**, 522–9.

Hurewicz, W. (1927). 'Normalbereich und Dimensionstheorie', *Math. Annln*, **96**, 736–64.

— (1941). 'On duality theorems', *Bull. Am. math. Soc.*, **47**, 562–3.

Hurwitz, A. (1896). 'Über die Zahlentheorie der Quanternionen', *Nachr. Ges. Wiss. Göttingen*, 313–34; *Mathematische Werke*, Vol. 2, Basel, 1933, pp. 303–30.

Illigens, E. (1899). 'Zur Weierstrass–Cantorischen Theorie der Irrationalzahlen', *Math. Annln*, **33**, 155–60.

Iseki, K. (1949). 'On definition of topological space', *J. Osaka Inst. Sci. Technol. (I)*, **1**, 97–8.

Jacobi, C. G. J. (1837). 'Zur Theorie der Variationsrechnung und der Differentialgleichungen', *Crelles Journal.*, **17**, 68–82.

— (1886). 'Vorlesungen über Dynamik, *Gesammelte Werke*, Vol. A, Berlin, 1884, p. 46.

Janiszewski, S. (1912). 'Sur les continus irreductibles entre deux points', *J. Éc. polytech.*, *Sér II*, **16**, 79–170.

Jeans, J. H. (1925). *The mathematical theory of electricity and magnetism*, 5th ed., Cambridge, Section 580.

Jech, T. J. (1973). *The axiom of choice*, Amsterdam–London.

Jeffreys, H. (1915). 'Certan hypotheses as to the internal structure of the earth and moon', *Mem. R. astr. Soc.*, **60**, 187–217 (especially pp. 211–3).

— (1924). 'On certain approximate solutions', *Proc. Lond. math. Soc. (2)*, **23**, 428–36; 'Certain solutions of Mathieu's equation', *ibid.*, 437–49; 'The modified Mathieu equation', *ibid.*, 449–54; 'Free oscillations of water in elliptical lake', *ibid.*, 455–76.

Jonquières, E. de (1890). 'Note sur le théorème d'Euler . . .', *C. r. hebd. Séanc. Acad. Sci., Paris*, **110**, 110–5, 169–73.

Jordan, C. (1866a). 'Des contours tracés sur les surfaces', *J. Math. pures appl. (2)*, **11**, 110–30.

— (1866b). 'Sur le deformation des surfaces', *J. Math. pures appl. (2)*, **11**, 105–9.

— (1882–7). *Cours d'analyse de l'école polytechnique*, 1st ed., Paris.

— (1887). *Cours d'analyse de l'école polytechnique*, Vol. 3, 1st ed., Paris, p. 593.

— (1892). 'Remarques sur les intégrales définis', *J. Math. pures appl. (4)*, 8, 69–99.

Jourdain, P. E. B. (1912). 'The development of the theories of mathematical logic and the principles of mathematics', *Q. Jl. pure appl. Math.*, **43**, 219–314 (especially p. 273).

— (1915). *Contributions to the founding of the theory of transfinite numbers*, Chicago–London.

Kelley, J. L. (1955). *General topology*, New York.

Kellogg, O. D. (1953). *Foundations of potential theory*, Dover — reprint (first published 1929).

Kemble, E. C. (1937). *The fundamental principles of quantum mechanics*, New York, Chap. III.

Kingman, J. F. C. and Taylor, S. J. (1966). *Introduction to measure theory and probability*, Cambridge.

Kirchhoff, G. (1882). 'Zur Theorie der Lichtstrahlen', *Sber. preuss. Akad. Wiss. (II)*, 641–69.

— (1883). *Vorlesungen über mathematische Physik*, Leipzig, p. 22.

Klein, F. (1871). 'Über die sogenannte nicht-Euklidische geometrie', *Math. Annln*, **4**, 573–625.

— (1872). *Vorlesungen über nicht-Euklidische Geometrie*, Berlin.

— (1873). 'Über dis ogennante nicht-Euklidische geometrie', *Math. Annln*, **6**, 112–45.

— (1882). *Über Riemann's Theorie der Algebraischen Funktionen*, Leipzig, p. iv.

— (1893a). 'Vergleichende Betrechtungen über neuere geometrische Forschungen', *Math. Annln*, **43**, 63–100. [English translation: *N.Y. Math. Soc. Bull.*, **2**, 215–49 (1893).]

— (1893b). *On Riemann's theory of algebraic functions and their integrals*, Cambridge. (Based on lectures delivered in 1880–1 and translated by F. Hardcastle in 1893.)

— (1909). *Elementarmathematik von hoheren Standpunkte aus*, Berlin.

— (1926–7). *Vorlesungen über die Entwicklung der Mathematik in 19 Jahrhundert*, 2 vols., Berlin.

— and Sommerfeld, A. (1897). *Über die Theorie des Kreisels*, Leipzig.

Kneser, A. (1900). *Lehrbuch der Variationsrechnung*, Braunschweig.

Knopp, A. (1918). 'Ein einfaches Verfahren zur Bildung stetiger nirgends differenzialbarer Funktionen', *Math. Z.*, **2**, 1–26.

Koebe, P. (1906). 'Herleitung der partielle Differentialgleichung der Potentialfunktion aus Integraleigenschaft', *Sber. berl. math. Ges.*, **5**, 39–42.

Kolmogorov, A. N. (1933). *Grundbegriff der Wahrscheinlichkeitsrechnung*. (English translation: *Foundations of the theory of probability*, New York, 1950, 1956).

König, H. (1953). 'Neue Begründung der Theorie der Distribution', *Math. Nachr.*, **9**, 129–48.

König, J. (1903). *Einleitung in der allgemeine Theorie der algebraischen Grössen*, Leipzig.

Korevaar, J. (1955). 'Distributions defined from the point of view of applied mathematics', *Amsterdam Proc. (A) Math. Soc.*, **58**, 368–89, 483–503, 663–74.

Köthe, G. (1966). *Topologische lineare Räume*, Berlin.

Kramers, H. A. (1926). 'Wellenmechanik und halbzahlige Quantisierung', *Z. Phys.*, **39**, 828–40.

Kronecker, L. (1869). 'Über systeme von Funktionen mehrer Variabeln', *Mh. K. Akad. Berlin*, 159–93; *Gesammelte Werke*, Leipzig, 1895, Vol. I, pp. 175–212, Vol. II, pp. 213–26.

— (1882). 'Grundzüge einer arithmetischen Theorie der algebraischen Grössen', *Crelles Journal.*, **92**, 1–122.

— (1887). 'Über der Zahlbegriff', *Crelles Journal*, **101**, 337–55.

— (1891). 'Sur le nombre des racines', *C. r. hebd. Séanc. Acad. Sci., Paris*, **113**, 1006–12.

Kummer, E. E. (1847). 'Zur Theorie der complexen Zahlen', *Crelles Journal*, **35**, 319–26.

Kuratowski, C. (1920). 'Sur la notion de l'ensemble fini', *Fundam. Math.*, **1**, 180.

— (1922a). 'Une méthode d'élimination des nombres infinis des raisonnements mathématiques', *Fundam. Math.*, **5**, 76–180. (especially p.89).

— (1922b). 'Sur l'opération A de l'analyse situé', *Fundam. Math.*, **3**, 182–99.

Lagrange, J.-L. (1760). 'Essai d'une nouvell méthode pour déterminer les maxima et les minima', *Oeuvres*, Vol. 1, Paris, 1867, pp. 335–62.

Laguerre–Verly, E. (1853). 'Sur la théorie des foyers', *Nouv. Annls Math.*, **12**, 57–66.

Laguna, T. de (1922). 'Point, line and surface, as sets of solids', *J. Phil.*, **19**, 393–461.

Lakatos, I. (1976). *Proofs and refutations*, Cambridge.

Lamb, H. (1920). *Higher mechanics*, Cambridge, p. 13.

Lamé, G. (1833). 'Sur la propagation de la chaleur ...', *J. Éc. polytech.*, **14**, 194–251.

— (1859). *Leçons sur les coordinées curvilignes*, Paris.

Landkof, N. S. (1972). *Foundations of modern potential theory*, Berlin–Heidelberg–New York.

Lasker, E. (1905). 'Zur Theorie der Moduln und Ideale', *Math. Annln*, **60**, 20.

Laugwitz, D. (1959). 'Eine Einführung der δ-Funktionen', *Sber. bayer. Akad. Wiss.*, **4**, 41–59.

— (1961). 'Anwendungen unendlich kleiner Zahlen, I, *Crelles Journal*, **207**, 53–60; 'II', *Crelles Journal*, **208**, 22–34.

Lebesgue, H. (1902). 'Intégrale longeur, aire', *Annls Math. pures appl. (3)*, **7**, 231–59.

— (1904). *Leçons sur l'intégration*, 1st ed., Paris.

— (1910). 'Sur l'intégral de Stieltjes . . .', *C. r. hebd. Séanc. Acad. Sci., Paris*, **150**, 86–8.

— (1911). 'Sur le non-applicabilité de deux domaines', *Math. Annln*, **70**, 166–8.

— (1912). 'Sur le problème de Dirichlet', *C. r. hebd. Séanc. Acad. Sci., Paris*, **154**, 335–7.

— (1924). 'Conditions de regularité . . . dans le problème de Dirichlet', *C. r. hebd. Séanc. Acad. Sci., Paris*, **178**, 349–54.

— (1928). *Leçons sur l'intégration*, 2nd ed., Paris.

— (1941). 'Les controverses de la théorie des ensembles et la question des fondements', *Les entretiens de Zurich*, Zurich, pp. 109–27.

Lebesgue, V.-A. (1841). 'Mémoire sur une formule de Vandermonde . . .', *J. Math. pures appl. (1)*, **6**, 17–35.

Lefschetz, S. (1926). 'Intersections and transformations of complexes and manifolds', *Trans. Am. math. Soc.*, **28**, 1–49.

— (1943). 'Existence of periodic solutions for certain differential equations', *Proc. natn. Acad. Sci. U.S.A.*, **29**, 29–32.

— (1963). *Differential equations—geometric theory*, 2nd ed., New York, pp. 366–70.

— (1967). 'Geometric differential equations', *Differential equations and dynamical systems* (Eds. J. K. Hale and J. P. la Salle), New York, pp. 1–14.

Legendre, A. M. (1976). 'Mémoire sur la manière de distinguer les maxima des minima dans le calcul des variations', *Histoire de l'Academie des Sciences*, Paris, 1788, pp. 7–37.

Leray, J. (1933). 'Étude des diverses équations intégrales . . .', *J. Math. pures appl.*, **12**, 1–82.

— (1945). 'Sur les équations et les transformations', *J. Math. pures appl.*, **24**, 201–40.

— and Schauder, J. (1934). 'Topologie et équations fonctionelles', *Annls scient. Éc. norm. sup., Paris*, **51**, 45–78.

Levi-Civita, T. (1892/3). 'Sugli infiniti ed infinitesimi attuali', *Atti Ist. veneto Sci.*, **4**(7), 1765–1815.

— (1917). 'Nozione di parallelismo in una varietà qualunque', *Rc. Circ. mat. Palermo*, **42**, 173–205.

Levinson, N. and Smith, O. K. (1942). 'A general equation for relaxation oscillations', *Duke math. J.*, **3**, 382–403.

Lewis, C. I. (1918). *A survey of symbolic logic*, Berkeley.

— and Longford, C. H. (1960). *A survey of symbolic logic*, Dover reprint.

L'Huillier, S. A. J. (1812). 'Théorèmes de polyedrométrie', *Ann. Math.*, **3**, 169–89.

Liapunov, A. M. (1892). *Problème générale de la stabilité du mouvement*, Princeton.

— (1898). 'Sur quelques questions qui se retachent au problème de Dirichlet', *J. Math. pures appl. (5)*, **4**, 241–311.

Lie, S. (1877). 'Theorie des Pfaff'schen Problems', *Arch. Math. Naturv.*, **2**, 338–79.

— (1886). 'Bemerkungen zu von Helmholtz' Arbeit . . .', *Ber. Verh. sächs Akad. Wiss.*, *Leipzig*, **38**, 337.

— (1890). 'Grundlagen der Geometrie', *Ber. Verh. Sächs Akad. Wiss.*, *Leipzig*, **42**, 355.

— (1893). *Theorie der Transformationsgruppen*, Vol. III, Part V, Leipzig, pp. 393–543.

Liénard, A. (1928). 'Étude des oscillations entretenues . . .', *Revue gén. Élect.*, **23**, 901–12, 946–54.

Lighthill, M. J. (1949). 'A technique for rendering approximate solutions to physical problems uniformly valid', *Phil. Mag. (7)*, **40**, 1179–201.

Lindelöf, E. L. (1903). 'Sur quelques points de la théorie des ensembles', *C. r. hebd. Séanc. Acad. Sci., Paris*, **137**, 697–700.

Lindemann, F. (1882). 'Uber die Zahl π', *Math. Annln*, **20**, 213–25.

Linstedt, A. (1883). 'Beitrag zur Integration der Differentialgleichungen der störingstheorie', *Mem. Acad. St. Pétersbourg*, **31**, No. 4, 1–20.

Liouville, J. (1832). 'Sur le calcul des différentielles à indices quelconques', *J. Éc. polytech.*, **21**(13), 71–162.

— (1834). 'Mémoire sur une formule d'analyse', *Crelles Journal*, **12**, 273–87.

— (1836). 'Sur le développement des fonctions . . .', *J. Math. pures appl. (1)*, **1**, 253–65.

— (1837). 'Sur le développement des fonctions . . .', *J. Math. pures appl. (1)*, **2**, 16–35.

— (1838). 'Sur la théorie de la variation des constants arbitraires', *J. Math. pures appl. (1)*, **3**, 342–9.

— (1856). 'Sur deux mémoires de Poisson', *J. Math. pures appl. (2)*, **1**, 1–6.

Lipschitz, R. (1864). 'De explications per series trigonometricas . . .', *Crelles Journal*, **63**, 296–308.

— (1876). 'Sur le possibilité d'intégrer complétement un système donné d'équations différentielles', *Bull. Sci. math.*, **10**, 149–59.

Listing, J. B. (1861). 'Der Census raümliche Komplexe', *Abh. Akad. Wiss. Göttingen*, **10**, 97–180.

Littlewood, J. E. (1953). *A mathematician's miscellany*, London, pp. 62–4.

Lomax, H. and Sluder, L. (1949). 'Downwash . . . behind a triangular wing . . .', *N.A.C.A. Technical Report*, No. 1803.

Looman, H. (1925). 'Über die Perronsche Integraldefinitionen', *Math. Annln*, **93**, 153–6.

Loš, J. (1955). 'Quelques remarques, théorèmes et problèmes sur les classes définissables d'algèbres', *Mathematical interpretation of formal systems*, Amsterdam, pp. 98–113.

Love, A. E. H. (1927). *A treatise on the mathematical theory of elasticity*, 4th ed., Cambridge, Chap. XIV.

Lowenheim, L. (1915). 'Über Möglichkeiten in Relativkalkül', *Math. Annln*, **76**, 447–70.

Lüroth, J. (1907). 'Über Abbildung von Mannigfaltigkeiten', *Math. Annln*, **63**, 222–38.

Luxembourg, W. A. J. (1962). *Lectures on A. Robinson's theory of infinitesimals and infinitely large numbers*, Pasadena.

Macaulay, F. S. (1913). 'On the resolution of a given modular system into primary systems', *Math. Annln*, **74**, 66–121.

— (1916). 'The algebraic theory of modular systems', *Camb. Tracts Math.*, No. 19.

MacLane, S. (1971). *Categories for the working mathematician*, New York –Heidelberg–Berlin.

Martineau, A. (1961). 'Les hyperfonctions de M. Sato', *Sem. Bourbaki, 13è année*, No. 214, 1–13.

Mathieu, E. (1868). 'Memoire sur le mouvement vibratoire d'une membrane de forme elliptique', *J. Math. pures appl.*, **13**, 137–204.

Maxwell, J. C. (1873). *Treatise on electricity and magnetism*, Oxford.

— (1892). *A treatise on electricity and magnetism*, 3rd ed., Vol. 1, Oxford, pp. 174–6.

Mayer, A. (1868). 'Über die Kriterion des Maximums and Minimums der einfachen Integrale', *Crelles Journal*, **69**, 238–63.

— (1878). 'Über das allgemeinste Problem der Variationsrechnung . . .', *Ber. Verh. sächs Akad. Wiss.*, *Leipzig, Math.-phys. Kl.*, **30**, 17–32.

— (1895). 'Die Lagrange'sche Multiplicatorenmethode . . .', *Ber. Verh. sächs Akad. Wiss.*, *Leipzig, Math.-phys. Kl.*, **47**, 129–44.

McAulay, A. (1898). *Octonions*, Cambridge.

McColl, H. (1878). 'The calculus of equivalent statements . . .', *Proc. Lond. math. Soc.*, **9**, 9–20, 177–86.

— (1879). 'The calculus of equivalent statements', *Proc. Lond. math. Soc.*, **10**, 16–28.

— (1880). 'The calculus of equivalent statements', *Proc. Lond. math. Soc.*, **11**, 113–121.

— (1906). *Symbolic logic and its application*, London.

McLachlan, N. W. (1947). *Theory and application of Mathieu functions*, Oxford.

Meltzer, B. (1962). *On formally undecidable propositions*, Edinburgh–London. (English translation of Gödel 1931b.)

Menger, K. (1923). 'Über die Dimensionalität von Punktmengen', *Mh. Math.*, **33**, 148–60.

— (1924). 'Über die Dimensionalität von Punktmengen', *Mh. Math.*, **34**, 137–61.

— (1926a). 'Zur Entstehung meiner Arbeit über Dimensions- und Kurventheorie', *Koninklijke Akademie von Wetenschappen te Amsterdam*, **29**, 1122 –4.

— (1926b). 'Zur Dimensions- und Kurventheorie', *Mh. Math.*, **36**, 411–32.

— (1926c). 'Zur Begrundung einer axiomatische Theorie der Dimension', *Mh. Math.*, **36**, 193–218.

— (1926d). 'Bericht über die Dimensionstheorie', *Jber. dt. MatVerein*, **35**, 113–50.

— (1927). 'Das Hauptprobleme über die dimensionalle Struktur der Raüme', *Koninklijke Akademie von Wetenschappen te Amsterdam*, **30**, 138–44.

— (1928). *Dimensionstheorie*, Leipzig.

— (1940). 'Topology without points', *Rice Inst. Pamph.*, No. XXVII, pp. 80–107.

Méray, Ch. (1869). 'Remarques sur la nature des quantités definis par la condition de servir de limites à des variables données', *Revue Socs. sav. sci. math.*, **4**, 280–9.

— (1887). 'Sur le sens qu'il convient d'attacher à l'expression "Nombre incommensurable"', *Annls scient. Ec. norm. sup.*, *Paris (3)*, **4**, 341–60.

Mie, G. (1893). 'Beweis der Integrierbarkeit gewöhnlichen Differential-gleichungensysteme nach Peano', *Math. Annln*, **43**, 553–68.

Mikusiński, J. (1948). 'Sur la méthode de généralisation de M. Laurent Schwartz . . .', *Fundam. Math.*, **35**, 235–9.

Milne, E. A. (1935). *Relativity, gravitation and world structure*, Oxford, pp. 341–7.

Milnor, J. (1963). 'Rapid course in Riemannian geometry', *Morse theory*, Princeton.

Minding, F. (1858). 'Über die Transformation, welche in der Variationsrechnung zur Nachweisung grössten oder kleinsten Werthe dienen', *Crelles Journal.*, **55**, 301–9.

Minkowski, H. (1909). 'Raum und Zeit', *Phys. Z.*, **10**, 104–11.

Minorsky, N. (1962). *Non-linear oscillations*, Princeton, Chap. 16.

Möbius, A. F. (1827). *Der barycentrische Calcul*, Leipzig.
— (1887). 'Über geometrische Addition und Multiplication', *Gesammelte Werke I*, Vol. 4, Leipzig, pp. 659–97.
Moigno, F. M. N. (1861). *Leçons de calcul différentiel et de calcul intégrale* (Ed.), Paris.
Molk, J. (1885). 'Sur une notion qui comprend celle de la divisibilité et sur la théorie générale de l'elimination', *Acta math.*, **6**, 1–166.
Moore, E. H. (1910). *Introduction to a form of general analysis*, Yale.
— and Smith, H. L. (1922). 'A general theory of limits', *Am. J. Math.*, **44**, 102–21.
Moore, R. L. (1935). 'A set of axioms for plane analysis situs', *Fundam, Math.*, **25**, 13–28.
Morse, M. (1925). 'Relations between the critical points of a real function of n independent variables', *Trans. Am. math. Soc.*, **27**, 345–96.
— (1934). *The calculus of variations in the large*, Providence, Rhode Island.
— (1969). *Critical point theory in global analysis and differential topology*, New York–London.
— (1973). *Variational analysis*, New York.
Moser, J. (1962). 'On invariant curves of area-preserving mappings of an annulus', *Nachr. Akad. Wiss. Göttingen, Math.-phys. Kl.*, **2a**, No. 1, 1–20.
— (1968). 'Lectures on Hamiltonian systems', *Mem. Am. math. Soc.*, **81**.
— (1973). *Stable and random motions in dynamical systems*, Princeton, 51–3.
Mostowski, A. (1964). *Sentences undecidable in formalized arithmetic*, Amsterdam.
Nagumo, M. (1950). 'Degree of mapping of manifolds . . .', *Osaka math. J.*, **2**, 105–18.
— (1951a). 'A theory of degree of mapping . . .', *Am. J. math.*, **73**, 485–96.
— (1951b). 'Degree of mapping . . .', *Am. J. Math.*, **73**, 497–511.
Netto, E. E. (1879). 'Beitrag zur Mannigfaltigkeiten', *Crelles Journal*, **86**, 263–8.
Neumann, C. (1870). 'Zur Theorie des logarithmischen und des Newtonischen Potentials', *Ber. Verh. sächs Akad. Wiss.*, *Leipzig*, **22**, 49–56, 264–321.
— (1877). 'Zur Theorie der logarithmischen und der Newton'schen Potential', *Math. Annln*, **11**, 558–66.
— (1884). *Vorlesungen über Riemann's Theorie der Abel'schen Integralen*, 2nd ed., Leipzig.
— (1887). *Vorlesungen über Potential und Kugelfunktionen*, Leipzig.
— (1901). 'Über die Methode des arithmetischen Mittels . . .', *Math. Annln*, **54**, 1–48.
Neumann, J. von (1923). 'Zur Einführung der transfiniten Zahlen', *Acta Litt. Scient. R. Univ. hung. Francisco-Josephina, Sect. Sci. Math.*, **1**, 199–208.
— (1925). 'Eine Axiomatisierung der Mengenlehre', *Crelles Journal*, **154**, 219–40; **155**, 128.
— (1928). 'Die Axiomatisierung der Mangenlehre' (Doctoral dissertation, Budapest, 1925), *Math Z.*, **27**, 669–752.
Neville, E. H. (1921). *Multilinear functions of direction* . . ., Cambridge.
Nicod, J. (1924). *Geometry and induction* (English translation), London, 1970, Part 1, Chap IV.
Nikodym, O. (1930). 'Sur une généralisation des intégrales de M. J. Radon', *Funam. Math.*, **15**, 131–79.
Noether, E. (1921). 'Idealtheorie in Ringbereichen', *Math. Annln*, **83**, 24–66.
Nuenhuis, A. (1952). *Theory of the geometric object*, Amsterdam.
Pal. J. (1912). 'Beweis des Lebesgue–Young'schen Sätzen', *Rc. Circ. mat. Palermo*, **33**, 352–3.
Pasch, M. (1882a). *Einleitung in die Differential- und Integralrechnung*, Leipzig.

— (1882b). *Vorlesgungen über die neuere Geometrie*, Leipzig (2nd ed., 1926).

Pauli, W. (1925). 'Über den Zusammenhang des Abschlusses der Elektronengruppen', *Z. Phys.*, **31**, 765–83.

— (1927). 'Zur Quantenmechanik des magnetischen Elektrons', *Z. Phys.*, **43**, 601–23.

Peano, G. (1887). *Applicazioni geometriche de calcolo infinitesimale*, Turin.

— (1889). *Arithmeticae principia, nova methodo exposita*, Turin.

— (1890a). 'Demonstration de l'intégrabilité des équations différentialles ordinaires', *Math. Annln*, **37**, 182–228.

— (1890b). 'Sur une courbe qui remplit toute une aire plane', *Math. Annln*, **36**, 157–60.

— (1891). 'Principii di logica matematica', *Riv. Mat.*, **1**, 1–10.

— (1894). 'Sui fondamenti della geometria', *Riv. Mat.*, **4**, 51–94.

— (1895–6). 'Saggio di calculo geometrico', *Atti Accad. Sci., Torino*, **31**, 952–75.

— (1895–1903). *Formulaire de mathématiques*, Turin; Vol. 1, 1895; Vol. 2, Part 1, 1897; Vol. 2, Part 2, 1898; Vol. 2, Part 3, 1899; Vol. 3, 1903; Vol. 4, 1903.

Peirce, B. (1881). 'Linear associative algebra', *Am. J. Math.*, **4**, 97.

Perrin, J. (1906). 'La discontinuité de la matière', *Revue du Mois*, **1**, 323–344.

— (1914). *Les atomes*, Paris. (English translation by D. Ll. Hammick. London, 1920.)

Perron, O. (1914). 'Über denn Integralbegriff', *Sber. heidelb. Akad. Wiss.*, **A16**, 1–16.

— (1915). 'Ein neuer Existenzbeweis für die Integrale der Differentialgleichung', *Math. Annln*, **76**, 471–84.

— (1923). 'Eine neue Behandlung der ersten Randwertaufgabe für $\Delta u = 0$', *Math. Z.*, **18**, 42–54.

Péter, R. (1951). *Rekursive Funktionen*, Budapest.

Picard, E. (1891a). 'Sur le nombre des racines communes à plusiers équations simultanées', *C. r. hebd. Séanc. Acad. Sci., Paris*, **113**, 356–8; *Traité d'analyse*, Vol. 1, Paris, pp. 123–7.

— (1891b). 'Sur la recherche du nombre des racines communes . . .', *C. r. hebd. Séanc. Acad. Sci., Paris*, **113**, 669–72.

— (1891c). 'Du nombre des racines communes . . .', *C. r. hebd. Séanc. Acad. Sci., Paris*, **113**, 1012–4.

— (1891–8). *Traité d'analyse*, 1st ed., Paris.

— (1893). 'Sur l'application des méthodes d'approximations . . .', *J. Math. pures appl. (4)*, **8**, 217–32.

— and Simart, G. (1897). *Théorie des fonctions algébriques*, Vol. 1, Paris, pp. 29–31.

Pieri, M. (1898). 'I principii della geometria di posizione', *Memorie Accad. Sci. Torino (2)*, **48**, 1–62.

Pincherle, S. (1880). 'Saggio di une introduzione alle teoria delle funzioni analitiche secondo i principii del Prof. C. Weierstrass', *G. Mat.*, **18**, 178–254, 317–57.

Poe, Edgar Allan (1845). 'The purloined letter', *The works of Edgar Allan Poe*, Vol. II, New York, 1884, pp. 389–416 (especially p. 407).

Poincaré, H. (1881). 'Mémoire sur les courbes définis par une équation différentielle', *J. Math. pures appl. (3)*, **7**, 375–422.

— (1882a). 'Théorie des groupes fuchsiens', *Acta Math.*, **1**, 1–62 (especially pp. 8 and 52).

— (1882b). 'Sur l'intégration des équations différentielles par les séries', *C. r. hebd. Séanc. Acad. Sci., Paris*, **94**, 577–8.

— (1886). 'Sur les intégrales irrégulières . . .', *Acta Math.*, **8**, 295–344.

— (1887a). 'Sur les hypotheses fondamentales de la geométrie', *Bull. Soc. math. Fr.*, **15**, 203–16.

— (1887b). 'Sur le probléme de la distribution électrique', *C. r. hebd. Séanc. Acad. Sci., Paris*, **104**, 44–6.

— (1890). 'Sur les équations aux dérivées partielles', *Am. J. Math.*, **12**, 211–94.

— (1892a). *Les méthodes nouvelles de la mécenique céleste*, Paris.

— (1892b). 'Sur l'application de la méthode de M. Lindstedt', *C. r. hebd. Séanc. Acad. Sci., Paris*, **114**, 1305–9.

— (1895a). 'Analysis situs', *J. Éc. polytech. (2)*, **1**, 1–122.

— (1895b). 'Sur la méthode de Neumann et le problème de Dirichlet', *C. r. hebd. Séanc. Acad. Sci., Paris*, **120**, 347–52.

— (1896–7). 'La méthode de Neumann . . .', *Acta Math.*, **20**, 59–142.

— (1899). 'Complément à l'analyse situs', *Rc. Circ. mat. Palermo*, **13**, 285–343.

— (1908). 'Le raisonnement mathématique', *Science et méthode*, Paris. (Translated by F. Maitland: *Science and method*, London, 1914.)

— (1912a). 'Pourquoi l'espace a trois dimensions', *Revue de metaphysique et de morale*, **20**, 483–504.

— (1912b). 'Sur un théorème de géometrie', *Rc. Circ. mat. Palermo*, **23**, 375–407.

— (1913). 'Mathematical creation', *The foundations of science* (Translated by G. Bruce Halstead), New York, p. 387.

Poisson, S. D. (1820), 'Sur la manière d'exprimer les fonctions par des séries de quantités périodiques', *J. Éc. polytech.*, **11** (18), 417–89.

— (1823). 'Sur la manière d'exprimer les fonctions par des séries de quantités périodiques', *J. Éc. Polytech.*, **11** (19), 145–62.

Pol, B. van der (1926). 'On relaxation oscillations', *Phil. Mag. (7)*, **2**, 978–92.

— (1927). 'Forced oscillations in a circuit with non-linear resistance', *Phil. Mag. (7)*, **3**, 65–80.

Poncelet, J.-V. (1822). *Traité des propriétés projectives des figures*, Paris (2nd ed., 1865).

Post, E. L. (1921). 'Introduction to a general theory of elementary propositions', *Am. J. Math.*, **43**, 163–85.

Rado, T. and Reichelderfer, P. V. (1955). *Continuous transformations in analysis*, Berlin.

Radon, J. (1913). 'Theorie und Anwendung der absolut additiven Mengenfuntionen', *Sber. Akad. Wiss. Wien (IIa)*, **122**, 1295–438.

Rayleigh, Lord (1877). *The theory of sound*, London (2nd ed. revised and enlarged, 1894).

— (1887). 'On the maintenance of vibrations . . .', *Phil. Mag. (5)*, **24**, 145–59.

— (1894). *Theory of sound*, 2nd ed., London, p. 78.

Rham, G. de (1931). 'Sur l'analysis situs des variétés à *n* dimensions', *J. Math. pures appl. (Sér. 9)*, **10**, 115–200.

— (1936). 'Relations entre le topologie et la théorie des intégrales multiples', *Enseign. math.*, No. 3–4, 213–28.

— (1955). *Variétés différentiables . . .*, Paris.

— and Kodaira, K. (1950). *Harmonic integrals*, Princeton.

Ricci, G. (1886). 'Sui parametri e gli invarianti della forme quadratiche differenziale', *Annali Mat. pura appl. (Ser. IIa)*, **14**, 1–11.

— and Levi-Civita, T. (1901). 'Méthodes de calcul différentiel absolu et leurs applications', *Math. Annln*, **54**, 125–201.

Riemann, B. (1847). 'Versuch einer allgemeiner Auffassung der Integration und Differentiation', *Gesammelte mathematische Werke*, No. XIX, New York, 1953, pp. 354–66.

— (1851). 'Grundlagen für eine allgemeine Theorie der Funktionen . . .', *Inaugural Dissertation*, Göttingen; *Gesammelte mathematische Werke*, No. XII, Leipzig, 1876, 1892, 1902.

— (1854). 'Über die Darstellbarkeit einer Function durch eine trigonometrische Reihe', *Gesammelte mathematische Werke*, No. II, 2nd ed., Leipzig, 1892.

— (1857a). 'Beiträge zur Theorie der durch die Gauss'sche Reihe darstellbaren Funktionen', *Gesammelte mathematische Werke*, No. XII, Leipzig, 1876.

— (1857b). 'Theorie der Abelschen Funktionen', *Crelles Journal*, **54**, 115–55.

— (1861). 'Commentatio mathematica . . .', *Gesammelte mathematische Werke*, No. XXII, Leipzig, 1876, pp. 391–404 (2nd ed., 1892; Supplement, 1902).

— (1867). 'Über die Hypothesen, welche der Geometrie zu Grunde legen', *Abh. Akad. Wiss. Göttingen, Math.-Phys. Kl.*, **13**. [Translated by W. K. Clifford, *Nature*, **8**, 14–17, 36–37 (1873).]

— (1876). 'Über die Darstellbarkeit einer Funktion durch eine trigonometrische Reihe' (read as his 'Habilitation' in 1854), *Gesammelte mathematische Werke*, No. XII, Leipzig (2nd ed., 1892). (French translation by L. Laugel, Paris, 1898.)

Riesz, F. (1909a). 'Stetigkeitsbegriff und abstrakte Mengenlehre', *Atti del IV Congresso Internationale di Mathematica (Roma, 1908)*, Vol. II, pp. 18–24.

— (1909b). 'Sur les opérations fonctionelles linéaires', *C. r. hebd. Séanc. Acad. Sci., Paris*, **149**, 974–7.

— (1910). 'Untersuchungen über Systeme integrierbaren Funktionen', *Math. Annln*, **69**, 449–97.

— (1925). 'Über die subharmonische Funktionen . . .', *Acta Litt. Scient. R. Univ. hung. Francisco-Josephina, Sect. Sci. Math.*, **2** (Part 2), 87–100.

— (1926). 'Sur les fonctions subharmoniques . . .', *Acta Math.*, **48**, 329–43.

— (1930). 'Sur les fonctions subharmoniques . . .', *Acta Math.*, **54**, 321–60.

Riesz, M. (1936). *Comptes rendus du congrès international des mathématiciens*, Oslo, Vol. 2, pp. 44–5.

— (1938). 'Intégrales de Riemann–Liouville et potentiels', *Acta. Litt. Scient. R. Univ. hung. Francisco-Josephina, Sect. Sci. Math.*, **9**, 1–42.

— (1940). 'Intégrales de Riemann–Liouville et potentiels', *Acta. Litt. Scient. R. Univ. hung. Francisco-Josephina, Sect. Sci. Math.*, **10**, 116.

— (1949). 'L'intégrale de Riemann–Liouville', *Acta Math.*, **81**, 1–223.

Robb, A. A. (1936). *Geometry of space and time*, Cambridge.

Robin, G. (1887). 'Distribution de l'électricité sur une surface fermée', *C. r. hebd. Séanc. Acad. Sci., Paris*, **104**, 1834–6.

Robinson, A. (1948). 'On source and vortex distributions . . .', *Q. Jl. Mech. appl. Math.*, **1**, 408–432.

— (1966). *Non-standard analysis*, Amsterdam.

— and Laurmann, J. A. (1956). *Wing theory*, Cambridge.

Robinson, R. M. (1937). 'The theory of classes', *J. Symbolic Logic*, **2**, 29–36.

Rubin, H. and Rubin, J. E. (1963). *Equivalents of the axiom of choice*, Amsterdam.

Russell, B. (1897). *An essay on the foundations of geometry*, Cambridge.

— (1903). *The principles of mathematics*, 1st ed., Vol. 1, Cambridge.

— (1908). 'Mathematical logic as based on the theory of types', *Am. J. Math.*, **30**, 222–62.

— (1922). *Our knowledge of the external world*, London, Chap. IV.

— (1937). *The principles of mathematics*, 2nd ed., London, pp. 278–86.

Sato, M. (1959). 'Theory of hyperfunctions', *J. Fac. sci. Tokyo Univ.*, **8**, 139–93.

Schapiro, P. (1969). *Théorie des hyperfonctions*, Berlin–Heidelberg–New York.

Schauder, J. (1930). 'Der Fixpunktsatz in Funktionalraümen', *Studia math.*, **2**, 170–9.

Scheefer, L. (1894). 'Zur Theorie der realen Funktionen', *Acta Math.*, **5**, 52.

Schlesinger, L. (1906). 'Sur certaines séries asymptotiques', *C. r. hebd. Séanc. Acad. Sci., Paris*, **142**, 1031–3.

— (1907). 'Über asymptotische Darstellungen . . .', *Math. Annln*, **63**, 277–300.

Schmidt, E. (1907a). 'Zur Theorie der linearen und nichtlinearen Integralgleichungen', *Math. Annln*, **63**, 433–76.

— (1907b). 'Zur Theorie der linearen und nichtlinearen Integralgleichungen', *Math. Annln*, **64**, 161–74.

— (1908). 'Zur Theorie der linearen und nichtlinearen Integralgleichungen', *Math. Annln*, **65**, 370–99.

Schmeiden, C. and Laugwitz, D. (1958). 'Eine Erweiterung der Infinitesimal-rechnung', *Math. Z.*, **69**, 1–39.

Schoenflies, A. (1906), 'Über die Möglichkeit einer projektiven Geometrie bei transfiniter (nicht archimedischer) Massbestimmung', *Jber. dt. MatVerein.*, **15**, 26–41.

Schouten, J. A. and Dantzig, D. van (1935). 'Was ist Geometrie?', *Abh. Sem. Vektor Tenzor anal. Moskau*, **2–3**, 15–48.

— and Haantjes, J. (1936). 'On the theory of the geometric object', *Proc. Lond. math. Soc. (2)*, **42**, 356–76.

Schröder, E. (1890–5). *Vorlesungen ber die Algebra der Logik*, Leipzig; Vol. 1, 1890; Vol. 2, 1891; Vol. 3, 1895.

— (1896). 'Über G. Cantor'sche Satze', *Jber. dt. MatVerein.*, **5**, 81–2.

Schur, F. (1902). 'Uber die Grundlagen der Geometrie', *Math. Annln*, **55**, 265–92.

Schwartz, H. A. (1870). 'Über einen Grenzübergang durch altirnirendes Verfahren', *Crelles Journal.*, **74**, 218–53.

— (1872). 'Zur Integration der Partielle Differenzialgleichungen', *Crelles Journal*, **74**, 218–53.

Schwartz, L. (1945). 'Généralisation de la notion de fonction ...', *Ann. Univ. Grenoble*, **21**, 57–74.

— (1948). 'Généralization de la notion de fonction et de dérivation', *Annls Télécommun.*, **3**, No. 4, 135–40.

— (1950). *Théorie des distributions*, 1st ed., Paris.

— (1966). 'Courants sur une variété', *Theorie des distributions*, nouvelle éd., Paris, pp. 312–400.

Scott, C. A. (1899). 'A proof of Noether's fundamental theorem', *Math. Annln*, **52**, 593–7.

Sebastião e Silva, J. (1945). *Sur une construction axiomatique de la théorie des distributions*, Lisbon.

— (1960). 'Sur la définition et le structure des distributions vectorielles', *Port. math.*, **19**, (1), 1–80.

Serret, J. A. (1854). *Cours d'algèbre supérieure*, Paris.

Sieger, B. (1908). 'Die Beugung einer ebener elektrischer Wellen ...', *Annln Phys.*, **27**, 626–64.

Sierpinski, W. (1918). 'L'axiome de M. Zermelo et son rôle dans la théorie des ensembles de l'analyse', *Bull. int. Acad. Sci. Lett. Cracovie. Cl. Sci. math. nat. Ser. A.*, 49–152.

— (1927). 'La notion dérivée comme base d'une théorie des ensembles abstraits', **97**, 321–37.

Signorini, A. (1912). 'Esistenza di un'estremale chiusa dentro un contorno di Whittaker', *Rc. Circ. mat. Palermo*, **33**, 187–93.

Sikorski, R. (1954). 'A definition of the notion of distribution', *Bull. Acad. pol. Sci., Cl. III., Math.*, 207–11.

Silberstein, L. (1914). *The theory of relativity*, London.

Skolem, T. (1920). 'Logisch-kombinatorische Untersuchungen ...', *Vidensk. Skrifter I, Math.-nat. Klasse*, no. 4.

— (1922). 'Einige Bemerkung zur axiomatischen Begründung der Mengenlehre', *Math. Kongressen Helsingfors.*, 217–32.

— (1923). 'Begründung der elementaren Arithmetik ...', *Vidensk. Skrifter I, Mat.-nat. Klasse*, No. 6, 302.

— (1929). 'Über einige Grundlagenfragen der Mathematik', *Skr, norske. Vidensk-Akad., Mat.-nat. Kl.*, No. 4.

Smith, H. J. S. (1894). *Report on the theory of numbers*, Vol. 1, Oxford, p. 93.

Smith, H. L. (1938). 'A general theory of limits', *Natn. Math. Mag.*, **12**, 371–9.

Sobolev, S. L. (1936). 'Méthode nouvelle à resoudre le problème de Cauchy', *Mat. Sbornik*, **1**, 39–72.

Somerville, D. M. Y. (1970). *Bibliography of non-Euclidean geometry*, New York (originally published 1911).

Staudt, J. von (1847). *Geometrie der Lage*, Nuremberg.

— (1856). *Beiträge zur Geometrie der Lage*, Nuremberg (2nd ed., 1857; 3rd ed., 1860).

Stegmann, F. L. (1854). *Lehrbuch der Variationsrechnung*, Cassel, p. 90.

Steiner, J. (1881–2). 'Systematische Entwicklung der Abhängigkeit geometrischen Gestalten von einander', *Gesammelte Werke*, Vol. 1, Berlin, pp. 229–460 (reprinted by Chelsea Publishing Company, New York, 1971).

Stieltjes, T. J. (1894). 'Récherches sur fractions continués, *Annls Fac. Sci. Univ. Toulouse*, **8**, 1–122.

— (1895). 'Récherches sur fractions continués', *Annls Fac. Sci. Univ. Toulouse*, **9**, 1–47.

Stoker, J. J. (1950). *Non-linear vibrations*, Appendix I, New York.

Stokes, G. G. (1854). 'A Smith's prize paper', *Cambridge University Calendar*.

Stolz, O. (1885). *Vorlesungen über allgemeine Arithmetik*, Leipzig, p. 82.

Stone, M. H. (1936). 'The theory of representations for Boolean algebra', *Trans. Am. Math. Soc.*, **40**, 37–111.

— (1937). 'Applications of the theory of Boolean rings to general topology', *Trans. Am. Math. Soc.*, **41**, 374–481.

Suzuki, D. T. (1934). *An introduction to Zen Buddhism*, Kyoto.

Sylvester, J. J. (1839), 'On rational derivation from equations of coexistence', *Phil. Mag.*, **15**, 428–35.

— (1841). 'Examples of the dialytic method', *Camb. math. J.*, **2**, 232–6.

Tait, P. G. (1867). *Elementary treatise on quaternions*, 1st ed., Oxford (2nd ed., 1873; 3rd ed., 1890).

— and Thomson, W. (1867). *Treatise on natural philosophy*, Cambridge.

Tannery, J. (1886). *Introduction à la Théorie des Fonctions d'une Variable*, Paris.

Tarski, A. (1925). 'Sur les ensembles finis', *Fundam. Math.*, **6**, 45.

Thiess, F. B. (1968). 'Analytic functionals in quantum field theory . . .', *J. Math. Phys.*, **9**, 305–15.

Thomae, K. J. (1880). *Elementare Theorie der analytischen Funktionen, einer komplexen Verändlichlen*, Halle, p. 112 (2nd ed., 1898).

Thompson, S. P. (1910a). *Calculus made easy: by F. R. S.*, London (2nd ed., 1914; 3rd ed., 1965).

— (1910b). *The life of William Thomson, Baron Kelvin of Largs*, Vol. II, London, p. 1070.

Thomson, Sir William, Baron Kelvin (1847). 'Sur une équation aux différences partielles . . .', *J. Math. pures appl.*, **12**, 493–6.

— (1848). 'Potential of surface S carrying charge Q', *Reprint of papers on electrostatics and magnetism*, No. XIII, London, 1884, pp. 139–43.

— (1869). 'On vortex motion', *Trans. R. Soc. Edinb.*, **25**, 217–60.

Todhunter, I. (1861). *A history of the progress of the calculus of variations during the nineteenth century*, Cambridge.

— (1865). *A history of the mathematical theory of probability*, Cambridge.

— (1871). *Researches in the calculus of variations*, Cambridge.

— (1886–93). *A history of the theory of elasticity* (edited and completed by M. Pearson), Vols. 1 and 2, Cambridge.

Tonelli, L. (1921). *Fondamenti di calcolo delle variazioni*, Vol. 1, Bologna.

— (1923). *Fondamenti di calcolo delle variazioni*, Vol. 2, Bologna.

Tukey, J. W. (1940), 'Convergence and uniformity in topology', *Ann. Math. Stud.*, **2**.

Turing, A. M. (1937). 'On computable numbers . . .', *Proc. Lond. math. Soc. (2)*, 42, 230–65; 43, 544–6.

Tychonoff, A. (1935). 'Ein Fixpunktsatz', *Math. Annln*, 111, 767–76.

Urysohn, P. (1922). 'Les multiplicités Cantoriennes', *C. r. hebd. Séanc. Acad. Sci., Paris*, 175, 440–2.

— (1925). 'Mémoire sur les multiplicités Cantoriennes', *Fundam. Math.*, 7, 30–137, 225–356.

— (1926). 'Mémoire sur les multiplicités Cantoriennes', *Fundam. Math.*, 8, 225–309.

Val, P. du (1964). *Homographies, quarternions and rotations*, Oxford.

Vallée-Pouissin, C. J. de la (1915). 'Sur l'intégrale de Lebesgue', *Trans. Am. math. Soc.*, 16, 435–501.

— (1916). *Intégrales de Lebesgue*, Paris.

Van der Corput, J. G. (1952). *Reports on asymptotic expansion*, National Bureau of Standards, Los Angeles, California.

— (1959a). 'Neutrices', *J. Soc. ind. appl. Math.*, 7, 253–79.

— (1959b). 'Introduction to the neutrix calculus', *J. Analyse math.*, 7, 281–399.

Van der Waerden, B. L. (1930). 'Ein einfaches Beispiel einer nicht differenzial-baren stetigen Funktion', *Math. Z.*, 32, 474–5.

— (1970). 'The foundations of algebraic geometry from Severi to André Weil', *Archs Hist. exact. Sci.*, 7, 171–80.

Veblen, O. (1904). 'A system of axioms for geometry', *Trans. Am. math. Soc.*, 5, 343–84.

— and Thomas, J. M. (1926). 'Projective invariants of affine geometry of paths', *Ann. Math.*, 27, 278–96.

— and Whitehead, J. H. C. (1932). 'The foundations of differential geometry', *Camb. Tracts Math.*, No. 29.

— and Young, J. W. (1910). *Projective geometry*, Vol. 1, Boston.

— — (1918). *Projective geometry*, Vol. 2, Boston.

Veronese, G. (1891). *Fondamenti di geometria*, Padua, p. 166.

— (1894). *Grundlage der Geometrie* (German translation), Leipzig.

Vietoris, L. (1927). 'Über den höheren Zusammenhand kompakten Räume . . .', *Math. Annln*, 97, 454–72.

Volterra, V. (1887). 'Sopra la funzioni che dipendono de altra funzioni', *Atti Accad. naz. Lincei Rc. (4)*, 3 (2), 97–105, 141–6, 153–8.

— (1913). *Leçons sur les fonctions de lignes*, Paris.

— (1936). *Theory of functionals* (lectures given in Madrid, 1925), Paris.

Wald, A. (1932). 'Die Widerspruchsfreiheit des Kollektivbegniffes der Wahr-scheinlichkeitsrechnung', *Ergebniss eines mathematisches Kolloquiums*, Vol. 3, Vienna, p. 6.

Walker, A. G. (1955). 'Connexions for parallel distributions', *Q. Jl Math.*, 6, 301–8.

— (1958). 'Connexions for parallel distributions', *Q. Jl. Math.*, 9, 221–31.

Wallmann, H. (1938). Lattices and topological spaces', *Ann. Math.*, 39, 112–26.

Wang, H. (1957). 'The axiomatization of arithemtic', *J. Symbolic Logic*, 22, 145–57.

Ward, G. N. (1955). *Linearized theory of steady high-speed flow*, Cambridge.

Wasow, W. (1955). 'On the convergence of an approximate method of M. J. Lighthill', *J. Rational Mechanics*, 4, 751–67.

Watson, G. N. (1922). *A treatise on the theory of Bessel functions*, Cambridge.

Weierstrass, K. (1870). 'Über die sogennante Dirichlet's Princip', *Gesammelte Werke*, Vol. 2, Berlin, 1895, pp. 49–54.

— (1876). 'Über continuirliche Funktion eines reellen Arguments die für keinen Werth des letzteren einen bestimmten Differential quotienten besitzen', *Abh. dt. Akad. Wiss. Berl.*, 680–93; *Gesammelte Werke*, Vol. 2, Berlin, 1895, pp. 71–4.

Weil, A. (1937). 'Sur les espaces à structure uniforme', *Actual. scient. ind.*, 551.

— (1946). *Foundations of algebraic geometry*, Providence, Rhode Island.

Wentzel, G. (1926). 'Eine Verallgemeinerung der Quantenbedingungen ...', *Z. Phys.*, **38**, 518–29.

Weyl, H. (1921). 'Über die neue Grundlagen krise der Mathematik', *Math. Z.*, **10**, 39–79.

— (1923). *Mathematische Analyse des Raumproblems*, Berlin.

— (1940). 'The method of orthogonal projection in potential theory', *Duke Math. J.*, **7**, 411–44.

— (1951). 'A half-century of mathematics', *Am. math. Mon.*, **58**, 523–53.

Whitehead, A. N. (1897). *Universal algebra*, Book IV, Cambridge.

— (1906). 'The axioms of projective geometry', *Camb. Tracts Math.*, No. 4.

— (1907). 'The axioms of descriptive geometry', *Camb. Tracts Math.*, No. 5.

— (1919). *Enquiry concerning the principles of natural knowledge*, Cambridge.

— (1922). *The principle of relativity with applications to physical science*, Cambridge.

— (1929). *Process and reality*, Cambridge. (Presented at Edinburgh in 1927–8 as Gifford lectures).

— and Russell, B. (1910). *Principia mathematica*, Cambridge.

Whitney, H. (1935). 'Differentiable manifolds', *Ann. Math. (2)*, **37**, 645–80.

Whittaker, E. T. (1902). 'On periodic orbits', *Mon. Not. R. astr. Soc.*, **62**, 186.

— (1914). 'On the general solution of Mathieu's equation', *Proc. Edinb. math. Soc.*, **32**, 75–80.

— (1903). 'On the partial differential equations of mathematical physics', *Math. Annln*, **57**, 333–55.

— (1904). *A treatise on the analytical dynamics of particles and rigid bodies*, Cambridge, pp. 376–8.

— (1937). *A treatise on the analytical dynamics of particles and rigid bodies*, 4th ed., Cambridge, pp. 386–9.

Whittaker, Sir Edmund (1953). 'A history of the theories of aether and electricity', *The modern theories, 1900–26*, London, Chap. II.

Whittemore, J. K. (1901). 'Lagrange's equation in the calculus of variations', *Ann. Math. (2)*, **2**, 130–6.

Wiener, N. (1924). 'Certain notions in potential theory', *J. Math. Phys.*, **3**, 24–51.

— (1926). 'On the representation of functions by trigonometrical integrals', *Math. Z.*, **24**, 575–616.

Wundheiler, A. (1937). 'Objekte, Invarianten und Klassifikation der Geometrie', *Abn. Sem. Vektor Tenzor anal. Moskau*, **4**, 366–75.

Young, L. C. (1933). 'On approximation by polygons in the calculus of variations', *Proc. R. Soc. (A)*, **141**, 325–41.

— (1969). *Lectures on the calculus of variations and optimal control theory*, Philadelphia–London–Toronto.

Young, W. H. (1905). 'On upper and lower integration', *Proc. Lond. math. Soc. (2)*, **2**, 52–66.

— (1910). 'On functions of bounded variation', *Q. Jl. Math.*, **42**, 54–85.

— and Young, G. C. '1915). 'On the reduction of sets of intervals', *Proc. Lond. math. Soc. (2)*, **14**, 111–30.

Zaremba, S. (1911). 'Sur le principe de Dirichlet', *Acta Math.*, **34**, 293–316 (especially p. 310).

Zariski, O. (1971). 'Algebraic geometry', *Encyclopaedia Brittanica, Macropaedia*, 15th ed., Vol. 7, pp. 1070–6.

Zermelo, E. (1894). 'Untersuchung zu Variationsrechnung', *Dissertation*, Berlin, p. 41.

— (1904). 'Beweiss dass jede Menge wohlgeordnet werden kann', *Math. Annln*, **59**, 514–6.

— (1908a). 'Neuere Beweis für die Möglichkeit einer Wohlordnung', *Math. Annln*, **65**, 107–28.

— (1908b). 'Untersuchungen über die Grundlagen der Mengenlehre', *Math. Annln*, **65**, 261–81.

Zorn, M. (1935). 'A remark on method in transfinite algebra', *Bull. Am. math. Soc.*, **41**, 667–70.

Index of Authors

General Index

313